TABLE OF CONTENTS

Preface	ix
Mathematics Teacher Educator Learning and Development: An Introduction *Barbara Jaworski*	1

Section 1: Challenges to and Theory in Mathematics Teacher Education

Chapter 1: The Challenge of Mathematics Teacher Education in an Era of Mathematics Education Reform *Martin A. Simon*	17
Chapter 2: Being Mathematical with and in Front of Learners: Attention, Awareness, and Attitude as Sources of Differences between Teacher Educators, Teachers and Learners *John Mason*	31
Chapter 3: Facing the Challenge of Educating Educators to Work with Practising Mathematics Teachers *Ruhama Even*	57
Chapter 4: Sociocultural Perspectives on Learning to Teach Mathematics *Merrilyn Goos*	75
Chapter 5: Meeting the Challenges of Mathematics Teacher Education through Design and Use of Tasks that Facilitate Teacher Learning *Orit Zaslavsky*	93
Chapter 6: Mathematics Teacher Educators' Learning from Research on Their Instructional Practices: A Cognitive Perspective *Olive Chapman*	115

TABLE OF CONTENTS

Section 2: Reflection on Developing as a Mathematics Teacher Educator

Chapter 7: Profound Awareness of the Learning Paradox (PALP): A Journey towards Epistemologically Regulated Pedagogy in Mathematics Teaching and Teacher Education — 137
Ron Tzur

Chapter 8: Becoming a Mathematics Teacher Educator: Processes and Issues — 157
Razia Fakir Mohammad

Chapter 9: Reflecting the Development of a Mathematics Teacher Educator and His Discipline — 177
Konrad Krainer

Chapter 10: A Quest for 'Good' Research: The Mathematics Teacher Educator as Practitioner Researcher in a Community of Inquiry — 201
Simon Goodchild

Section 3: Working with Prospective and Practising Teachers: What We Learn; What We Come to Know

Chapter 11: Knowledgeable Teacher Educators and Linking Practices — 223
Christer Bergsten and Barbro Grevholm

Chapter 12: The Elementary Mathematics Methods Course: Three Professors' Experiences, Foci, and Challenges — 247
Amy Roth McDuffie, Corey Drake and Beth Herbel-Eisenmann

Chapter 13: Tools for Learning about Teaching and Learning — 265
Pat Perks and Stephanie Prestage

Chapter 14: What to Teach and How to Teach It: Dilemmas in Primary Mathematics Teacher Education — 281
Victoria Sánchez and Mercedes García

Chapter 15: Caring Relations in the Education of Practising Mathematics Teachers — 299
Paola Sztajn

Chapter 16: Trust and Respect: A Path Laid While Walking — 315
A.J. (Sandy) Dawson

The Mathematics Teacher as a Developing Professional

Edited by

Barbara Jaworski
Loughborough University, UK

and

Terry Wood
Purdue University, West Lafayette, USA

SENSE PUBLISHERS
ROTTERDAM / TAIPEI

 from the Library of Congress.

ISBN 978-90-8790-550-7 (paperback)
ISBN 978-90-8790-551-4 (hardback)
ISBN 978-90-8790-552-1 (e-book)

Published by: Sense Publishers,
P.O. Box 21858, 3001 AW Rotterdam, The Netherlands
http://www.sensepublishers.com

The cover picture is a compilation of scenes from the Western Pacific, provided by Sandy Dawson. The first one shows five mentors from (left to right) American Samoa, Palau, Palau, Marshall Islands and American Samoa together with the model of a Yapese meeting house which they built as part of a cultural sharing activity. They are wearing the tradition Yapese lei around their necks. They used materials brought by the Yapese mentors – bamboo wood and string made from the husks of coconuts – and were guided in their construction by the Yapese mentors. Sandy's chapter [16] emphasises relationships between local mentors and incoming mathematics teacher educators that encourage fruitful approaches to mathematics teaching development centring on respect and trust.

Printed on acid-free paper

All rights reserved © 2008 Sense Publishers

No part of this work may be reproduced, stored in a retrieval system, or transmitted in any form or by any means, electronic, mechanical, photocopying, microfilming, recording or otherwise, without written permission from the Publisher, with the exception of any material supplied specifically for the purpose of being entered and executed on a computer system, for exclusive use by the purchaser of the work.

THE MATHEMATICS TEACHER EDUCATOR AS A L

The International Handbook of Mathematics Teacher Education

Series Editor:

Terry Wood
Purdue University
West Lafayette
USA

This *Handbook of Mathematics Teacher Education*, the first of its kind, addresses the learning of mathematics teachers at all levels of schooling to teach mathematics, and the provision of activity and programmes in which this learning can take place. It consists of four volumes.

VOLUME 1:
Knowledge and Beliefs in Mathematics Teaching and Teaching Development
Peter Sullivan, *Monash University, Clayton, Australia* and Terry Wood, *Purdue University, West Lafayette, USA* (eds.)
This volume addresses the "what" of mathematics teacher education, meaning knowledge for mathematics teaching and teaching development and consideration of associated beliefs. As well as synthesizing research and practice over various dimensions of these issues, it offers advice on best practice for teacher educators, university decision makers, and those involved in systemic policy development on teacher education.
paperback: 978-90-8790-541-5, hardback: 978-90-8790-542-2, ebook: 978-90-8790-543-9

VOLUME 2:
Tools and Processes in Mathematics Teacher Education
Dina Tirosh, *Tel Aviv University, Israel* and Terry Wood, *Purdue University, West Lafayette, USA* (eds.)
This volume focuses on the "how" of mathematics teacher education. Authors share with the readers their invaluable experience in employing different tools in mathematics teacher education. This accumulated experience will assist teacher educators, researchers in mathematics education and those involved in policy decisions on teacher education in making decisions about both the tools and the processes to be used for various purposes in mathematics teacher education.
paperback: 978-90-8790-544-6, hardback: 978-90-8790-545-3, ebook: 978-90-8790-546-0

VOLUME 3:
Participants in Mathematics Teacher Education: *Individuals, Teams, Communities and Networks*
Konrad Krainer, *University of Klagenfurt, Austria* and Terry Wood, *Purdue University, West Lafayette, USA* (eds.)
This volume addresses the "who" question of mathematics teacher education. The authors focus on the various kinds of participants in mathematics teacher education, professional development and reform initiatives. The chapters deal with prospective and practising teachers as well as with teacher educators as learners, and with schools, districts and nations as learning systems.
paperback: 978-90-8790-547-7, hardback: 978-90-8790-548-4, ebook: 978-90-8790-549-1

VOLUME 4:
The Mathematics Teacher Educator as a Developing Professional
Barbara Jaworski, *Loughborough University, UK* and Terry Wood, *Purdue University, West Lafayette, USA* (eds.)
This volume focuses on knowledge and roles of teacher educators working with teachers in teacher education processes and practices. In this respect it is unique. Chapter authors represent a community of teacher educators world wide who can speak from practical, professional and theoretical viewpoints about what it means to promote teacher education practice.
paperback: 978-90-8790-550-7, hardback: 978-90-8790-551-4, ebook: 978-90-8790-552-1

Section 4: Synthesis

Chapter 17: Development of the Mathematics Teacher Educator and Its
Relation to Teaching Development   335
Barbara Jaworski

PREFACE

It is my honour to introduce the first *International Handbook of Mathematics Teacher Education* to the mathematics education community and to the field of teacher education in general. For those of us who over the years have worked to establish mathematics teacher education as an important and legitimate area of research and scholarship, the publication of this handbook provides a sense of success and a source of pride. Historically, this process began in 1987 when Barbara Jaworski initiated and maintained the first Working Group on mathematics teacher education at PME. After the Working Group meeting in 1994, Barbara, Sandy Dawson and I initiated the book, *Mathematics Teacher Education: Critical International Perspectives,* which was a compilation of the work accomplished by this Working Group. Following this, Peter de Liefde, while at Kluwer Academic Publishers, proposed and advocated for the *Journal of Mathematics Teacher Education.* In 1998 the first issue of the journal was printed with Thomas Cooney as editor of the journal who set the tone for quality of manuscripts published. From these events, mathematics teacher education flourished and evolved as an important area for investigation as evidenced by the extension of JMTE from four to six issues per year in 2005 and the recent 15th ICMI Study, *The professional education and development of teachers of mathematics.* In preparing this handbook it was a great pleasure to work with the four volume editors, Peter Sullivan, Dina Tirosh, Konrad Krainer and Barbara Jaworski and all of the authors of the various chapters found throughout the handbook.

Volume 4, *The Mathematics Teacher Educator as a Developing Professional,* skillfully edited by Barbara Jaworski, focuses on the professionalization of mathematics teacher educators. This aspect of mathematics teacher education has only recently been realized as an important area for investigation and development. This volume is fourth in the series and is an excellent culmination to the handbook.

Terry Wood
West Lafayette, IN
USA

REFERENCES

Jaworski, B., & Wood, T. (Eds.). (2008). *International handbook of mathematics teacher education: Vol. 4 The mathematics teacher educator as a developing professional.* Rotterdam, the Netherlands: Sense Publishers.

Wood, T. (Ed.), Jaworski, B., Krainer, K., Sullivan, P., & Tirosh, D. (Vol. Eds.). (2008). *International handbook of mathematics teacher education.* Rotterdam, the Netherlands: Sense Publishers.

BARBARA JAWORSKI

MATHEMATICS TEACHER EDUCATOR LEARNING AND DEVELOPMENT

An Introduction

This chapter serves as an introduction to Volume 4 as a whole. The stated aim of the volume is to open up the practice of mathematics teacher educators to scrutiny and critique: locating our practices internationally, identifying issues both local and global, and seeing ourselves as practitioners alongside teacher practitioners with whom we work. The chapter begins with a brief discussion of mathematics teacher educator knowledge, learning and development, locating this both historically and theoretically. It goes on to address the three sections: Challenges to and theory in mathematics teacher education; Reflection on developing as a mathematics teacher educator; Working with prospective and practising teachers: what we learn; what we come to know; and presents a short account of each chapter. The chapter ends with a vision of teacher educator practice for the future.

Mathematics teacher educators are professionals who work with practising teachers and/or prospective teachers to develop and improve the teaching of mathematics. They are often based in university settings with academic responsibilities. The qualities required of teacher educators are in many respects the same as those required of mathematics teachers. They need to know mathematics, pedagogy related to mathematics, mathematical didactics in transforming mathematics into activity for learners in classrooms, elements of educational systems in which teachers work including curriculum and assessment, and social systems and cultural settings with respect to which education is located. In addition they need a knowledge of the professional and research literature relating to the learning and teaching of mathematics, knowledge of theories of learning and teaching, and knowledge of methodologies of research that inquires into learning and teaching in schools and educational systems. In many cases mathematics teacher educators have been mathematics teachers themselves and bring a profound professional experience to their work with teachers. Such experience brings with it credibility: teachers can see that teacher educators have themselves grappled with the practical realities of classroom settings and the systemic demands that teachers face. Educators by the very nature of their roles in higher education do not know intimately the students with whom teachers work or the particularities of the schools where teaching takes place. Teachers are the ones with profound knowledge in these areas, although they might reasonably expect that educators

B. Jaworski and T. Wood (eds.), The Mathematics Teacher Educator as a Developing Professional, 1–13.
© 2008 Sense Publishers. All rights reserved.

have past experiences that enable an empathetic appreciation of the day to day issues teachers face. In a synthesis of the theoretical and the practical, MTE's develop knowledge of how teaching and new practices are learned, how such learning occurs, and the associated pitfalls and this parallels teachers knowing and sensitivity to students learning. The diagram in Figure 1 represents the relationships I have just sketched.

From the perspective of this diagram which is necessarily simplistic, educators draw on their knowledge in A and B to promote growth of knowledge in B and C. Location of their own knowledge in B ensures a recognition that educators are learning as well as teachers. However, the direction of influence from A to B and C is not the only one we need to consider: we need to be aware of the mutual and reciprocal influences of knowledge in practice of educators and teachers in B and C, a complexity that is only implicit in this diagram. The knowledge of *prospective* teachers is also located in B and C and, for prospective teachers, both educators and practising teachers are educators. This diagram serves to introduce

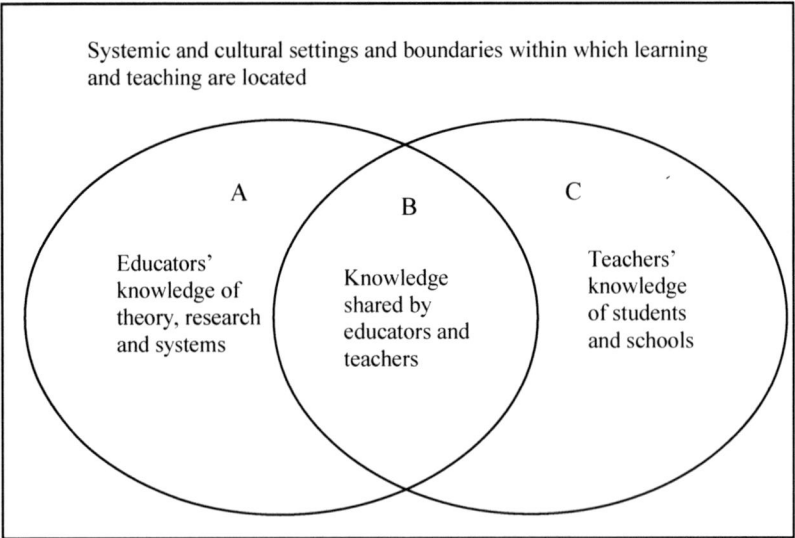

Figure 1. Knowledge in teacher education

the complexity of factors and issues that this volume addresses. I return to this diagram in Chapter 18, seeking to synthesise from the collection of chapters.

Teacher educators have responsibility towards development of knowledge in teaching. This might be seen as developing the knowledge of teachers – teaching teachers – and certainly this is part of the picture. After six years of editing the *Journal of Mathematics Teacher Education* (JMTE), I have read a considerable number of papers from teacher educators who have researched their own programmes of teaching teachers, and reported the outcomes. Such papers report from programmes for prospective teachers and from programmes for practising teachers; programmes that take several years to complete, or short courses or summer institutes; programmes that encourage teacher reflection, teachers' involvement in research, and extensive use of technology. In most such papers, findings report outcomes for teachers from engagement in the programme and raise issues for teachers or for teacher education more generally. Some papers advise other teacher educators as to approaches that can achieve certain outcomes or of where to be aware of particular issues. A very few papers reflect critically on the teacher education process, on what teacher educators themselves learn from engaging in teacher education, through reflecting on their own practice, and through research into the programmes they design and lead. And even fewer papers report on the learning of the teacher educator or on programmes designed to educate educators. Authors in this last category are highly represented within the chapters of this volume.

In 1999, in a chapter in an edited volume on *Mathematics Teacher Education* (Jaworski, Wood and Dawson), Sandy Dawson wrote as follows:

> There seems to be a culture of mathematics inservice education hinted at in the programmes described earlier in the book. Reading between the lines of those descriptions one could argue that this manifestation of inservice culture seems to have the following basic principle: there is something wrong with mathematics teaching world-wide and we, as mathematics educators, must fix it. Many mathematics teachers have bought into this culture. Such teachers seem to be seeking new ways to fix their practice. But this places mathematics teachers in a relationship of co-dependence with mathematics teacher educators. Mathematics teachers need someone to fix them, and mathematics teacher educators need someone to fix. The two groups seem made for each other. (p. 148)

The heavy irony here was deliberate. It suggested that many programmes were designed to pass on knowledge and expertise from the educator to the teacher.

In the same volume, Terry Wood and I synthesized themes and issues from papers in the volume, and reported that many papers claimed to ground programmes in a *constructivist* theory of knowledge and learning. Cutting through the complexities, such perspectives eschewed a transmissive approach to teaching and sought centrality of the learner in the learning process: as with mathematics teaching in classrooms, so with the teaching of mathematics teachers (see pages 138-139). Papers spoke of "'challenging teachers; conceptions' and providing opportunities for teachers to reflect on and possibly change these conceptions" (Amit & Hillman, 1999. p. 18). Other papers spoke of teachers: "He considers that materials allow ... the construction of pupils' own reasoning" (Serrazina & Loureiro, 1999, p. 53). Carter and Richards (1999, p. 69) spoke of their inquiry approach being "framed in a constructivist epistemology", indicating a focus on "sense making, building up of one's ideas". Irwin and Britt (1999, p.91) set out in their teacher education programme to test a suggestion that "if teachers were to develop a new teaching methodology compatible with a constructivist view of learning, the professional development itself should be constructivist in nature". On the face of it, such programmes might be seen as countering what Dawson expresses as "designed to pass on knowledge and expertise from the educator to the teacher". The philosophy was rather to engage teachers in activity through which they could develop their knowing in the domains detailed above.

The book referred to above resulted from a working group of 10 years' duration at yearly Psychology of Mathematics Education (PME) conferences (1986-1996). The mathematics education literature in those years was theoretically dominated by constructivist theory, evidenced by seminal work including Cobb, Wood and Yackel, 1990; Confrey, 1990; Davis, Maher & Noddings, 1990; Glasersfeld, 1987; Steffe and Thompson, 2001. It seemed unsurprising, therefore, at the time, that so many papers presented a constructivist perspective on teaching-learning at both classroom and teacher education levels. What *is* perhaps surprising was that very few papers at this time challenged the over-riding constructivist perspective or offered alternative ways to theorise teacher education. While conceptualising learning through constructivist theory, teaching and teacher education was premised on educators creating programmes in which teachers would in a sizeable variety of ways encourage their pupils' construction of mathematics. The pedagogies that this should involve were the focus of many of the programmes reported; inculcating or fostering such pedagogies was the aim of the programmes. The successes of, and/or issues raised by such pedagogies were largely reported without a concomitant questioning of the underlying rationale to which the irony of Dawson above draws attention.

Almost a decade later, JMTE still receives many papers in the same vein. However, there has been a shift. One obvious difference is that constructivism has moved from a largely cognitive, psychological focus to take into account social contextual and institutional factors, often relating to the work of Vygotsky (e.g. Cobb, 1994; Confrey 1995). In parallel, sociocultural theories, rooted in the work of Vygotsky and followers have become better known and understood in mathematics education, with a challenge, implicit or explicit, to constructivism (e.g., Forman & Ansell, 2001; Lerman 1996), and social, cultural, political and policy issues have become more evident in the mathematics education literature (e.g., Cooper & Dunne, 2000; Vital & Valero, 2003; Zevenbergen, 2001), perspectives of teacher educators have moved into more social frames (e.g., Blanton, Westbrook & Carter, 2005; Goos, Galbraith & Renshaw, 1999) with recognition also of the wider influences of system and society (Krainer, 2006; Lachance & Confrey, 2003).

While theoretical issues such as these are relatively obvious, the main difference that I would point to is less easy to articulate. It concerns a shift in tone and nuance in the ways educators write about educating teachers. There is less of a surety of models of practice that educators promote with teachers and much more a sense of uncertainty. With this uncertainty comes, almost paradoxically, a strength of purpose, new ways of speaking about mathematics teacher education, and new paradigms of practice. These build on notions of reflection for both teachers and teacher educators, on teacher-as-researcher and simultaneously educator-as-researcher positions, and on growing recognitions of epistemology, of complexity and the importance of not trying to oversimplify (e.g., Brown & Coles, 2000; Davis, Simmt & Sumara, 2006; Potari & Jaworski, 2002; Steinbring, 1998).

The chapters in this volume have all been written, by invitation, for this volume. So, with regard to historical development, they are state of the art. Some authors have written before about teacher educator development; others are writing about it explicitly for the first time, albeit drawing on a wide experience of research in and critical analysis of teacher education programmes. An aim of this volume is to open up the practice of mathematics teacher educators to scrutiny and critique: locating our practices internationally, identifying issues both local and global and seeing ourselves as practitioners alongside the teacher practitioners with whom we work. Roles of teacher educators and the dilemmas, issues and anomalies we face are central to this volume.

The volume is organised into four sections.
- Challenges to and theory in mathematics teacher education;
- Reflection on developing as a mathematics teacher educator;
- Working with prospective and practising teachers: what we learn; what we come to know;
- Synthesis.

CHALLENGES TO AND THEORY IN MATHEMATICS TEACHER EDUCATION

The six chapters in this section take a broad focus within the discipline of mathematics teacher education, focusing particularly on challenges to, or within, the discipline or on particular theoretical perspectives, or both. Where they draw on particular programmes or initiatives, it is largely to exemplify more general theories, issues or perspectives. Theoretical perspectives differ across these papers.

The first chapter, written by Martin Simon, takes a critical look at the challenges facing mathematics teacher education and hence mathematics educators. Coming from a cognitive constructivist perspective, and focusing on "courses and workshops for teachers in which teacher educators aim to promote particular mathematical and pedagogical concepts, skills and dispositions", Martin highlights four key areas of research-based knowledge that he sees as currently insufficient for teacher education efforts that promote envisaged reforms. He roots his arguments in perspectives of teachers and educators that are *perception-based* or *conception-based*. In the former, perceptions of how things are for the teacher or educator dominate what is possible in fostering the learning of others. In the latter, teachers or educators struggle to address how other learners develop concepts and to know the prior concepts on which new knowledge is to be built. The rationale here is psychological drawing on Piaget's concepts of assimilation and reflective abstraction. The chapter opens up the conceptual frame in mathematics teacher education and challenges teacher educators to be aware of psychological (under)currents in their development of educational programmes and design of research.

From a psychological perspective, cognitive and affective, John Mason focuses on three *As, attention, awareness* and *attitude,* in addressing learners', teachers' and teacher educators' interactions with each other and with their substance of learning. He uses a metaphor of human psyche as *chariot* to write about the human body (chariot), senses (the horses), cognition/awareness (the driver), imagery (the reins) and so on. He draws attention to the act of *drawing attention to*, recognising that, in any learning-teaching moment, attention of learner and teacher may be quite different signifying different awarenesses. A teachers' awareness *in discipline* can allow the teacher to direct students' attention to key aspects of the action in which they engage through a careful choice of tasks and tools; in doing so the teacher has to be aware of students' awareness in the action of engaging with mathematics. The complexity of these states of awareness is compounded when we consider the teacher educator working with teachers to draw-attention-to/bring-to-awareness aspects of discipline including those of attention and awareness. John refers to these educative stances as second and third order disciplines.

Ruhama Even writes about the education of educators of practising teachers with particular reference to a unique programme, *Manor*, in Israel. She draws attention to both the lack of attention in the literature to the education of teacher educators and to the (concomitant) lack, internationally, of such education programmes. Three problematic aspects are defined as: almost no research on the education of mathematics teacher educators; the ill-defined nature of the field of educating practicing mathematics teachers; and a lack of information on the practice of mathematics teacher educators working with practicing teachers. Ruhama uses the term "knowtice" to capture theoretically a unique blend of knowledge and practice, a pragmatic view, in the educative process using the Manor programme as a paradigmatic example. She traces elements of conceptualization, recruitment and considerations of curriculum and practice, the latter both in seminars for prospective teacher educators in educative settings and in practice in settings involving activity with practicing teachers. The meta-layers of educators educating educators are clearly distinguished. The chapter ends with a focus on future needs with attention to key research questions relating to the nature of the field, its knowledge base and the kinds of practices it encompasses.

Merrilyn Goos's chapter focuses on the use of sociocultural theories to analyse, explain and promote development in mathematics teacher education. Within a Vygotskian framework, she focuses particularly on the work in Valsiner in three developmental "zones", Zone of Free Movement (ZFM), Zone of Promoted Action (ZPA) and Zone of Proximal Development (ZPD), and refers briefly to studies in mathematics teacher education that have used such theory in their analyses. She draws on her own research to offer analyses of the learning of two beginning teachers using the inter-relationships in these zones. Relationships between the zones enable a holistic addressing of teachers' settings, actions and beliefs and ways in which these relationships might change over time across a variety of contexts. Merrilyn extends zone theory to situations and roles of the mathematics teacher educator, suggesting that zone theory can offer a means to analyse, explain and promote the activity of the mathematics teacher educator researcher.

MATHEMATICS TEACHER EDUCATOR LEARNING

The focus of Orit Zaslavsky's chapter is on the design and use of tasks to promote learning of teachers and concomitantly that of teacher educators, relating implicitly to the influence of design theory in the learning sciences (e.g. Kelly 2003; Wood and Berry, 2003). Her chapter begins with consideration of seven broad themes that address qualities and kinds of competence and knowledge that mathematics teacher education seeks to promote in prospective and practising teachers in a broad sense. Within this panorama she identifies the role of carefully designed tasks to mediate between on the one hand *facilitating teacher learning* and on the other *researching teacher practice and knowledge*. The seeking for, design and use of tasks which have the aim of promoting teachers' learning is a formative process for the teacher educator from which knowledge and awareness develop. Orit draws on her own experience in designing a task for teachers that went through several metamorphoses or iterative cycles through research into its use and the feeding back of knowledge gained in practice to the design process. The increasing sophistication of the task mirrored the growth of understanding of the educators, analysed with respect to the seven themes. The chapter ends by considering the demands on teacher educators as facilitators of teacher learning.

The final chapter in this section, by Olive Chapman, offers a review of a selected set of papers focusing on teacher educators' programmes for prospective teachers that take a cognitive perspective. The review addresses what we can learn as teacher educators from what is reported in this research, categorising the range of factors under broad headings of characteristics of instructional practices, and characteristics of learning outcomes. These instructional practices and learning outcomes include both *teacher working with classroom student as learner of mathematics* and *teacher educator working with prospective teachers as learners of mathematics teaching*. Olive points out that the reports from these studies present the outcomes of the research as sources of learning for those reading the reports rather than for those writing the reports. In other words, teacher educators as researchers take mainly an *outsider* position in reporting their research; only a few reflect the insider position of teacher educator learning and its impact on their practice.

REFLECTION ON DEVELOPING AS A MATHEMATICS TEACHER EDUCATOR

Consideration of *insiders* and *outsiders* takes us neatly into the second section of this volume. Here the focus is overtly on the mathematics teacher educator as an insider researcher developing practice through research in and on practice. The chapters here reflect personal journeys in mathematics teacher education through which the teacher educators themselves develop their professional practice.

The first chapter, by Ron Tzur, picks up an earlier story (Tzur, 2001) of Ron's own development as a Mathematics Teacher Educator and expands it theoretically to conceptualise and characterise a construct called PALP (Profound Awareness of the Learning Paradox). The learning paradox (Bereiter, 1985) concerns the need for a conceptual structure at least as complex as the concept to be learned, posing a serious challenge for the intuitive teaching of reform-minded teachers. PALP leads mathematics teachers or teacher educators to construct tasks that might seem counter-intuitive to those lacking such awareness, but which result in learning of desired concepts by the students or teachers who are the focus of the tasks. Through three cases in which he worked with teachers in designing tasks for students, Ron charts his own developing awareness of addressing the learning paradox, and synthesises five 'capacities' that serve as goals for mathematics teacher educators' development.

Razia Fakir Mohammad writes about her own personal development as a teacher educator in Pakistan, working with teachers in Pakistani schools and in developmental courses at her university. Starting from her personal philosophy for education, a belief in co-learning as an approach to working with teachers, Razia describes episodes from her work that raise issues for her moral and ethical stance and create dilemmas in her practice. We see the Pakistani context itself with its hierarchies of respect and authoritarian school systems offering major challenges to her work as a teacher educator. Through her interactions with teachers in which she struggles with her own aims and beliefs related to the teachers themselves and the students they have responsibility to teach, Razia charts a personal growth of understanding for dealing with contradictory forces and living with the outcomes. She ends with a theoretical synthesis to generalise this process.

Konrad Krainer also offers a personal account of his development as a mathematics teacher educator, juxtaposing it with a model of growth and development within a mathematics education discipline. Using a model of seven nested domains with mathematics at the centre, he traces both his own development and a (historical) development of research in mathematics education. Here we see the focus changing from early studies related to mathematical content to current focuses on the roles of teacher educators and development of mathematics learning and teaching within school and educational systems. The growth of individuals, mathematics teachers and teacher educators, can have only a minor influence so long as perspectives remain at local levels. Konrad offers a powerful thesis that development will only be sustainable with an attention to the wider systems, structures and populations within the societies that mathematics serves. As long as mathematics remains unknown, elitist, separatist, feared by significant members of society and education, it can not become accessible to all, nor bring its powerful potential for communication to fruition.

Simon Goodchild speaks of achieving "good research". This sits alongside aims for achieving *better* mathematics teaching and learning. Simon's own biographical development shows a progression from research characterised as "data extraction" – extracting and analysing data from research settings with minimal involvement of the participants in those settings, the teacher and students – to developmental research in which all participants are engaged fundamentally in the research and influenced by this involvement to improve practice. As an example, he focuses on a particular large-scale research project in Norway in which didacticians and teachers worked together to inquire into practices and processes of teaching mathematics in school classrooms and the associated learning outcomes for pupils. Simon uses this project to illustrate key facets of a developmental research process and relates these to principles in critical theory through which emancipation and empowerment are theoretical goals and practical gains.

WORKING WITH PROSPECTIVE AND PRACTISING TEACHERS: WHAT WE LEARN; WHAT WE COME TO KNOW

In this section the focus moves more overtly into the practice field, albeit with a continuing theoretical dimension. Here we find commentary on programmes and practices and on the perspectives and rationalisations that underpin them. Practices include programmes for prospective or practising teachers and address forms of knowledge, professional competencies and dilemmas of practice that challenge teacher educators in their professional lives.

Christer Bergsten and Barbro Grevholm identify a key concept, which they call the 'didactic divide', that separates teachers' learning in theoretical contexts based on their university study from that in practically-based contexts relating to classrooms and pupils. Their chapter traces in the literature a progression from competency models relating to teacher knowledge and competency, through recognition of the interactivity of teachers and teacher educators in teacher education programmes to consideration of the practices of teachers educators and their associated knowledge. The complexity of educational processes is emphasised throughout with attention to the problems of fragmentation and desire for holistic approaches. The chapter presents a range of approaches through which power differentials are reduced and learning of pupils, prospective teachers, practicing teachers and teacher educators take place within mutually sustaining environments that present opportunities to overcome the didactic divide.

Amy Roth McDuffie, Corey Drake and Beth Herbel-Eisenmann report on their collaboration as a threesome in reflecting on and analysing their design and teaching of mathematics methods courses for prospective elementary teachers. These are typical of such courses across the United States and thus fit a cultural stereotype, subject to institutional and systemic factors in the US educational system. The three MTEs compared and contrasted their design and thinking about the courses, reflecting on both the material of the courses and the motivational thinking that underpinned activity and action. As well as presenting a detailed account of what they do, how they do it and the associated reasons related to research on prospective elementary teacher development, the three MTEs acknowledge their own learning through collaborative preparation of this chapter. Particularly they characterise the developmental process in learning to teach as "teaching as learning in practice" and speak of using inquiry approaches throughout their practice with prospective teachers. Thus as they encourage prospective teachers to learn through inquiry into the learning of their students of mathematics in classroom settings, the three MTE's inquire into their own practice and become more knowledgeable about their activity and its development.

The chapter by Pat Perks and Stephanie Prestage focuses on tools for learning, including the learning of pupils, prospective teachers and their own learning as teacher educators at secondary level. In these three layers, learning in an inner layer generates the tools for the next outer layer. Within a Vygotskian theoretical frame, these authors recognise the teacher/educator role in promoting scientific concepts and simultaneously nurturing spontaneous concepts. Tasks and associated activity are designed to encourage generation of labels to describe key elements of pedagogy which can then be available for use in the classroom or university seminar. Pat and Stephanie reflect on their own use of tools, particularly the tool of writing, in which they encapsulate and synthesise concepts that are in process of formation.

Victoria Sánchez and Mercedes Garcia use the idea of 'dilemmas' in education to reflect on and analyse their design and development of a teacher education programme for prospective primary teachers. They define a *dilemma* as denoting a potential action and opting for a practical strategy to manage inconsistencies between beliefs and practice. The dilemmas they negotiate take them deeply into consideration of three layers of design and development in their programme: 'given' by the institution versus 'chosen' by the teacher educator; adapting from other teacher education programmes versus building on research; and problems to be handled versus problems for research. In discussing these three dilemmas, they delve deeply into the literature in mathematics teacher education, searching what others have written for insights into and support for their own trajectories of thinking. We see in their account a broad consideration of theory and practice internationally as it is represented in this literature. Finally they come through the account of design and development of a programme, and the many choices it has encompassed to reflection on their own learning as educators in a mathematical context for primary teachers. This returns to theoretical perspectives in their account of design particularly those of community of practice, extending now to community of inquiry as they examine their own learning through inquiry in their developmental research process.

Paola Sztajn writes about 'caring' relations. Drawing on the work of Nel Noddings, she analyses the role of teacher educator as *carer* for the teachers with whom he or she works. Importantly, she makes clear that the caring relationship, for its success, depends on the foundational human quality of reciprocity between the carer and the cared-for: this acknowledges an overt recognition, by the cared-for, of the caring approach such that a reciprocal caring emerges. Offering three examples from stages in her own development, working with practising teachers in a variety of settings and contexts, Paola shows how caring relationships can differ and how the nature of the (reciprocal) relationship affects the outcomes of the mutual process. The apparent contradiction of carer putting aside own goals to facilitate goals of the cared-for resulting in achievement of the carer's own goals is carefully analysed.

The final chapter in this section is written by Sandy Dawson who, following an extensive career as mathematics teacher educator, has recently spent eight years working with teachers, teacher educators and the people of the island territories widely spread across the Pacific Ocean. Building on earlier work with First Nations communities in Canada, Sandy discusses his educative approach on the building of mutual respect and trust between peoples. The chapter draws on episodes from this

MATHEMATICS TEACHER EDUCATOR LEARNING

work to show how peoples from very different cultures come together to learn mathematics and teaching mathematics, and how Sandy as a teacher educator came to know and understand how to generate trust and respect in relation to new and strange customs. The chapter introduces readers to the practice of the "wisdom circle" and the theory of "subtle" as underpinning the growth of knowledge in overtly conscious cultural settings. This chapter, together with the ones from Razia Fakir Mohammad and Paula Sztajn, emphasises the social, contextual and affective sides of teacher education and the complex functionings of, and challenges for, a mathematics teacher educator alongside teachers as partners in the educative process.

SYNTHESIS

The final section of the volume is one of synthesis and includes just one chapter which I have written. Here, I have taken on the demanding and (for me) exciting challenge of synthesising from the richness and diversity of ideas as I have seen them to emerge through my reading and re-reading of the chapters in this volume. In these chapters we see both reinforcing commonality and differences in ways of seeing issues, particularly in terms of theoretical perspectives. It has seemed important to me to emphasise the complexity within our field and ways in which the theoretical perspectives help us to explain and tackle issues and tensions. The theme of the final chapter, that emerges from this process, is development of the mathematics teacher educator and its relation to teaching development. It sees teachers and educators both as learners and researchers in practice and suggests that co-learning between teachers and educators can be seen as the way ahead for developing mathematics teacher education practice. I see this volume not only as the first collection of writings concerning the mathematics teacher educator as a developing professional but also as an inspiration for further seriously focused work in this area.

REFERENCES

Amit, M., & Hillman, S. W. (1999). Changing mathematics instruction and assessment: Challenging teachers' conceptions. In B. Jaworski, T Wood, & S. Dawson (Eds.), *Mathematics teacher education: Critical international perspectives* (pp. 17–25). London: Falmer Press.

Blanton, M. L., Westbrook, S., & Carter, G. (2005). Using Valsiner's zone theory to interpret teaching practices in mathematics and science classrooms. *Journal of Mathematics Teacher Education, 8,* 5–33.

Brown, L., & Coles, A. (2000). Complex decision-making in the classroom: The teacher as an intuitive practitioner. In T. Atkinson & G. Claxton (Eds.), *The intuitive practitioner: On the value of not always knowing what one is doing* (pp. 165–181). Buckingham: Open University Press.

Carter, R., & Richards, J. (1999). Dilemmas of constructivist mathematics teaching: Instances from classroom practice. In B. Jaworski, T Wood, & S. Dawson (Eds.), *Mathematics teacher education: Critical international perspectives* (pp. 69–77). London: Falmer Press.

Cobb, P. (1994). Where is the mind? A coordination of sociocultural and cognitive constructivist perspectives. *Educational Researcher, 23,* 13–23.

Cobb, P., Wood, T., & Yackel, E. (1990). Classrooms as learning environments for teachers and researchers. In R. B. Davis, C. Maher, & N. Noddings (Eds.), *Constructivist views on teaching and learning mathematics* (pp. 125–146). Journal for Research in Mathematics Teacher Education, Monograph Series, No. 4. Reston, VA: National Council of Teachers of Mathematics.

Confrey, J. (1990). What constructivism implies for teaching. In R. B. Davis, C. Maher, & N. Noddings (Eds.), *Constructivist views on teaching and learning mathematics* (pp. 107-124). Journal for Research in Mathematics Teacher Education, Monograph Series, No. 4. Reston, VA: National Council of Teachers of Mathematics.

Confrey, J. (1995). How compatible are Radical Constructivism, Sociocultural Approaches, and Social Constructivism? In L. P. Steffe & J. Gale (Eds.). *Constructivism in education*. Hillslade, NJ: Lawrence Erlbaum Associates.

Cooper, B., & Dunne, M. (2000). *Assessing children's mathematical knowledge: Social class, sex and problem solving*. Buckingham, UK: Open University Press.

Davis, B., Simmt, E., & Sumara, D. (2006). Mathematics-for-teaching: The cases of multiplication and division. In J. Novotná, H. Moraová, M. Krátká, & N. Stehlíková (Eds.), *Proceedings of the 30th Conference of the International Group for the Psychology of Mathematics Education* (Vol. 2, pp. 385–392). Prague, Czech Republic: Charles University.

Davis, R. B., Maher, C., & Noddings, N. (Eds.)·(1990) *Constructivist views on teaching and learning mathematics* (pp. 125–146). Journal for Research in Mathematics Teacher Education, Monograph Series, No. 4. Reston, VA: National Council of Teachers of Mathematics.

Dawson, S. (1999). The enactive perspective in teacher development: 'A path laid while walking'. In B. Jaworski, T Wood, & S. Dawson (Eds.), *Mathematics teacher education: Critical international perspectives* (pp. 148–162). London: Falmer Press.

Forman, E., & Ansell, E. (2001). The multiple voices of a mathematics classroom community. *Educational Studies in Mathematics, 46*, 115–142.

Glasersfeld, von E. (1987). Learning as a constructive activity. In C. Janvier (Ed.), *Problems of representation in the teaching and learning of mathematics* (pp. 3–17). Hillsdale, NJ: Lawrence Erlbaum.

Goos, M., Galbraith, P., & Renshaw, P. (1990). Establishing a community of practice in a secondary mathematics classroom. In L. Burton (Ed.), *Learning mathematics: From hierarchies to networks* (pp. 36–61). London: Falmer Press.

Irwin, K. C., & Britt, M. S. (1999). Teachers' knowledge of mathematics and reflective professional development. In B. Jaworski, T Wood, & S. Dawson (Eds.), *Mathematics teacher education: Critical international perspectives* (pp. 91–101). London: Falmer Press.

Jaworski, B., & Wood, T. (1999). Themes and issues in inservice programmes. In B. Jaworski, T Wood, & S. Dawson (Eds.), *Mathematics teacher education: Critical international perspectives* (pp. 125–147). London: Falmer Press.

Jaworski, B., Wood, T., & Dawson, S. (1999). *Mathematics teacher education: Critical international perspectives*. London: Falmer Press.

Kelly, A (Ed.) (2003). Theme issue: The role of design in educational research. *Educational Researcher, 32*.

Krainer, K. (2006). How can schools put mathematics in their centre? Improvement = content + community + context. In J. Novotná, H. Moraová, M. Krátká, & N. Stehlíková (Eds.), *Proceedings of the 30th Conference of the International Group for the Psychology of Mathematics Education* (Vol. 1, pp. 84–89). Prague, Czech Republic: Charles University.

Lachance, A., & Confrey, J. (2003). Interconnecting content and community: A qualitative study of secondary mathematics teachers. *Journal of Mathematics Teacher Education, 6*, 107–137.

Lerman, S. (1996). Intersubjectivity in mathematics learning: A challenge to the radical constructivist paradigm? *Journal for Research in Mathematics Education, 27*, 133–150.

Potari, D., & Jaworski, B. (2002). Tackling complexity in mathematics teaching development: Using the teaching triad as a tool for reflection and analysis. *Journal of Mathematics Teacher Education, 4*, 351–380.

Serrazina, L., & Loureiro, C. (1999). Primary teachers and the using of materials in problem solving in Portugal. In B. Jaworski, T Wood, & S. Dawson (Eds.), *Mathematics teacher education: Critical international perspectives* (pp. 49–58). London: Falmer Press.

Steffe, L. P., & Thompson, P. W. (2000). Interaction or intersubjectivity? A reply to Lerman. *Journal for Research in Mathematics Education, 31*, 191–209.

Steinbring, H. (1998). Elements of epistemological knowledge for mathematics teachers. *Journal of Mathematics Teacher Education, 1*, 157–189.

Vital, R., & Valero, P. (2003). Researching mathematics education in situations of social and political conflict. In A. J. Bishop, M.A. Clements, C. Keitel, J. Kilpatrick, F. K. S. Leung (Eds.), *Second international handbook of mathematics education* (pp. 545–592). Dordrecht, the Netherlands: Kluwer Academic..

Wood, T. & Berry, B. (2003). What does "Design Research" offer mathematics teacher education? Journal of Mathematics Teacher Education, 6, 3, pp 195-199.

Zevenbergen, R. (2001). Language, social class and underachievement in school mathematics. In P. Gates (Ed.), *Issues in mathematics teaching* (pp. 38–50). London: Routledge/Falmer.

Barbara Jaworski
Mathematics Education Centre
Loughborough University

SECTION 1

CHALLENGES TO AND THEORY IN MATHEMATICS TEACHER EDUCATION

MARTIN A. SIMON

1. THE CHALLENGE OF MATHEMATICS TEACHER EDUCATION IN AN ERA OF MATHEMATICS EDUCATION REFORM

In countries engaged in promoting major reform of mathematics teaching practice, mathematics teacher education faces enormous challenges. Two issues critical to the success of reform-oriented teacher education are the adequacy of the knowledge base and the sufficiency of the human infrastructure. Both of these issues are important in thinking about the education of mathematics teacher educators. My examination of the knowledge base focuses on re-conceptualization of mathematics teaching, identification of key pedagogical understandings, and understanding teachers' conceptions. My point of reference is, for the most part, reform efforts in the United States. However, the issues that I raise are likely to be relevant in other countries engaged in mathematics education reform.

Many countries are currently involved in movements to reform the teaching of mathematics. These reform efforts tend to be oriented towards increasing conceptual learning, problem solving, and effective mathematical communication for all students (e.g., National Council Teachers of Mathematics, 2000). Perhaps the greatest obstacle to these reforms is that the teachers in their respective countries were educated under the traditional system of mathematics instruction. Thus, their view and understanding of mathematics, mathematics learning, and mathematics teaching are incompatible with those needed for implementing the envisioned reforms. The challenge for mathematics teacher education[1] is significant. The conceptual and cultural shift needed for traditionally educated teachers to participate effectively in envisioned reforms is profound and requires effective and sustained interventions. Success is not guaranteed. The challenges of teacher education in a reform context translate into a set of questions for mathematics teacher educator education. What do mathematics teacher educators need to know and who is prepared to conduct effective teacher education in line with reform goals?

In this chapter, I will focus on two issues critical to the success of reform-oriented teacher education: the adequacy of the knowledge base and the sufficiency of the human infrastructure. My examination of the knowledge base is not meant to be exhaustive; it is focused on issues of teacher conceptualization of mathematics,

---

[1] Throughout this chapter *mathematics teacher education* will refer to both the preparation of prospective teachers and the professional development of practising teachers, unless otherwise noted.

mathematics learning, and mathematics teaching. My point of reference is, for the most part, reform efforts in the United States. However, the issues that I raise are likely to be relevant in other countries. The intended mathematics education reform in the United States is extensive incorporating major changes in mathematics, mathematics teaching, and mathematics learning. It represents a commitment to equal opportunity for mathematics education for all students. Finally, it is important to note that the need for change is motivated in part by national concern for the capability of the US workforce.

TWO CATEGORIES OF MATHEMATICS TEACHER EDUCATION

Teacher professional development efforts can be sorted into two categories, those with process goals only and those that have content and process goals. Highlighting the former category are programmes based on the Japanese Lesson Study model (e.g., Fernandez, 2005; Yoshida, 2001) and programmes focused on teacher inquiry or teacher research (e.g., Dana, & Yendol-Silva, 2003). The basis of these programmes is that the engagement of teachers in inquiry-based, reflective practices in the context of professional support and communication structures can support the ongoing professional development of mathematics teachers. There are no a priori learning goals for teachers involved in these programmes (other than learning the processes of inquiry, reflection, etc.). The second category of professional development consists of courses and workshops for teachers in which teacher educators aim to promote particular mathematical and pedagogical concepts, skills, and dispositions.

In an era of reform, Category One programmes are useful, but not sufficient. Lesson Study in Japan is not meant to be a vehicle for major instructional reform. Rather, it is an opportunity for professional growth within an established system. A goal of teacher re-conceptualization of mathematics, mathematics learning, and mathematics teaching requires targeted interventions in which knowledge developed through mathematics education research can be promoted through innovative teacher education efforts. In this chapter, I focus on demands on the knowledge base for such efforts.

DEMANDS ON THE KNOWLEDGE BASE

I will discuss four areas of research-based knowledge that is currently insufficient for teacher education efforts that promote envisioned mathematics education reforms. I have attempted to arrange them in an order in which loosely the first is foundational to the other three, the second is foundational to the third and fourth, and so on. The suggested order does not preclude simultaneous progress being made in several categories.

1. Re-conceptualization of Mathematics Teaching

In order for any type of instruction to be effective, the knowledge to be learned must be clearly identified. For mathematics teacher education, it is knowledge about teaching (and learning and mathematics) that must be made explicit. Unfortunately, in the US and in a number of other countries, the vision and principles of teaching towards which the reform is aimed have not been well articulated. Rather, reform efforts focus on a commitment to remove direct instruction as the predominant form of instruction and on the promulgation of particular tools and classroom structures. Given that telling and showing students what they are to learn has been deemed ineffective, what is lacking is a clear sense of how to promote learning of mathematical concepts. Let me treat these points in greater detail.

Two areas of work in mathematics education led to the current mathematics reform: empirical research revealing that the majority of students were weak in conceptual understanding and problem solving and empirical and theoretical studies suggesting that human learners do not simply take in the understandings of teachers, but rather develop understandings on the basis of their prior knowledge and current goals and activity. These findings led to a rejection of direct instruction as the primary means for promoting conceptual understanding and problem solving and commitment to more active involvement of the learner in classroom mathematics. This movement was accompanied by an increase in the use of a number of instructional tools (e.g., non-routine problems, manipulatives, software environments, calculators) and organizational structures (e.g., collaborative groups, classroom discussions). These changes have had positive effects in some situations. Students have a broader range of mathematical representations to work with; they often see more diverse solution strategies; they have more opportunities to solve problems and communicate mathematical ideas. Nonetheless, the proliferation of these tools and structures provides no guidance as to how to design and carry out lessons that promote the learning of particular mathematical concepts. (Why this is not troubling to more mathematics educators and mathematics teacher educators is beyond the scope of this chapter). The lack of articulation of new visions/principles of teaching means that the goal of mathematics teacher education is under specified.

2. Identification of Key Pedagogical Understandings

As I asserted above, effective mathematics teacher education requires specification of the learning that it aims to promote. The lack of well-articulated models of teaching hampers the process of specifying goals. However, there is more to this issue. Currently, the identification of goals for teacher education courses, to the extent it is done explicitly, is generally done by teacher educators in the context of their practices and is not the focus of theoretical and empirical research reports. A review of articles in the *Journal of Mathematics Teacher Education* since its inception turned up very few articles on this subject. Some articles focus on process goals such as developing reflective practitioners (e.g., McDuffie, 2004).

Hiebert, Morris and Glass (2003) focus on learning to teach from practice. Within this broad objective, they identify specific requisite dispositions and skills.

Literature that focuses more on specific learning includes reports of fostering teachers' understanding of students thinking (e.g., Crespo, 2000). Schifter, Bastable, & Russell (cf., 1999) have developed teacher education curricula targeted at developing knowledge of students' thinking as teachers learn particular mathematics and reflect on related teacher interventions. The Cognitively Guided Instruction Project (cf., Carpenter, Fennema, Franke, Levi, & Empson, 1999) focused on providing teachers with research-based information on students' solution strategies.

What is missing almost entirely from this literature is an articulation of key pedagogical concepts[2] that might be promoted in teacher education. What is it that we want teachers to understand about teaching and learning? What are the key concepts that are fundamental to mathematics teaching that is consonant with current reform goals? To both demonstrate what it might mean to articulate pedagogical concepts and to engage the discussion of such concepts, I offer four examples of pedagogical concepts that have emerged in the context of research done by our research team. For each, I point to related research, if available, and I attempt to articulate the understandings (concepts) that would be the learning goal for the teachers involved.

2.1. Understanding of classroom norms and their negotiation.

Research. Yackel, Cobb, and Wood (1991) brought to the attention of the mathematics community the power of viewing mathematical classrooms in terms of the norms operating in the classroom. They identified particular norms and the process by which the norms were negotiated in a second grade mathematics classroom. Subsequently, a number of research projects have contributed in this area (cf., Edwards, 2007; McNeal & Simon, 2000). At this time, I am unaware of research that specifically looks at the development of teachers' conceptions of classroom norms and the negotiation of norms.

Articulation of the concept. Every classroom can be seen as having its own set of norms. These include specific expectations of the students and the teacher and what it means to be an effective member of the classroom community. Sociomathematical norms (Yackel & Cobb, 1996) include what counts as a mathematical justification, what counts as a different solution to a problem, and how mathematical validity is determined in the classroom community. Classroom norms (social and sociomathematical) are negotiated not imposed. Negotiation of norms is both explicit (discussion of expectations for behaviour) and implicit (i.e., affected by a chain of interaction). The negotiation of norms is an ongoing process, and takes place whether or not the teacher is aware of the process. Understanding

---

[2] "Pedagogical concepts" refers to concepts of mathematics teaching and learning and is meant to be distinct from mathematical concepts.

the notion of classroom norms, how they are negotiated, and the teacher's role in the negotiation process are key understandings for teachers that can contribute to the teacher's ability to establish a classroom community that supports conceptual learning, problem solving, and mathematical communication. The nature of the classroom sociomathematical norms also affects students' learning about the nature of mathematics.

2.2. Understanding Assimilation.

Research: Informed by the theoretical work of Cobb, Yackel, & Wood (1992), Simon et al, (2000) conducted an empirical study that proposed a construct for characterizing the perspective of many teachers participating in the mathematics education reform in the US, a *perception-based perspective*. From a perception-based perspective teachers see mathematics as existing as part of the external world, accessible to all. They consider mathematical relationships to be clearly displayed in certain situations, and therefore each person will see the same mathematical relationships in a situation chosen for such clarity. Finally, they take understanding to be paramount and consider understanding to result from first-hand experience with these carefully selected situations. Simon et al (2000) contrasted a perception-based perspective with what they considered to be a more powerful perspective for mathematics education, a *conception-based perspective*. A conception based perspective is grounded in the concept of assimilation (Piaget, 1970) and is the basis for the concept articulated in this section.

Articulation of the concept: Students extant knowledge/understanding determines what students perceive and understand, and the resources they bring to learning situations. Thus, students' perceptions of learning situations are structured by their current conceptions. (Students do not necessarily see and understand tasks, manipulatives, and other representations in the same way as their teachers do.) Mathematics is the result of how humans structure their experience. Teaching mathematics involves understanding students' extant conceptions and engineering situations in which the students can use their extant conceptions to build more advanced conceptions.

2.3. Understanding what is involved in learning a new mathematical operation.

Research: I know of no formal research on this pedagogical concept. The articulation of this concept derives from my teacher education practice and builds on the constructs of assimilation and conception-based perspective. It is part of a larger issue of what it means to conceptualize a new mathematical object. Many teachers approach teaching students a new operation by teaching *about* the operation. Thus, they teach *about* multiplication to students who have no concept of multiplication. They do not problematize the issue of helping students, who do not see the world as examples and non-examples of the operation, to do so.

Articulation of the concept: The name of an operation must label for the students a commonality that *they* perceive in a set of situations. Multiplication involves the perception of a commonality among a set of problem situations. For the student who does not see that commonality, there is no operation called multiplication[3]. However, one does not apprehend commonality of mathematical relationships the way one perceives commonality of colour. Rather, one can only see commonality in what one does with the situations (how one organizes the quantities, solves the problem, etc.). It is only when a student observes that what I did in this problem about the cost of 5 candy bars is "the same" as what I did (or would do) in this problem about 7 boxes of pencils, that she begins to have something to label as multiplication – that commonality.

4.4. Understanding the difference between reflective abstraction and empirical learning.

Research: Piaget (2001) postulated reflective abstraction as the process by which higher-level mental structures are developed from lower-level structures. In Simon, (2006), I discussed the distinction between reflective abstraction and *empirical learning processes*[4], situations in which students can observe a pattern of inputs and outputs without any understanding of why the pattern is generated.

Articulation of the concept: A mathematical concept is a *learned anticipation of the logical necessity of a particular pattern or relationship* (Simon, 2006). Understanding the logical necessity of a particular pattern or relationship is generated through reflective abstraction and not through empirical learning processes. Empirical learning processes result in knowing *that* something is true (i.e., that a pattern exists), not the anticipation of the logical necessity of a mathematical relationship.

The above pedagogical concepts are examples of ideas proposed as important goals for teacher education. They represent only one potential contribution in an important inquiry and discourse in mathematics teacher education.

3. Understand Teachers' Conceptions

By applying the concept of *assimilation* (discussed above) to teacher learning, we can claim that what teachers perceive in their classrooms, what they perceive in teacher education experiences, and how they make sense of what they perceive in

---

[3] Most operations are themselves abstractions that related sub-operations (e.g., quotitive and partitive division). Development of the sub-operations precedes abstraction of the commonality among them.

[4] I use the term "empirical learning process" rather than Piaget's term, "empirical abstraction," to indicate a category that is broader than Piaget's. Piaget (1978) used "empirical abstraction" (also called "physical" or "simple" abstraction) to refer to an inductive process that leads to abstraction of properties of physical objects or the material aspects of a physical action. In contrast, I use empirical learning processes to also include processes that are empirical in nature, but not based on physical objects or actions.

both types of situations is a function of their current conceptions. How well do we understand teachers' conceptions? In an era of reform, it is imperative that we understand the conceptions of teachers as they progress from traditional concepts of mathematics, teaching, and learning to concepts that are more compatible with the goals of the reform.

I use the construct of *perception-based perspective* (Simon et al, 2000), discussed above, to illustrate this issue. The construct derived from analyses of data from teaching experiments in teacher education classes with a combined group of prospective and practising teachers and from case studies with individuals from that group. The construct was a way to make sense of several different subsets of data that were initially puzzling.

One basic principle in our research is that teachers' thinking is coherent.[5] That is, what they think and do make sense from *their* perspectives. Leatham (2006) described a similar stance:

> In addition, the sensible system framework assumes that what one believes influences what one does, adopting Rokeach's (1968) description: "All beliefs are predispositions to action" (p. 113). This assumption does not imply, however, that an individual holding a belief must be able to articulate that belief, nor even be consciously aware of it. With this perspective, "beliefs cannot be directly observed or measured but must be inferred from what people say, intend, and do – fundamental prerequisites that educational researchers have seldom followed" (Pajares, 1992, p. 207). In order to infer a person's beliefs with any degree of believability, one needs numerous and varied resources from which to draw those inferences. You cannot merely ask someone what their beliefs are (or whether they have changed) and expect them to know or know how to articulate the answers.

Based on the assumption of coherence, if some piece of our data (i.e., a teacher action, explanation, question, etc.) does not seem to us as researchers to be a reasonable action, it means that we do not yet understand the teacher's perspective. Based on this principle, we developed a methodology, a variation of case study, called *accounts of practice* (Simon & Tzur, 1999) in which we explicate the teacher's perspective from the researcher's perspective.

Example: Perception-based Perspective. The power of the construct *perception-based perspective* is in its ability to explain how teachers make sense of teacher education opportunities and classroom experiences. Further, during design of teacher education opportunities, a construct of this type can lead to anticipation of teachers' responses and contribute to identifying appropriate goals for teacher learning. In this example I describe four different situations for which this construct proved useful.

---

[5] Researchers who collect data on "expressed beliefs" might have reason to argue for a lack of coherence in some situations. Our data collection is restricted to observations of practice and specific questions about what is to be or what was observed. Within those parameters our principle of coherence in teachers' thinking makes sense.

In teacher education programmes with which I have been involved, as in many other programmes, two of the important components of the programme are mathematical learning experiences for teachers, in which the teachers are mathematics students engaged in an inquiry mathematics class, and opportunities for teachers to conduct and analyze task-based interviews with students. The construct, perception-based perspective is useful for understanding some of teachers' responses to these situations.

Through the experiences of learning mathematics in an inquiry classroom, teachers often find that they *understand* for the first time areas of mathematics that they studied years before. As a result, they are committed to providing experiences of these types for their students. Towards this end, they often attempt to use the problems and/or representations that were used with them to teach their students. Of course this is problematic; their students do not approach the lessons with the same understandings that the teachers did. However, when we use the construct of perception-based perspective to think about this phenomenon, it makes sense. For the teachers, the mathematical concepts were contained in and transparently portrayed by the mathematical tasks and representations. They had the experience of *seeing* the relationships[6] in the tasks and representations. They considered their mathematical breakthrough to have been afforded by the opportunity for firsthand experience with the particular tasks and representations.

A second situation is teachers conducting and analyzing task-based interviews. In the resulting discussions, teachers often make insightful comments about the student's thinking. We also have conversations about opportunities for assessing students' conceptions in classrooms, where in-depth interviews are seldom a possibility. However, we find that this work seldom carries over to teachers' classrooms. Teachers' assessment of students' understandings, when practiced, is in service of knowing what students still need to learn, not, as we intend, what students have to build on and with. Again the construct of perception-based perspective is helpful. If the mathematical relationships are universally perceivable and accessible through particular tasks and representations, teaching does not need to involve attention to prior conceptions. The teacher's focus is on identifying those mathematical situations that will transparently display the concept to the student.

Another situation that was illuminated by the perception-based perspective construct involved the concept of number. The teachers and prospective teachers in our study had a great deal of difficulty thinking about how children develop a concept of number; they found it impossible to consider that children could not *see* number in the world around them (Simon et al, 2000). For the teachers, five was a property of 5 objects. They could consider that the children did not know the word "five" for identifying the number of objects, but they could not fathom that the children did not see the five-ness of the set of objects. From a perception-based

---

[6] Cobb (1989) pointed out that the "subjective experience" of *seeing what is true* is an essential part of doing and learning mathematics.

perspective, not seeing number is not an option; number exists in a set of objects and is there to be seen.

Finally, this situation of our use of the perception-based perspective construct is from a case study of Ivy (Heinz et al, 2000), a sixth-grade teacher (11-year old students). Ivy was committed to promoting her students' understanding of the canonical long-division algorithm. She gave the students problems and asked them to solve them using base-ten blocks. Most of them were able to do so. She then asked for students to describe their block solutions to the problems. As a student, reported the steps of his solution, Ivy put up the corresponding step in the algorithm. Of course, the students did not report the blocks solution in a way that emphasized the isomorphism, so Ivy asked leading questions and paraphrased liberally to be able to record parallel steps. Even with these interventions, Ivy's students did not see the relationship between the block solutions and the algorithm. Ivy's only recourse was to repeat and refine what she was doing. From a perception-based perspective, Ivy considered the juxtaposition of the block solutions and the algorithm to clearly portray the relationship between the two. She had no way to think about why this was not so for her students. She was not able or inclined to think about how what she understood afforded seeing these relationships, how what she had abstracted about the blocks solution affected what she saw in the algorithm and vice versa.

This example was meant to illustrate the power of having an empirically-based, theoretical construct that explains teachers' thinking across a variety of situations. When such a characterization of teachers' conceptions is contrasted with a more powerful conception, it contributes to understanding the nature of the learning that would need to be fostered in order to engender a significant change in the teachers' thinking. How to promote a change in teachers' assimilatory structures is a significant challenge for research in mathematics teacher education. The first step is research that illuminates these structures.

4. Promoting Teachers' Development of Important Pedagogical Concepts

This section is included to complete a discussion, although it is beyond the scope of recent and current work in the field. As the knowledge base for teacher education focused on reform becomes more adequate, that is, progress is made in articulating models of teaching, identifying key pedagogical concepts, and understanding the conceptions of teachers in transition (the areas discussed above), it becomes possible and important to study processes by which and contexts in which teachers can develop important pedagogical concepts. Studies of learning and teaching of specific concepts are the most difficult to do, because they involve not just the collection and analysis of evidence *that* learning took place, but also the collection and analysis of evidence of how the learning took place. Focusing on the process of change is difficult. Even in the area of mathematics learning, a domain in which the knowledge base is richer than in the learning of mathematics pedagogy, little progress has been made in understanding the processes through which learners develop new concepts.

Summary Comment: Impact on the Development of Mathematics Teacher Educators

The inadequate knowledge base is a serious impediment in the development of mathematics teacher educators. In the US teacher educators are being prepared in doctoral programmes without the conceptual frameworks that they require in order to work with prospective and practising teachers. Thus, the goals of their work and the developmental process that they endeavour to support and promote are under-defined. New teacher educators are likely to learn about activities that are considered to be educative for prospective and practising teachers. However, a focus on activities and broad goals (e.g., listening to students, accepting multiple solutions) is not sufficient for the important work of mathematics teacher education.

INFRASTRUCTURE FOR REFORM-ORIENTED MATHEMATICS TEACHER EDUCATION AND RESEARCH ON TEACHER EDUCATION

The ultimate goal of research on teacher education in a reform context is knowledge of how to effect large scale change. However, because the context of reform is one in which educators were originally educated through a traditional approach, there exists major problems of building a human infrastructure to carry out large-scale reform-oriented mathematics teacher education and research on large-scale efforts. Let me unpack this claim.

I work from the following set of assumptions. Each assumption is taken as evident by some, and not by others, who have key roles in the overall enterprise of improving mathematics instruction.

1. Having a deep understanding of mathematics is necessary, but not sufficient, to make one an effective mathematics teacher.

2. Being an effective mathematics teacher is necessary, but not sufficient, to be an effective mathematics teacher educator.

3. To be an effective researcher in mathematics teacher education one needs a deep understanding of relevant mathematics, mathematics learning, mathematics teaching, mathematics teacher education, and research in mathematics education and mathematics teacher education.

How do we get a sufficient number of well-prepared mathematics teacher educators and researchers in mathematics teacher education to have the capacity to do large-scale work? There is an aspect of the chicken-egg problem here. The traditional system generally does not prepare people for participation in the reform effort. The educational system would need to be changed to do so. However, until appropriately prepared mathematics educators are in place, the system will not change. Only through a well-funded organized effort can the infrastructure be slowly built up. This requires a commitment to ongoing development of the research base, a commitment to intensive education of mathematics educators, an intensive effort to prepare effective mathematics teachers to be teacher educators, and a cautious approach to upscaling of teacher education efforts.

The above discussion also raises the question of whether mathematics education doctoral programmes adequately prepare their candidates to do quality research in mathematics teacher education. Mathematics teacher education is more difficult and complex than mathematics education, because it subsumes all of the latter. Likewise, research in mathematics teacher education is more difficult and complex than research in mathematics education. Many recent doctoral graduates have gone into research in mathematics teacher education, probably in part because of its compatibility with their teaching responsibilities in mathematics teacher education. How prepared are they to do this work that is critical to progress of the reform movement? Do they have the preparation to do quality empirical and theoretical work in this area?

In a reform context, there are two additional factors that complicate the development of the infrastructure.[7] First, many of those involved in decision-making with respect to mathematics education or the regulation and funding of the overall enterprise do not really understand the reform and what it would take to make it happen on a large scale. Their understandings were honed under the traditional system. Second, large scale educational improvement is needed *now*, particularly for underserved segments of the population. These efforts cannot be put off until the research base and the infrastructure are more adequate. Of course efforts are made to use cutting-edge knowledge in these efforts. However, these efforts are hampered by the knowledge and infrastructure deficiencies discussed above. A further concern is that there is often a conflation of these efforts with the reform efforts that are ultimately needed and an unfortunate competition for funding priorities. This competition takes the form of *do we fund large-scale efforts targeting underserved populations now, or do we fund work that will progressively build the knowledge base and infrastructure?*

CONCLUDING REMARKS

In this chapter, I have tried to describe challenges facing mathematics teacher education in a context of mathematics education reform. In doing so, I have tried to point out prominent holes in the current knowledge base. Although research in mathematics education is a young field with significant problems and challenges, it is far ahead of research in mathematics teacher education in articulating the knowledge to be learned, mapping concepts in the domain, and understanding the conceptions of learners. I use my perspectives as a researcher in mathematics education to assist my consideration of research in mathematics teacher education.

---

[7] Of course, there are many other problems not discussed in this chapter.

The particular challenges of mathematics teacher education in a context of reform involve what Ball (1990) has called "breaking with experience." Research that is the engine of change has demands upon it that are not found in a system that endeavours to recreate itself. Developing mathematics teacher educators who are well prepared to prepare teachers for a new approach to mathematics education is a significant challenge as well.

REFERENCES

Ball, D. L. (1990). Breaking with experience in learning to teach mathematics: The role of a preservice methods course. *For the Learning of Mathematics, 10*, 10–16.

Carpenter, T., Fennema, E., Franke, M., Levi, L., & Empson, S. (1999). *Children's mathematics: Cognitively guided instruction.* Portsmouth, NH: Heinemann.

Cobb, P. (1989). Experiential, cognitive, and anthropological perspectives in mathematics education. *For the Learning of Mathematics, 9*, 32–42.

Cobb, P., Yackel, E., & Wood, T. (1992). A constructivist alternative to the representational view of mind in mathematics education. *Journal for Research in Mathematics Education, 23*, 2–33.

Crespo, S. (2000). Seeing more than right and wrong answers: Prospective teachers' interpretations of students' mathematical work. *Journal of Mathematics Teacher Education 3*, 155–181.

Dana, N., & Yendol-Silva, D. (2003). *The reflective educator's guide to classroom research: Learning to teach and teaching to learn through practitioner inquiry.* Thousand Oaks, CA: Corwin Press.

Edwards, J. (2007) The language of friendship: Developing sociomathematical norms in the secondary school classroom. In *Proceedings of the Fifth Congress of European Research in Mathematics Education (CERME 5).* Spain, European Society for Research in Mathematics Education (ERME).

Fernandez, M. (2005). Exploring lesson study in teacher preparation. In H. Chick & J. Vincent (Eds.), *Proceedings of the 29th Annual Conference of the International Group for the Psychology of Mathematics Education, Vol 2* (pp. 305–312). Melbourne, Australia.

Heinz, K., Kinzel, M., Simon, M. A., & Tzur, R. (2000). Moving students through steps of mathematical knowing: An account of the practice of an elementary mathematics teacher in transition. *Journal of Mathematical Behavior, 19*, 83–107.

Hiebert, J., Morris, A. and Glass, B. (2003). Learning to learn to teach: an "experiment" model for teaching and teacher preparation in mathematics. *Journal of Mathematics Teacher Education 6*, 201–222.

Leatham, K. (2006). Viewing mathematics teachers' beliefs as sensible systems. *Journal of Mathematics Teacher Education, 9*, 91–102.

McDuffie, A. R. (2004). Mathematics teaching as a deliberate practice: an investigation of elementary pre-service teachers' reflective thinking during student teaching. *Journal of Mathematics Teacher Education 7*, 33–61.

McNeal, B. & Simon, M. (2000). Mathematics culture clash: Negotiating new classroom norms with prospective teachers. *Journal of Mathematical Behavior, 18*, 475–509.

National Council of Teachers of Mathematics (2000). *Principles and standards for school mathematics.* Reston, VA: Author.

Pajares, M. F. (1992). Teachers' beliefs and educational research: Cleaning up a messy construct. *Review of Educational Research, 62*, 307–332.

Piaget, J. (1970). *Genetic epistemology.* New York: Columbia University Press.

Piaget, J. (1978). *Success and understanding.* London and Henley: Routledge and Kegan Paul.

Piaget, J. (2001). *Studies in reflecting abstraction.* Sussex, England: Psychology Press.

Rokeach, M. (1968). *Beliefs, attitudes, and values: A theory of organization and change.* San Francisco: Jossey-Bass.

Schifter, D., Bastable, V., & Russell, S. (1999). *Making meaning for operations.* Parsippany, NJ: Dale Seymour Publications.

Simon, M. A. (2006). Key developmental understandings in mathematics: A direction for investigating and establishing learning goals. *Mathematical Thinking and Learning, 8*(4), 359–371.

Simon, M. & Tzur, R. (1999). Explicating the teacher's perspective from the researchers' perspective: Generating accounts of mathematics teachers' practice. *Journal for Research in Mathematics Education, 30*, 252–264.

Simon, M., Tzur, R., Heinz, K., Kinzel, M., & Smith, M. (2000). Characterizing a perspective underlying the practice of mathematics teachers in transition. *Journal for Research in Mathematics Education, 31*, 579–601.

Yackel, E. & Cobb, P. (1996). Sociomathematical norms, argumentation, and autonomy in mathematics. *Journal for Research in Mathematics Education, 27*(4), 458–477.

Yackel, E., Cobb, P., & Wood, T. (1991) Small-group interactions as a source of learning opportunities in second-grade mathematics. *Journal for Research in Mathematics Education, 22*(5), 390–408.

Yoshida, M. (2001). American educators' interest and hopes for lesson study (jugyokenkyu) in the U.S. and what it means for teachers in Japan. *Journal of Japan Society of Mathematical Education, 83*(4), 24–34.

Martin Simon
New York University
Department of Teaching and Learning

JOHN MASON

2. BEING MATHEMATICAL WITH AND IN FRONT OF LEARNERS

*Attention, Awareness, and Attitude
as Sources of Differences between
Teacher Educators, Teachers and Learners*

My aim in this chapter is to probe beneath the surface of the, to me self-evident stance, that teachers cannot 'do' the learning for learners. It is convenient, for the differences I wish to highlight, to use the term learner to refer to students of mathematics being taught by teachers who themselves may or may not be influenced by teacher educators. Of course, at any time both teachers and teacher educators may be learning mathematics and more. I shall develop and use the constructs and discourse of attention, awareness and attitude to draw attention (sic) to differences between the experiential states of teacher educators, teachers and learners and between the kinds of actions they initiate in their respective communities of practice. The purpose of these distinctions is to highlight the developmental process of what I call mathematical being, which is the basis for and which informs being mathematical with and in front of others.

RATIONALE

Despite attempts to the contrary, teachers cannot make learners learn, nor can they do the learning for their learners. Authors of texts may provide tasks, exposition and examples intended to promote learning; and teachers may select tasks and interact with learners working on the tasks. However, none of these can ensure learning, and when conceived of as attempting to provide guarantees, are doomed to failure or at best, partial success. It seems patently clear that what teachers can do for learners, indeed perhaps the only thing they can actually do for learners, is to direct learners' attention. This they do by a variety of overt and covert means such as pointing, highlighting, and stressing, and by more or less subtle questioning and listening during activity arising from work on tasks. By their presence teachers can indirectly help learners maintain their attention where it might wander or when they might otherwise give up. Through their 'being' they can influence learners, especially when they are being mathematical with and in front of their learners.

When teaching mathematics, attention is directed to mathematical objects, relationships, properties, and reasoning. It can also be directed to manifestation of the presence of pervasive mathematical themes, and to forms of mathematical thinking, in the form of heuristics, use of natural sense-making powers (Mason & Johnston-Wilder, 2006) and habits of mind (Cuoco, Goldenburg, & Mark, 1996) and didactical tactics,[1] and how those choices are informed by reference to pedagogical constructs which together comprise distinctions, assumptions, beliefs and theories about teaching and learning. Where the choices being attended to match choices being made, there is consonance between what is practiced and what is 'preached': not only 'walking the talk' but 'talking the walk'.

When working with teacher educators, attention can be drawn not only to mathematics and to mathematical pedagogy, but also to theories and constructs, and to ways of working with both teachers and learners through different modes of interaction consistent with those theories and constructs, thus 'practicing what is being preached'. In each situation, attention may be drawn to actions which are being carried out with lesser or greater awareness, so that in the future the person might become aware of a possibility of choosing actions according to the circumstances (Mason, 1998). Overt actions as observed by others, implicitly or explicitly, form practices. By engaging with and in those practices, individuals become members of a community of practice (Lave & Wenger, 1991; Wenger, 1998) which may be local rather than global (Winbourne & Watson, 1998).

However, nothing can be ensured or guaranteed. It is perfectly possible to display behaviour without generating it from understanding of underlying structure. Thus a learner can carry out procedures without being able to justify, explain or reconstruct them, or, unhappily, without being able to modify them appropriately in changed situations. Similarly a teacher can enact pedagogical strategies and didactic tactics without be able to justify, explain or modify them appropriately, and even without being able to articulate them. Again similarly, a teacher educator can provide scholarly discourse about strategies and tactics without consistency between the theory being expounded and the behaviour being displayed, and without flexible response to changing situations.

METHODS

A word about my method of enquiry is in order. This chapter is not based on what is traditionally known as 'empirical studies', in the sense of collecting data through extra-spective observation of other people, followed by analysis consisting of classification. Rather, it draws upon my sometimes systematic observations of my own experience with myself and when working with others. These provide the background for contemplation of that experience, leading to the identification of phenomena, and attempts to find some informative organisation for them. The phenomena consist of similarities perceived by me in my interpretations of my observations of disparate events. Theoretical constructs are then used to organise, characterise, and account for these phenomena (Hewitt, 1994; Mason, 1991, 2002). As far as this exposition goes, the significant 'data' consist of events and incidents which come to your mind while reading, as resonated or triggered by what is read.

---

[1] I use didactic tactics to refer to topic or concept specific pedagogical actions, reserving pedagogic strategies for actions which can be used in several or many different topics.

However, to begin with phenomena would, for the purposes of this chapter, involve me in recording a large number of incidents which would exhaust both the space available and the reader, before getting to any analysis. Therefore I have elected instead to use only a few phenomena to illustrate points from the emerging analysis.

My analysis is based upon three aspects of human experience which I have been pondering for a very long time: the structure and role of attention, the nature of awareness (conscious and unconscious, explicit and implicit), and the influence of attitude (including intention and all aspects of affect) as manifestation of will. What makes mathematics education so frustrating yet intriguing is that people's behaviour is, despite myths to the contrary, rarely rational (in the sense of logically deduced from beliefs or dispositions; see Bennett, 1964; Nørretranders, 1998). Nor is behaviour ever simple, so rationalising observed behaviour by recourse to theories is at best an approximation, and attempts to locate cause-and-effect relationships are generally doomed.

The whole of my analysis, what could be called the theoretical framework for this chapter, is based on the ancient metaphor 'human psyche as chariot' which can be found in several of the Upanishads and which was developed by Gurdjieff, (1950, pp. 1192–1200). Plato (Republic II 488ff see Hamilton and Cairns, 1961, pp. 724–725) uses the setting of the crew of a ship to portray much the same idea. Among other elements in the extended analogy, there is an owner, a driver, the chariot itself, horses, reins and shafts. The chariot can be taken to refer to the body, whose general state of repair needs looking after if the chariot is to function. In the psyche this includes keeping techniques and skills fluent through rehearsal. The horses can be seen as representing the senses, and more generally as affect including emotions and feelings. These steeds provide the motive energy, but are prone to seeking grass and other delights along the side of the road. They therefore need directing. The driver does this by means of the reins attached to the harnesses, and the response of the chariot is governed by the shafts. The reins correspond to mental imagery, which is the means humans use to guide and direct their energies towards their goals. The shafts correspond to habits manifested through selves which transform motive energy (motivation) from the horses into patterns of behaviour (movement of the chariot). The owner (will) gives the driver general instructions, and the driver (cognition, awareness, though these terms need careful elaboration) provides the direction and guidance needed by the horses. Although I am confident that the western psychological tripartite division of the human psyche into enaction, affect and cognition is based on the Upanishad chariot metaphor, I have been unable to trace the origins of this division in western psychological literature. Study of attention (as the manifestation of will) is a relatively recent phenomenon although introduced by James (1890/1950). Piaget (1954/1981) held similar views of affect as the engine of action.

I have omitted from the title, and from this discussion, specific singling out of the enactive component of the psyche, the behaviour, not because it is unimportant, but because it is enough to work here on the three As. It is not, as far as I know, part of the Upanishads to consider the charioteer and owner as members of a social milieu influenced by the ever-evolving practices of the community, and there is insufficient space here to develop this aspect either. The three aspects of the human psyche being discussed here, attention, awareness and attitude, are human constructs: I attempt to delineate and discern components of an interactive complex which we call generally, 'being human', and specifically, 'being mathematical'. Some authors choose to stress the social as the source of influence on the individual, and in extreme forms, see the social as the only pertinent entity for educational purposes. Some authors choose to stress the individual as the agent responding to forces from its social and physical environment. Neither extreme is able to account fully for the rich tapestry of experience, in which the individual and the environment are engaged in a mutually evolving autopoesis or self-construction (Maturana & Varela, 1972; see also Cobb, 1995 and James, 1890/1950).

ATTENTION

Over many years it has become clearer and clearer to me that attention is the central concern for and of teachers. It came about because in every workshop I did for some 25 years, someone would ask at the end "where is your accent from?" I finally woke up to the fact that people were spending time in sessions attending not to what I was saying or asking, but to my accent. This led me to questions such as 'what are learners attending to?', with the assumption that if they and the teacher are attending to different things, then there is likely to be miscommunication at best because of little common ground, no matter what is taken-as-shared.

Attention is the manifestation of will, of intention. It is not a thing, but its influence can be inferred, though certainly not observed, in others. To a significant but not complete extent, it *is* observation: it is the medium through which observation takes place. In a certain sense 'you are your attention', or 'you are where your attention is': thus we have habits of speech such as "give me your attention", "thank you for your attention", and the more sarcastic "are you with us?". The military command "attention!" is intended to startle people into a heightened state of wakefulness. These uses all signal the centrality of attention in human experience. Of significance with attention is not only what is attended to, but the nature or structure of that attention, as will be developed shortly. How people attend and how they indicate what they are attending to and in what ways, is culturally based, but directed by personal propensities and interests.

In order to be sensitive to what learners are attending to, it is essential to be aware oneself not only of what is being attended to, but also how. Without this it is impossible to make sense of what learners are experiencing or reporting about that experience. What is more, attention is not an all-or-nothing experience: it has both a macro and a micro structure.

Macro Structure of Attention

Drawing on a variety of ancient sources, on my own experience, and on the results of exercises tried out with others, I began my studies with the macro-structure of attention (Mason, 1998). It is common experience that you can attend to more than one thing at once; it may be rapid serial succession as in a serial-computer operating system, and it may be in parallel; that is a matter for neuroscience to discern. The experience itself often seems like effectively parallel awareness. For example, on the one hand, while reading this text you can form an image of where you are likely to partake of your next meal, and you may also be able to imagine the walls around you painted in stripes or polka dots of different colours. You can recite a poem, nursery rhyme, or other familiar word sequence while at the same time revisiting mentally a holiday or other pleasant context. On the other hand, many people find that having a radio or television playing makes it difficult to think, especially when trying to read signs in an unfamiliar city while driving. Attention therefore can have multiple loci or be multiply directed.

A more subtle aspect of location is the experience of where attention is based. With a little work it is possible to become aware, for example, of being slightly outside yourself, perhaps looking from the side, or from a bit behind, or from a bit in front, or from inside the back or front of your head. Sometimes people speak of 'being up front' or 'laid back', frozen metaphors which capture something of the location of attention. Mostly we are unaware of the location of the 'search beam' which is our attention (Harding, 1961).

Attention can be expansive, like a broad beam of light, taking in a wide variety of impressions, and it can be narrow, as if confined like binoculars or a microscope. Independently of its breadth it can be sharply focused, especially on something specific or confined, but it can also be fuzzy, indistinct and encompassing: there can be a general 'sense' of what is being attended to without concern about details. Thus at the macro level, attention can vary in multiplicity, locus, focus and sharpness.

Learners' macro attention is usually caught up by their current action. Often they lose track of overall goals. Short-term memory and cognitive load are terms which have been used to describe the same phenomena. The same applies to teachers and to educators: immediate action may absorb so much attention that there is little if any left over for keeping track of what is happening, even for noticing sufficiently sharply so that moments become salient and so can be re-entered later during reflection. Thus the experience may be all of a blur.

When Bruner (1986, pp. 75–76) talks about a teacher 'being consciousness for two' he is describing ways in which a teacher's macro attention can keep in mind aspects of a situation which a learner lets go of because of the demands made on their attention by a local task. Thus someone who has not internalised addition or multiplication facts needs support of an external consciousness when exploring number patterns so that sufficient attention can be focused on calculation without losing track of the purposes of the calculation. By contrast, someone who needs little or no attention to carry out mental arithmetic does not need an external 'consciousness for two' to keep track and so can function more independently.

JOHN MASON

Micro Structure of Attention

Beyond these macro qualities, attention also has micro qualities, which are conveniently described in terms of how you are attending rather than to what you are attending. Informed by Eastern thinking (Bennett, 1966) I find it useful to distinguish five different ways of attending, which I call the structure of attention. These are holding wholes, discerning details, recognising relationships, perceiving properties and reasoning on the basis of agreed properties. There is a remarkable alignment of these ways of attending with descriptions of levels of geometric thinking (van Hiele, 1986) with the SOLO taxonomy (Biggs & Collis, 1982) and with the 'onion' model of understanding (Pirie & Kieren, 1994), but there is no space here to elaborate on these connections. The principle differences are that I see the different structures of attention, the different ways of attending, as following one another in rapid succession and in no particular order, rather than as sufficiently stable to be classified as 'levels' or 'layers' of reasoning, understanding or thinking.

Take for example the equation

$$\frac{\sqrt{3}+\sqrt{2}}{(\sqrt{3}-\sqrt{2})(5+2\sqrt{6})} = 1.$$

A first impression might be that it is very complicated. You might or might not actually be aware of components, but just that there are several square roots. The right hand side may attract attention because of its simplicity, leading to an affective response of disbelief or surprise, but it might actually be overlooked due to the absorbing or distancing complexity of the left hand side. When attention is allowed to play over the symbols, a relationship between the numerator and denominator might come to the fore. Some people are tempted into multiplying out the denominator, while others notice that one of the factors in the denominator is the conjugate of the numerator, so multiplying numerator and denominator by the numerator will eliminate the first bracket in the denominator. Some people might notice that the $\sqrt{6}$ is the product of $\sqrt{2}$ and $\sqrt{3}$ and so experience a slight ray of hope that the answer might indeed be 1. Others might go further and experience an association with the famous paper by Dedekind (1912) challenging people to prove that $\sqrt{6}$ does actually equal $\sqrt{2} \times \sqrt{3}$.

Associations, connections and links which 'come to mind' can be accounted for through either or a mixture of two processes: *metonymic* triggering and *metaphoric* resonance. These grammatical terms were re-introduced by Jakobsen (1951) and exploited by Lacan (1985) and Lakoff and Johnson (1980). Metonymies are associations which play on the surface of meaning, usually highly idiosyncratic and often based on syntactic elements such as homonyms. They are grammatical constructions in which one thing is referred to by some association such as a property or a specific part (literally, 'change of name'). The verb *trigger* is used with metonymies because the links are largely within the affect-emotional domain, very rapid and often not even conscious. By contrast, *metaphor*, which means literally 'carrying across' involves describing one thing in terms of another. Underneath every metaphor there is an extended analogy (originally a mathematical term for proportion), which is why metaphor is concerned with structure and meaning. The verb *resonance* is used to indicate the structural, semantic basis of how metaphors bring things to mind.

One form of attending is gazing, not really focused on anything in particular, yet taking in the whole. An assonant description is *holding wholes*. These 'wholes' may be parts of other entities which have been discerned. Thus you can gaze at a geometric diagram, waiting for it to 'speak to you' or at some particular part of a diagram. The experience is of 'waiting for things to come to mind'. Similarly you can gaze at a collection of symbols, and you can hold a situation in your mind and allow your subconscious to work away at it. You can gaze at the expression above without focusing on particular components. This may last only a few microseconds before you make distinctions, or discern details.

Making distinctions, which can be assonantly referred to as *discerning details*, is a structure of attention, a way of attending. Elements or aspects which might serve as useful sub-wholes are distinguished and identified, often through 're-cognition' either metonymically or metaphorically. Discerning details is neither algorithmic nor logically sequential. It doesn't happen all at once. You don't 'suddenly discern details', and then stop discerning details; rather, your attention is caught by some detail, allowing you to distinguish some aspect from some other aspect, and these distinctions participate in and contribute to subsequent attending. You may return briefly to gazing, whether at the detail or at some larger part. Discerning details applies to hearing and sensation as well as to sight. Most often attention darts around like a butterfly, settling for a brief moment, and then moving on. One of the things that has to be learned is to direct your own attention (using will) rather than be subject to will-o'-the-wisp movement. As learners of mathematics, people need to become familiar with worthwhile details to look out for; as teachers of mathematics people need to become familiar with ways of attracting and holding attention on worthwhile details; as teacher educators people need to become familiar with ways of drawing teachers' attention to worthwhile details, to ways of directing learner attention, as well as to integrate these with pedagogical constructs and theories.

An excellent example is afforded by the use of a simple partly shaded diagram to direct attention to the ways in which it is possible to change both the unit used to compare two things, and what constitutes the whole. In Figure 1, below, find something which is two-fifths of something else; something which is three-fifths of something else; something which is five-thirds of something else; something which is five-halves of something else (Thompson, 2002). The task develops from there. But many people are so inured to seeing the whole of the figure as the only whole that they balk at the third and fourth parts.

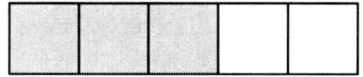

Figure 1. Partly shaded diagram

It is sometimes hard to imagine that you have not always discerned all the details of which you are currently aware, but if you put yourself in an unfamiliar situation you may become aware that experts are attending to details un-noticed by novices. This 'awareness' is of course vital for teachers and teacher educators. A significant if not major part of learning is discerning important details or aspects not previously discerned, including what it is that constitutes 'important'. Thus the algebra expert detects errors without even checking the calculations, and spots possible avenues of enquiry invisible to novices who are not distinguishing the same sorts of things. In the surds-fraction earlier, an expert discerns numerator, denominator, equal signs, the solitary 1, the repetition of the square-roots of 3 and 2 and the single square-root of 6 and immediately recognises some relationships.

Recognising of relationships between discerned elements is often an entirely automatic development from discerning details, but it is very difficult to be aware of a relationship between two or more terms when those terms are embedded in a mass of symbols or geometrical objects and your attention is not discerning at that level. Recognising relationships refers to specific relations between specific elements. In the fraction strip, it is the relations between 2, 3 and 5 which are crucial. In the surds-fraction someone might, for example, detect the presence of more than one square root of 3, or see the numerator and one term in the denominator as differing by a sign. A relationship can come to mind in its entirety, or there can be a coagulation of disparate discerned elements. The feature of relationships here is that they are embedded in the particular.

There is a subtle, but potentially important shift from recognising specific relationships between specific elements, to *perceiving properties*. When you are aware of a possible relationship and you are looking for elements to fit it, you are perceiving a property. Particular relations are seen as instances of general properties or abstract concepts.

Mathematical reasoning proceeds only when learners identify properties that can be used as the basis for reasoning. Without this, learners tend to dredge up everything they know, even when it has not yet been proved formally, or when it leads to circular reasoning. Formal reasoning identifies properties as either axioms or as 'already deduced', and then proceeds to use only them in further reasoning. This is *reasoning on the basis of specified properties*.

The significance of these subtly different forms of attention lies in the disparity of comprehension which can take place when teacher and learners are attending either to different things, or to the same thing(s) but in different ways. Thus if learners are trying to discern the elements that a teacher is relating together, they are unlikely to appreciate the relationship as well; if learners are recognising relationships between specific elements but not perceiving them as properties, then again they may not appreciate the import of what is being said and done.

A novice teacher observing a lesson may miss all sorts of subtle details evident to a more experienced teacher or educator. In particular, they may be unaware of what learners are attending to, or how, and oblivious to what the teacher may be attending to, or how. A teacher caught up in attending to learners' responses may not be attending to the negative (or positive) effects of the implicit choices they are making in how they interact with the learners. Similarly, an educator caught up in attending to teachers' responses may not be aware that their own behaviours may be influencing those responses in unexpected or unintended ways.

Overt and explicit directing of attention ("look here", "can you see ...?") are very local and ephemeral phenomena. There is no reason to expect that such direction will alert learners to the possibility of attending similarly in the future. Indeed, the refinements of the structure of attention introduced earlier serve as a reminder that even if learners are attending to the same thing as the teacher, they may not be attending in the same way.

AWARENESS

The word aware is usually used in connection with some degree of consciousness, so the word awareness then picks up that association. However, one of Caleb Gattegno's many insights is that as organisms, we have many 'awarenesses' which lie below the surface of consciousness (Gattegno, 1973, 1987, 1988). For example, somatic functioning such as breathing, perspiration, and heart-rate, not to say digestion and lymphatic defences all take place without our being conscious of them. Functionings like these which are controlled by the soma deserve to be considered awarenesses because they are dynamically responsive rather than mechanically determined. There are also internalised, habituated or automatic functionings to which we no longer need to attend which are also dynamically responsive and deserve to be thought of as 'awarenesses'. Awareness may be restricted in the degree to which the individual is specifically and explicitly aware and in what they could articulate about it. Thus a child may coordinate pointing and number-word-speaking without articulating anything to do with 'counting', nor even 'I am coordinating pointing and speaking'; a learner may draw axes, plot points and sketch straight lines through specified points without articulating anything to do with 'graphing'.

Awarenesses are influenced by and also influence affective responses, such as dislike (or stronger) of surds in the surd fraction. Finally there are functionings which arise through the use of human powers for making sense (literally as well as figuratively), which are developed and extended through social interaction, and which can also be observed as actions. For example, actions such as stressing and consequently ignoring different parts of the fraction strip, or treating the surd equation as an object to manipulate are functionings driven by awareness. Awarenesses form the backbone of disciplined modes of mathematical enquiry whether at the level of algebra, geometry, topology and so forth or at more refined levels associated with specific topics. Lakoff and Nunez (2000) exploit this in their conjecture that most mathematical concepts are based on bodily sensations and movement.

Gattegno (1987, 1988) proposed that disciplines such as mathematics or physics arise when people become aware of the fact that they have actions that they can carry out, explicit functionings, which he called awarenesses, even though they may be implicit and below the level of consciousness. Thus a discipline arises when people become aware of awarenesses which enabled or guided them to act in certain situations (Mason, 1998). The second order awareness means that they can begin to act upon and with the base awarenesses, to study them as phenomena, and, in Vygotskian terms, explicitly choose to make use of them. As people become aware of their below-the-surface functionings which are activated by the tasks they undertake and by the activity in which they engage and participate, they develop competence in and a sense of the associated discipline. Thus, the discipline of counting emerges from becoming aware of and making use of actions such as pointing, speaking, coordinating pointing and speaking, and speaking a particular word-sequence; the discipline of fractionating emerges from relating different wholes via a common unit.

Awarenesses trigger actions without our having to initiate them consciously, and often without any conscious knowledge at all. Since the term conscious is itself ambiguous, referring both to being awake rather than asleep, and to being mindful rather than oblivious or functioning in automatic mode (Claxton, 1984, 1997; Langer, 1997), I shall use the terms explicit awareness to refer to awareness which could be articulated, and implicit awareness to refer to awareness that is either not yet or no longer readily articulable. The importance of Gattegno's insight is that when we say that we are 'aware' of something, we obscure the origins of that 'conscious perception', and we overlook a myriad of functionings of which we are not explicitly aware but yet which comprise the awarenesses of our organism as a whole, since they initiate action of which we only later become explicitly explicitly aware.

Trying to distinguish between explicit and implicit awareness is fraught with difficulty. For example, it has been suggested that a multitude of experiments point to the fact that consciousness trails behind action and affect (Mandler, 1989). Nørretranders (1998) goes further and claims that it is an illusion that we (consciously) make choices at all. Rather, choices made in the moment, on the fly, as distinct from reasoned or debated choices, are made below the surface of awareness-attention-consciousness. Consciousness then makes up a narrative which gives itself the 'starring' role. Choices involving reactions or responses to stimuli are actually made on the basis of past experience before we become aware of them as choices. It is more convenient therefore to follow Gattegno and to use awareness to refer to the sensitivities of the organism to detect change and to respond to that change through action. Those actions may be well rehearsed habits or functionings, or may be cobbled together from fragments of such actions to suit the situation.

The significance of awarenesses is that development as conceived by Vygotsky (Valsiner, 1988, p. 334) lies in transformation of a person's ability to act in some way 'in himself', into an ability to act that way 'for himself'. Thus development, in Vygotskian terms, refers to becoming able to make a conscious, explicit choice, rather than simply being triggered or cued into action. Put another way, development has to do with the growth of conscious control and participation, not the extension of facility with a particular technique or procedure. When people refer to the zone of proximal development (ZPD) as imminent changes in behaviour (solving problems or doing tasks with the help of a relative expert) they misunderstand Vygotsky's intentions. What they are referring to really is a zone of proximal behaviour which is in some sense a projection of the ZPD into the enactive component of the psyche (Mason, Drury, & Bills, 2007).

Vygotsky's perspective is compatible with the proposal by Gattegno (1987) that 'only awareness is educable, and I added 'only behaviour is trainable' and 'only emotion is harnessable' in line with the Upanishad image of the chariot. Gattegno meant that functionings can be integrated (through subordination of attention), but that participation in a discipline (such as mathematics, or more particularly synthetic geometry or elementary algebra) occurs when you become aware (explicitly) of those awarenesses (functionings). Similarly, Vergnaud (1997) observed what he called theorems-in-action: children acting as if they knew a theorem or a technique such as the commutativity of addition or multiplication, even when they patently could not articulate it, nor was there any evidence that they were aware of it at more than a functional level.

Development, that is, education of awareness, comes about when implicit theorems-in-action become explicit, open for questioning and possible justification or refutation, and made use of in new ways or in new situations in which they might not have come to mind previously. This is what some people refer to as 'applications' and what behavioural psychologists referred to as transfer (Detterman & Sternberg, 1993; Lobato, 2006). Situated cognition stresses that 'knowing' arises in and because of a particular situation. It tried to account for transfer and its absence by describing learning as situated. However, the same issue then arises in terms of how situativity extends and broadens to encompass previously unfamiliar situations. Ference Marton (see Marton & Booth, 1997; Marton & Pang, 2006; Marton & Trigwell, 2000) sees learning a concept as becoming aware of aspects of examples which can be varied while still remaining examples. These are referred to as dimensions of (possible) variation which provides a more precise formulation of situativity than is usually available (Marton 2006). All of these articulations are, at heart, concerned with educating awareness, implicitly and explicitly. Thinking in terms of attention and 'what comes to mind' in a situation, whether triggered metonymically or resonated structurally through metaphor, highlights the importance of ways in which metonymy and metaphor can be supported and strengthened, and more generally, how noticing an opportunity to act can be developed (Mason, 2002).

Core Mathematical Awarenesses

Taking the perspective described here means that learners need to experience and integrate into their functioning various actions which lie at the core of mathematical topics and themes. Setting tasks, concomitant activity, accumulated experience and working on that experience is the surface structure of mathematical classrooms. Below the surface lies the essential purpose of the tasks, activity and experience, for at the heart of mathematics lie core awarenesses. Thus coordinating the number names experienced as a form of 'counting poem' with pointing, together with discerning a 'unit' are the actions which underpin the awareness which constitutes counting; the action of discerning a repeated unit lies at the core of measuring as well as counting; the action of combining and breaking up underpins addition and subtraction; the actions of scaling and of repeating underpin multiplication; the projection of motion along a path into motion in two (usually orthogonal) directions and the coordination of action in two directions to form a path underpins graphing; the action of selection from a range of possibilities underpins probability, and so on.

Each mathematical topic is based on core actions which learners can carry out under instruction (otherwise they cannot be expected to make much sense of the topic!). These actions constitute the core awarenesses around which the topic is built. A learner needs to encounter the use of previously familiar actions, possibly in a new form or with extensions or variations; an effective teacher needs to be aware of these awarenesses; an effective teacher educator needs to be aware of how to locate and bring to the surface these core awarenesses, derived from the literature and from experience. Tzur (Chapter 8, this *Handbook*, this volume). provides one articulation of how this can be done.

If teachers are not aware of relevant core awarenesses, then they are not in a position to choose appropriate tasks, nor to choose appropriate pedagogical strategies and didactic tactics. For example, partitioning a rectangle into equal pieces with vertical divisions, or with horizontal divisions is a core action which underpins the awareness of fractions as operators on objects. An associated pedagogical awareness is the fact that many learners have a predisposition for vertical partitions. Many require continued exposure to horizontal ones as well so that they develop the flexibility to choose for themselves, which is essential for using the rectangular area model for fraction operations (Kyriakides, 2006).

One of the classic interventions used by relative experts to enculturate novices into particular practices is often referred to as *scaffolding and fading* (Brown, Collins, & Duguid, 1989). A teacher repeatedly uses a particular prompt or question with learners, and then begins to use less and less direct prompts or meta-questions such as "what question am I going to ask you?" or "what did you do last time in this sort of a situation?", until the teacher need only rarely if at all remind learners of the prompt: the prompt has been internalised and become a spontaneous action (Mason, 1991). At first astonished by the question, learners become aware of the actions they were carrying out 'in themselves' as being available to carry out 'for themselves'. The prompts become internalised and available for use, at least in situations in which the prompt comes to mind, by the learners themselves. Requisite awarenesses are integrated into functioning so that they are likely to come to mind when required, in the form of actions to choose to carry out. This is perhaps the paradigmatic manner in which learners develop (in the sense of Vygotsky), although sometimes prompts and questions will come to mind even without being subjected to a process of deliberate scaffolding and fading. Here the propensities and other attitudinal components, together with the sensitivities to notice (even if below the surface of explicit awareness) play a major role in whether and how readily awarenesses trigger actions, or in other words, the extent to which someone has 'learned'.

AWARENESS AND ATTENTION

Awareness and attention are closely related. Someone may be attending to something in a particular way but unaware explicitly of the what or the how. This makes it particularly difficult for teachers and educators to work out what the people they are working with are attending to and aware of. For example, when you express a relationship in words, it is often highly ambiguous as to whether you are expressing a generality, that is, perceiving a property that can hold in different situations, or whether you are forced by language to express yourself in generalities, but in fact you are referring to the foci of your attention, namely specific elements and relationships between them (Mason, Drury, & Bills, 2007). When people are caught up in action, in the 'doing', they may be completely oblivious to relationships that are present, although some of these relationships may be enabling the action to be performed.

For example, if a learner says

> to subtract one fraction from another [meaning 'the ones I am looking at'] I make the bottoms the same and then adjust the tops and subtract the second from the first

the words are necessarily general. They may be thinking generally, but they may be thinking and attending only to the particular. Even if they are thinking generally, they may not be aware of what they are saying *as* a generality. This is what Noss and Hoyles (1996) are pointing to by using the label *situated abstraction*. If the learner had said

> I multiply the 4 by the 5 and the 10 by the 2 and adjust the 3 by multiplying by 5 as well, and the 7 by 2, giving me 15 minus 14 equals 1 so the result is 1/20

then there is stronger indication that the learner may be dwelling in the particular, but in fact they may also be exemplifying a generality of which they are aware. Thus a learner may have no sense of generality at all, and be dwelling entirely in the particular; there may be a vague but inchoate sense of generality; there may be an 'almost articulable' or 'tip of the tongue' sense of generality; there may be a strong sense of generality as a procedure, or as a procedure based on a generative appreciation of how the procedure works; there may be varying (and even unstable) degrees of confidence, varying and unstable degrees of facility, and various assumptions about the *range of permissible change* in the variation permitted by the generality (Mason, Drury, & Bills, 2007).

Notice that the ambiguity inherent in interpreting someone's utterances as statements of generality or of particularity applies at every level. A teacher working with learners needs to be aware of the potential gap between what learners appear to be saying and how they are actually attending; the educator working with teachers needs to be aware of the potential gap between what teachers appear to be saying about their teaching or about their learners, and how they are actually attending. As is well known, there are often gaps between espoused, intended and enacted beliefs about teaching and learning (Cooney, 1985) even when the researcher's own interpretations of these is taken into account (see Taylor & Dirkx, 2002). Educators need to be cognisant of their awareness when drawing on pedagogical constructs and didactical tactics so as to be alert and sensitive to how teachers might be attending to and interpreting what is being said: as generality or as particularity.

It is important to point out that the process of emerging awareness is not a monotonic process of constant improvement. Articulating your current sense of something of which you are only beginning to become aware is often fragmentary to the point of incoherence. In the context of beginning algebra, Nicolina Malara and GianCarlo Navarra (2003) referred to this as 'algebabble' by analogy with young children's babble prior to bursting into speech; Janet Ainley (1999) called it 'emergent algebra'. More generally it has been described as a cyclic process of manipulating confidence-inspiring objects, getting-a-sense of something, and struggling to articulate it (Floyd, Burton, James, & Mason, 1981; Mason & Johnston-Wilder, 2006).

The whole point of tasks and emergent activity is that learners become aware of mathematical concepts, techniques, themes, heuristics, ways of thinking, and the use of their own powers in making sense of mathematics and making mathematical sense of phenomena. The more explicit their awareness, the more likely it is that tasks, activity and intervention will be informed and guided by teacher/educator awareness of core elements of mathematical thinking. If attention is taken up by the interaction, there is little room for meta-awareness and hence the use of pedagogical strategies which will direct learner/teacher attention to useful objects and in useful ways.

Attitude

Attitude is another much used term which covers a multitude of meanings. Its etymological origins lie in the word *aptitude* in the sense of 'having the quality of likeliness or appropriateness'. Here it is taken as a synonym for the affective or emotional component of the psyche, the horses in the chariot metaphor described earlier. Strongly attitudinal or affective components include a veritable alphabet of terms: alignments, assumptions, beliefs, desires, dispositions, likes and dislikes, orientation, perspective, stance, *weltanschauung*, wishes, and so on. Attitude includes not only where the force comes from (as in the horses; see also Piaget, 1954/1981) but also the way in which that force is processed through the individual's selves which are active in the situation, producing reactions (automated or habituated actions) and responses (freshly formed or chosen actions). Of course those 'selves' develop through, and because of, interaction in the social milieu.

Selves

Following a long and ancient tradition extending back at least to Plato, I find it helpful to see people not as individuals, but as assemblages of competing selves (Bennett, 1964; Minsky, 1986). Each 'self' is a cluster of typical emotional states and associated habituated or automatic behaviours, with concomitant awarenesses which come to the surface. Different selves are dominant at different times. For example, the 'person' who leads a seminar is not the 'person' who buys petrol on the way home; the person who fills their briefcase with work they imagine they will do at home is not the person who rejects doing homework once at home: they may have the same outward body, but the gestures, postures, voice tones, vocabulary, way of relating to people, things that are attended to and thoughts that arise may sometimes be quite different in nature. Some people find 'multiple selves' disturbing and prefer to see these as aspects of one self, referred to as 'I'. But detailed investigation into the source of the word 'I' when it is uttered reveal that most often it is a habituated grammatical construction with no referent. Its use is primarily a form of social reassertion of the currently dominant self (Bennett, 1964; Mason, 2002). James (1890/1950) made the point nicely: "if from the one point of view I am one self, from another I am quite as truly many" (p. 202, quoted in Bullough, 2005, p. 146).

Each self therefore can be thought of as an interconnected and self-amplifying collection of propensities, behaviours, indeed a mini psyche. The term *attitude* then functions as a label for motive properties of whichever self is dominating at any given moment. Teachers are well aware that learners invoke different selves in different lessons, and that to connect with a learner so that an experience has some significance requires sensitivity to the learners' currently dominant selves, including invitations and prompts to access a more appropriate self. Adolescents particularly are engaged in the process of discovering that they can have some control, can exercise some choice about which self they bring into play, and that this can have an enormous impact on what they get from lessons. Similarly, educators are well aware that teachers invoke different selves at different times, with the result that sometimes there is eager identification and take up of proposals, and other times hostility or outright rejection. Where teachers cannot imagine themselves taking up an idea, or where learners cannot imagine themselves enacting a proposal, reluctance or rejection are likely outcomes.

When people are together in social groupings there is a social-amplification-attenuation effect. Sometimes there is a sufficient degree of alignment so that it is reasonable to speak of a 'group attitude'. But it is also possible for some people to react against proposals from the outside, for a range of reasons, and so act as a brake on social alignment. As Jaworski (2006) has pointed out, a functioning community of inquiry will experience a 'critical alignment' which is not necessarily a feature of a community of practice. Teachers sensitive to these subtle movements are better able to nudge and cajole, to massage a group into effective mathematical work. Educators sensitive to these subtle movements are better able to work effectively with groups of teachers.

JOHN MASON

SIMILARITIES AND DIFFERENCES IN DIFFERENT STATES

The terms *attention, awareness* and *attitude* provide a useful vocabulary for discerning similarities and differences between the experiential states of people acting as learners, teachers, and teacher educators. People rarely function solely in a single mode: many learners are remarkably aware of pedagogic choices; many teachers work on mathematics for themselves as well as with learners; many teacher educators work on mathematics and on teaching , both for themselves as well as with others. The fact that human beings are complex organisms embedded in a multitude of environments, coupled with the elaboration of aspects of attention, awareness and attitude suggests to me that teachers who are not learning about their learners, and who are not challenging themselves mathematically cannot expect to remain fresh and effective for long. Conversely, teacher educators who are not challenging themselves mathematically as well as pedagogically similarly cannot expect to remain fresh and effective for long. For teacher educators this means working on mathematics alone and with others, working on their own pedagogy by using pedagogical constructs to elaborate their practice as well as using their practice to elaborate constructs; for teachers this means working on mathematics alone and with others as well as working on their pedagogical practices and didactic tactics. In short, this can be re-cast as educating awareness, sensitising attention and enriching attitudes to mathematics, to learning mathematics and to teaching mathematics.

From the foregoing it is evident that the aims and intentions of learners, teachers and teacher educators are different (see also (Chapter 8, this *Handbook*, this volume). Learners' primary concern is functioning in relevant mathematical ways as they work on tasks and become aware of heuristics, themes, strategies and forms of mathematical thinking, as well as encountering and re-constructing for themselves important concepts and techniques. Learners may gradually become explicitly aware of some of these aspects of learning mathematics, but their main concern is that these come to mind when needed so that satisfactory responses to tasks can be achieved. Indeed, most learners act as if they have a theorem-in-action regarding learning, or, put another way, their side of the implicit didactic contract (Brousseau, 1997) is to complete the tasks they are set as best they can. Somehow students assume (rarely explicitly or consciously) that task-completion will produce the learning expected. In this sense, they naturally tend to dwell in the particular, at least until someone, or something in the practices in which they are embedded, draws them out of the particular to reflect upon and access encompassing generalities. Sometimes a task-completion orientation goes so far as obtaining answers by whatever method is available, including copying.

The implicit underlying theory of learning seems to be that learning happens as a result of doing and handing in work. Teachers who focus on assigning tasks, marking learner attempts, and giving answers, with various forms of worked examples and explanations, connive to reinforce this simplistic perspective on learning. By contrast, provoking learners in the use of their own natural powers and familiar actions so as to meet new challenges, and then drawing attention to what learners have achieved, and how, directs their attention to what really matters, and affords opportunities for learners awareness to be educated (Mason and Johnston-Wilder, 2006). From a similar perspective, Tzur (Chapter 8, this *Handbook*, this volume) focuses on the activity-effect relationship experienced in carrying out tasks as the core feature which requires reflection and attention for effective learning to take place. Thus teacher intentions and aims are that learners' attention be drawn to experiences in which powers and heuristics have been used effectively, and mathematical themes, concepts and techniques encountered. Teacher educator aims and intentions are that teachers become aware of ways in which learner attention can be directed appropriately to pertinent experiences. This of course requires that teachers themselves become aware of the powers, themes, heuristics which comprise mathematical thinking, as well as both the concepts and techniques themselves, and the psycho-social components of educating awareness and training behaviour associated with them. In short, then, learners are exposed to teachers who are being mathematical with and in front of their learners, supported and amplified by being in the presence of teacher educators who themselves are being mathematical and who are able to draw attention to what comprises that mathematical being. To be effective, teachers need to develop 'awareness-in-discipline', and teacher educators need to develop 'awareness-in-counsel' (Mason, 1998). This unusual label is used to emphasise the states of self-awareness which are necessary in order to be able to function at all the requisite levels, similar to the states needed by effective counsellors.

A narrow perspective on the role of tasks and the nature of learning is likely to lead to a constricted disposition towards mathematical thinking, seeing it as the obtaining of answers through the use of memorised procedures. Positive dispositions, self-image, and confidence in self as a mathematical thinker develop from meeting challenge (Dweck, 2000). Attempts to support learner confidence by giving them simple tasks merely reinforces their self-image and does nothing to open them up to other possibilities (Houssart, 2004; Prestage, Watson, & de Geest, 2007; Watson, Prestage, & de Geest, 2004).

Where generality is achieved by varying several parameters, including changing contexts, the learner experiences what Treffers (1987) referred to as *horizontal mathematisation* and has many of the characteristics of metonymy and syntactic or surface learning (Marton & Saljö, 1984). *Vertical mathematisation* arises when attention is directed from relationships to properties, and from properties to characteristic, defining or generative properties which can function as axioms from which all other relevant consequences flow. It has many of the characteristics of metaphor and semantic or deep learning (Marton & Saljö, 1984). Both flows are essential to learning and to developing positive dispositions towards mathematical enquiry and use.

Faced with a mathematical task, learners seek answers. To do this they call upon, usually without reflection or awareness, practices into which they have been initiated if not enculturated through engaging in the social practices of the mathematics classroom in particular and of schooling in general, and through interacting with their teachers and their peers. To teach effectively and efficiently requires working differently. Attention is not on answers but on the approach, the forms of thinking employed. Teachers attend to where and how learners are attending, and choose their interactions with learners so as to direct attention in what the teacher trusts will be useful ways. This accounts for the extensive literature on misconceptions (how to direct learner attention so that classic misconceptions or construals will be circumvented or else corrected through exposure to cognitive dissonance or surprise). Teachers also attend to learners' motivational-affective state, as is evident in their habitual reactions and their considered responses in lessons at different times of the day and in response to learners' all too explicitly expressed states of boredom, confusion and frustration. They seek ways to attract learner attention so as to engage them, and to connect to learners' natural curiosity and desire to explain phenomena.

However, teachers too are practitioners. They carry out actions with greater or lesser explicit awareness. They develop functionings or practices, ways of coping, many of which become habitual and below the surface of explicit awareness, as indeed they must if they are to survive the complex interactions with the energies of children and adolescents. As such they are similar to learners but in a different domain. As well as functioning mathematically, teachers function pedagogically and didactically. To become mathematical, that is, to enter and engage in the discipline of mathematics, it is necessary to become aware of the awarenesses which enable and direct functionings which can be construed as mathematical (Gattegno, 1987; Mason, 1998). For example, explicitly retreating from an overly complex task or situation by trying particular or special cases, building up particular cases so as to see through them to an encompassing generality, or making use of a theme such as invariance-in-the-midst-of-change or doing-and-undoing, begin to become 'second-nature' or automatic as they are integrated into functioning. Brown and Coles (2000) give a vivid description of how learners took over the phrase 'same and different' and turned it into a verb which they explicitly initiated as part of the practice of their classroom when working on new ideas. The same process of integration can be accomplished with any sensitivity to notice, any structure of attention, any awareness (Mason, 1999, 2002).

Distinctions such as that drawn by Skemp (1976) between instrumental and relational understanding, by Marton and colleagues (e.g., Marton & Saljö, 1984) between a surface and a deep approach to learning, between procedural and conceptual knowledge, and between syntactic and semantic appreciation all divert attention away from the richly complex experience of 'having pertinent possible actions come to mind' and towards a single spectrum of technique and comprehension. Valsiner (1988) notes that Vygotsky was trying to direct attention away from what the learner can already do (which is what most tests attempt to discover) to what the learner can soon choose to do unaided but can currently accomplish with the guidance of a more-experienced other. This is where learning, growth, and transformation are imminent, what Vygotsky meant by 'proximal development', and hence where the teacher can be of most value. Tzur (Chapter 8, this *Handbook*, this volume) sees the participatory stage of forming a new (to the learners) concept (through becoming aware of an activity-effect relationship) as a significant step towards learners being able to initiate similar activity themselves, which is the development of importance to Vygotsky.

For example, a focus on tasks and their results avoids the complexity of learning as a transformation of awareness, of what is spontaneously attended to and discerned, of the ways that attention is structured, and of mathematical themes, heuristics and powers which come to mind through metonymic triggering and metaphoric resonance. Such a focus concentrates on task-completion and the accumulation of procedures for every situation, eventually putting an excessive burden on memory and leaving the learner vulnerable to forgetting and to inflexibility in the face of unfamiliar challenges. By contrast, maintaining complexity when working with learners is more likely to provide them with sufficiently rich experiences that their awareness is educated, their attention honed and sharpened, and their attitudes enriched.

Teachers develop functional awareness that enables them to cope with the exigencies and dynamics of a classroom full of will-possessing learners who are themselves full of desires and concerns, interests and dislikes, with varying degrees of coordination between their affect, their behaviour and their intellect. Analogously with learners, as teachers become more and more explicitly aware of their implicit functionings or awarenesses which promote learning, they are in a better position to make choices, to respond sensitively rather than automatically to whatever situation emerges. This response covers the Shulman range from content knowledge to pedagogic subject knowledge and beyond (Shulman, 1987, p. 6).

Becoming aware not simply of the fact of different ways of intervening, but of the fact of subtle sensitivities which guide or determine choices between types and timings of interventions, is what is involved in becoming an effective teacher, a 'real teacher' (Mason, 1998). Such explicit awarenesses give rise to a discipline of teaching, which I refer to as *awareness-in-discipline* to distinguish it from *awareness-in-action* (*op cit*). The 'in-discipline' is stressing that the actions being brought from implicit to explicit awareness are in and of the discipline of mathematics rather than actions which constitute the discipline. This 'second-order' discipline that is emerging is really the discipline of mathematics education as experienced by a teacher of mathematics.

Someone acting as mentor to a novice teacher who is teaching a lesson while being observed by the mentor and-or by a teacher educator provides an ideal setting for exploring the complexity of interacting awarenesses. There are mirrored parallels between the novice teacher working with the learners, aware of possibilities not yet accessible to the learners, the mentor working with the novice teacher in the same way, and the teacher educator working with the mentor. In order to make sensible and suitable choices, each must also be aware of the world occupied by the other. Thus the mentor is aware of the mathematics being taught as well as the pedagogical strategies and didactic tactics being used; the teacher educator is aware of the mathematics, the strategies and also the constructs and theories which inform and underpin the strategies. Each must choose how much to try to bring to the surface for those with whom they are working. It is no wonder that teaching and learning are so complex, and perhaps it is even surprising that so much learning actually does take place in connection with attempts at teaching! (see also Jaworski, 2007).

The structures of awareness become even more complex when teachers move into mentoring and then into teacher education. When teachers first make this move, they tend to try to describe what they think they used to do when teaching continuing to see themselves as teachers rather than mentors. Over a period of time they discover that this is insufficient for pre-service or in-service work with prospective or practicing teachers. Something more is required. They become aware that there are useful ways of working with people who are themselves learning to teach, many of which parallel ways of working with learners on mathematics, but which alter subtly with the change of focus from the discipline of mathematics to the second-order discipline of teaching mathematics. In the perspective of teacher educator being developed here, the effective teacher educator aims to direct attention so that participants' attention is drawn out of the actions of doing mathematics and also out of the actions of teaching mathematics, so that awarenesses become explicit. In this way, individuals and their social milieu may serve to educate that awareness, and thus inform actions in the future. This forms a third-order discipline.

WORLDS OF EXPERIENCE AND MODES OF ACTION

One way of speaking about the range of structured attention, awarenesses, and attitudes, and structured attention of people in different situations is in terms of occupying or dwelling in different worlds. The image of living in different worlds is of course ancient, probably as old as human beings, as evidenced by shamanic traditions and more modern religions. Nelson Goodman (1978) used the image to consider how people individually and collectively create the worlds they occupy, in a manner consistent with a Buddhist perspective. Varela, Thompson and Rosch (1991) used a similar image in similar ways, but focusing on the role of embodiment in in-forming and pro-forming such construction. Jerome Bruner (1966) proposed three modes of representation (enactive, iconic and symbolic) and developed the importance of the imaginative as different from the world of material experience (Bruner, 1986). If re-presentation is interpreted as indicating the nature of a world of experience, then his three modes can be thought of as three worlds: physical or material, mental, and symbolic, and by metaphoric extension, manipulative, imagistic and intuitive, and as-yet-only-partially-coherent. Thus familiar and expressive symbols can be manipulated, so for such a person, these occupy a manipulative, metaphorically enactive world, whereas ideas which are as-yet abstract and not well connected occupy a symbolic world as yet out of reach in terms of articulation. This is the basis for frameworks such as *manipulating – getting – a-sense-of – articulating* developed from experience with Bruner's ideas in the Open University Centre for Mathematics Education (Floyd et al., 1981, see also Mason & Johnston-Wilder, 2006). David Tall (2004) has latterly been developing a cognate but slightly different version in terms of conceptual (embodied), proceptual (symbolic) and axiomatic (formal) worlds.

BEING MATHEMATICAL WITH LEARNERS

In my own case, I became explicitly aware of mathematical thinking processes such as specialising and generalising, conjecturing and convincing through seeing the film *Let Us Teach Guessing* (Pólya, 1965). It released a style of teaching I had experienced at school which promoted interaction with learners, prompting them to undertake the thinking. My awareness expanded to the second-order discipline which was manifested in the writing of *Thinking Mathematically* (Mason, Burton, & Stacey, 1982). However it took me several years of promoting 'mathematical investigation sessions' at Open University Summer Schools before I realised explicitly that many very mathematical tutors were not themselves spontaneously prompting learners to specialise and generalise, to conjecture and convince. Thus arose elements of the third-order discipline described above. I then embarked on developing ways to engage people in mathematical thinking, to draw them out of the actions forming their activity so that they became aware of those actions as potential actions to invoke for themselves, and ways of working with colleagues on working in this way with teachers. Finding myself in a distance teaching institution forced me to articulate to myself where in a face-to-face context I might have felt the need to engage with participants directly, but it also freed me from feeling that I had to be present and interacting for learning to take place. These different awarenesses combined with my propensity to see phenomena as metaphorical. For example, I saw the triad of enactive–iconic–symbolic not simply as three modes of (re)presentation, but more metaphorically as the necessary components of developing expertise: doing, imagining and symbolising and as constituting three worlds of experience (Mason, 1980).

Because attention is 'where we are', the form and structure of our attention not only determines but *is* the mental world we occupy. The 'we' in this case is led by a cluster of competing selves under the leadership of the currently dominant self, so that there is a particular self in its own characteristic world of attention and awareness. What we are aware of, implicitly as well as explicitly determines the fine-grain details of that world of experience. Our attitude in all of its complexity determines what actions will come to mind through a combination of metonymic triggers and metaphorical resonance, whether automatically or as conscious choices.

Alerted by a sensitivity to notice opportunities to initiate actions, learners, teachers and teacher educators sometimes have the chance to participate in making a choice. Among other things, such choices include a mode of interaction: whether to initiate action (say something, do something), to respond (to wait expectantly, to listen carefully and try to enter the world of the speakers) or to mediate (to influence activity merely through being present and having contributed to a rich milieu of ways of working, access and use of resources, and so on). This is what is meant by 'being mathematical with, and in front of' learners.

The world of the learner is essentially to do with locally focused attention to the tasks and consequent activities, but it needs to reach out to the underlying awarenesses. Teaching is about directing learner attention, and not only being consciousness for two or more, but scaffolding and fading support for learners so that eventually they take over the initiative for themselves. It is about being aware of what learners are not yet aware of and finding ways to prompt them to become aware. Educating teachers is about directing attention to practices and choices, constructs and theories which can inform choices when teaching. Again it is about being aware of what teachers are not yet explicitly aware of, and prompting relevant shifts of attention. It is must be made clear that there is *no* assumption that the teacher educator 'knows' what the teacher needs to be come aware of, for most often what practitioners need at any level, learner, teacher or teacher educator, are prompts which provoke them to become explicitly aware of what they are currently at best implicitly aware of, but which may be evidenced in their practices, their desires, or their aims.

REFERENCES

Ainley, J. (1999). Doing algebra-type stuff: Emergent algebra in the primary school. In O. Zaslavsky (Ed.), *Proceedings of the 23rd Conference of the International Group for the Psychology of Mathematics Education* (Vol. 2, pp. 9–16). Haifa, Israel: Psychology of Mathematics Education.

Bennett, J. (1964). *Energies: Material, vital, cosmic*, London: Coombe Springs Press.

Bennett, J. (1966). *The dramatic universe*. (Vols. 1–4). London: Routledge.

Biggs J., & Collis K. (1982). *Evaluating the quality of learning: The SOLO taxonomy*. New York: Academic Press.

Bullough, R. V. (2005). Being and becoming a mentor: School-based teacher educators and teacher educator identity. *Teaching and Teacher Education, 21*, 143–155.

Brousseau, G. (1997). *Theory of didactical situations in mathematics: Didactiques des mathématiques, 1970-1990*. N. Balacheff, M. Cooper, R. Sutherland, & V. Warfield, (Trans.), Dordrecht: Kluwer Academic.

Brown, J. S., Collins A., & Duguid, P. (1989). Situated cognition and the culture of learning. *Educational Researcher, 18*, 32–42.

Brown, L., & Coles, A. (2000). Same/different: A 'natural' way of learning mathematics. In T. Nakahara & M. Koyama (Eds.), *Proceedings of the 24th Conference of the International Group for the Psychology of Mathematics Education*, (Vol. 2, pp. 153–160). Hiroshima, Japan: Psychology of Mathematics Education.

Bruner, J. (1966). *Towards a theory of instruction*. Cambridge, MA: Harvard University Press.

Bruner, J. (1986). *Actual minds, possible worlds*. Cambridge, MA: Harvard University Press.

Claxton, G. (1984). *Live and learn: An introduction to the psychology of growth and change in everyday life*. London: Harper & Row.

Claxton, G. (1997). *Hare brain, tortoise mind: Why intelligence increases when you think less*. London: Fourth Estate.

Cobb, P. (1995). Continuing the conversation. *Educational Researcher, 24*, 25.

Cooney, T. J. (1985). A beginning teacher's view of problem solving. *Journal for Research in Mathematics Education, 16*, 324–336.

Cuoco, A., Goldenburg, P., & Mark, J. (1996). Habits of mind: An organizing principle for mathematics curricula. *Journal of Mathematical Behavior, 15*, 375–402.

Dedekind, R. (1912). Stetigkeit und irrationale zahlen. [Continuity and irrational numbers] (4th, ed.). Braunschweig: Friedr. Vieweg & Sohn.

Detterman, D., & Sternberg, R. (Eds.) (1993). *Transfer on trial: Intelligence, cognition, and instruction*, Norwood, MA: Ablex.

Dweck, C. (2000). *Self-theories: Their role in motivation, personality and development*. Philadelphia: Psychology Press.

Floyd, A., Burton, L., James, N., & Mason, J. (1981). *EM235: Developing mathematical thinking*. Milton Keynes, UK: Open University Press.

Gattegno, C. (1973). *In the beginning there were no words: The universe of babies.* New York: Educational Solutions.
Gattegno, C. (1987). *The science of education Part 1: Theoretical considerations.* New York: Educational Solutions.
Gattegno, C. (1988). *The mind teaches the brain* (2nd ed.). New York: Educational Solutions.
Goodman, N. (1978). *Ways of world making.* London: Harvester.
Gurdjieff, G. (1950). *All and everything.* London: Routledge & Kegan Paul.
Hamilton, E. & Cairns, H. (Eds.) (1961). *Plato: The collected dialogues including the letters.* Bollingen Series LXXI. Princeton: Princeton University Press.
Harding, D. (1961). *On having no head: Zen and the re-discovery of the obvious.* London: Arkana (Penguin).
Hewitt, D. (1994). *The principle of economy in the learning and teaching of mathematics.* Unpublished doctoral dissertation, Open University, Milton Keynes, UK.
Houssart, J. (2004). *Low attainers in primary mathematics: The whisperers and the maths fairy.* London: Routledge.
Jakobson, R. (1951). *Fundamentals of language* (2nd ed.). Den Hague, the Netherlands: Mouton de Gruyter.
James, W. (1890/1950). *Principles of psychology*, (Vol. 1). New York: Dover.
Jaworski B. (2006). Theory and practice in mathematics teaching development: Critical inquiry as a mode of learning in teaching. *Journal of Mathematics Teacher Education, 9*, 187–211.
Jaworski, B. (2007). Developmental research in mathematics teaching and learning: Developing learning communities based on inquiry and design. In P. Lillejdahl (Ed.), *Proceedings of the Canadian Mathematics Education Study Group* (pp. 3–16). Vancouver: Simon Fraser University.
Kyriakides, A. O. (2006). Modelling fractions with area: The salience of vertical partitioning. In J. Novotná, H. Moraová, M. Krátká, & N. Stehlková (Eds.), *Proceedings of the 30th Conference of the International Group for the Psychology of Mathematics Education* (Vol. 4, pp. 17–24). Prague: Charles University.
Lacan, J. (1985). Sign, symbol and imagery. In M. Blonsky (Ed.), *On signs* (pp. 203–209). Oxford: Blackwell.
Lakoff, G., & Johnson, M. (1980). *Metaphors we live by.* Chicago: University of Chicago Press.
Lakoff, G., & Nunez, R. (2000). *Where mathematics comes from: How the embodied mind brings mathematics into being.* New York: Basic Books.
Langer, E. (1997). *The power of mindful learning.* Reading, PA: Addison-Wesley.
Lave, J., & Wenger, E. (1991). *Situated learning: Legitimate peripheral participation.* Cambridge: Cambridge University Press.
Lobato, J. (2006). Alternative perspectives on the transfer of learning: History, issues, and challenges for future research. *Journal of the Learning Sciences, 15*, 431–449.
Malara, N., & Navarra, G. (2003). *ArAl Project. Arithmetic pathways towards favouring pre-algebraic thinking.* Bolgona: Pitagora.
Mandler, G. (1989). Affect and learning: Causes and consequences of emotional interactions, in D. McLeod & V. Adams (Eds.), *Affect and mathematical problem solving: A new perspective* (pp. 3–19). London: Springer-Verlag.
Marton, F. (2006). Sameness and difference in transfer. *Journal of the Learning Sciences, 15*, 499–535.
Marton, F., & Booth, S. (1997). *Learning and awareness.* Hillsdale, NJ: Lawrence Erlbaum.
Marton, F., & Pang, M. (2006). On some necessary conditions of learning. *Journal of the Learning Sciences, 15*, 193–220.
Marton, F., & Saljö, R. (1984). Approaches to learning. In F. Marton, D. Hounsell, & N. Entwistle (Eds.), *The experience of learning.* Edinburgh: Scottish Academic Press.
Marton, F., & Trigwell, K. (2000). 'Variatio est Mater Studiorum'. *Higher Education Research and Development. 19*(3), 381–395.
Mason, J. (1980). When is a symbol symbolic? *For the Learning of Mathematics, 1*(2), 8–12.

JOHN MASON

Mason, J. (1991). Epistemological foundations for frameworks which stimulate noticing. In R. Underhill (Ed.), *Proceedings of Psychology of Mathematics Education-North America 13* (Vol. 2, pp. 36–42). Christiansburg, VA: Christiansburg Printing Company.

Mason J. (1998). Enabling teachers to be real teachers: Necessary levels of awareness and structure of attention. *Journal of Mathematics Teacher Education, 1*, 243–267.

Mason, J. (1999). The role of labels for experience in promoting learning from experience among teachers and students. In L. Burton (Ed.), *Learning mathematics: From hierarchies to networks* (pp. 187–208). London: Falmer Press.

Mason, J. (2002). *Researching your own practice: The discipline of noticing*. London: RoutledgeFalmer.

Mason, J., & Johnston-Wilder, S. (2006). *Designing and using mathematical tasks* (2nd ed.). St. Albans, UK: Tarquin.

Mason, J., Burton L., & Stacey K. (1982). *Thinking mathematically*. London: Addison-Wesley.

Mason, J., Drury, H., & Bills, E. (2007). Explorations in the zone of proximal awareness. In J. Watson & K. Beswick (Eds.), *Mathematics: Essential research, essential practice: Proceedings of the 30th annual conference of the Mathematics Education Research Group of Australasia* (Vol. 1, pp. 42–58). Adelaide: MERGA.

Maturana, H., & Varela, F. (1972). *Autopoesis and cognition: The realization of the living*. Dordrecht: the Netherlands: Reidel.

Minsky, M. (1986). *The society of mind*. New York: Simon and Schuster.

Nørretranders, T. (1998). *The user illusion: Cutting consciousness down to size*. (J. Sydenham Trans.). London: Allen Lane.

Noss, R., & Hoyles, C. (1996). *Windows on mathematical meanings: Learning cultures and computers*. Mathematics Education Library, Dordrecht, the Netherlands: Kluwer Academic.

Piaget, J. (1954/1981). *Intelligence and affectivity: Their relationship during child development*. T. Brown & C. Kaegi (Trans. & Ed.). Palo Alto, CA: Annual Review.

Pirie, S., & Kieren, T. (1994). Growth in mathematical understanding: How can we characterise it and how can we represent it? *Educational Studies in Mathematics, 26* (2-3), 165–190.

Pólya, G. (1965). *Let Us Teach Guessing*, (film) Washington: Mathematical Association of America.

Prestage, S., Watson, A., & de Geest, E. (2007). *Building learning in mathematics: Deep progress in mathematics*. London: Continuum International.

Shulman, L. S. (1987). Knowledge and teaching: Foundations of the new reform. *Harvard Educational Review, 57*, 1–14.

Skemp, R. (1976). Relational and instrumental understanding. *Mathematics Teacher, 77*, 20–26.

Tall, D. (2004). Thinking through three worlds of mathematics. In M. Høines & A. Fuglestad (Eds.), *Proceedings of the 28th Conference of the International Group for the Psychology of Mathematics Education* (Vol. 4, pp. 281–288). Bergen, Norway: Bergen University College.

Taylor, E., & Dirkx, J. (2002). The relationship between espoused and enacted beliefs about teaching adults: A photo-elicitation perspective. In *Proceedings of the 21st annual meeting Canadian Association for the Study of Adult Education* 275–281. Toronto: OISE.

Thompson, P. (2002). Didactic objects and didactic models in radical constructivism. In K. Gravemeijer, R. Lehrer, B. van Oers, & L. Verschaffel (Eds.), *Symbolizing, modelling, and tool use in mathematics education* (pp. 191–212). Dordrecht, the Netherlands: Kluwer Academic.

Treffers, A. (1987). *Three dimensions: A model of goal and theory description in mathematics education*. Dordrecht, the Netherlands: Reidel.

Valsiner, J. (1988). *Developmental psychology in the Soviet Union*. Brighton, UK: Harvester.

van Hiele, P. (1986). *Structure and insight: A theory of mathematics education*. London: Academic Press.

Varela, F., Thompson, E., & Rosch, E. (1991). *The embodied mind: Cognitive science and human experience*. Cambridge, MA: MIT Press.

Vergnaud, G. (1997). The nature of mathematical concepts. In T. Nunes and P. Bryant (Eds.) *Learning and teaching mathematics: An international perspective*. London: Psychology Press.

Watson, A., De Geest, E., & Prestage, S. (2004). *Deep progress in mathematics: The improving attainment in mathematics project*. Oxford: Department of Education, Oxford University.

Wenger, E. (1998). *Communities of practice: Learning, meaning and identity*, Cambridge: Cambridge University Press.

Winbourne, P., & Watson, A. (1998). Participation in learning mathematics through shared local practices. In A. Olivier & K. Newstead (Eds.), *Proceedings of the 22nd Conference of the International Group for the Psychology of Mathematics Education* (Vol. 4, pp. 177–184). Stellenbosch, SA: Psychology of Mathematics Education.

John Mason
Centre for Mathematics Education
Open University

RUHAMA EVEN

3. FACING THE CHALLENGE OF EDUCATING EDUCATORS TO WORK WITH PRACTISING MATHEMATICS TEACHERS[1]

This chapter centers on the challenge of educating mathematics educators to work with practising teachers. The first part of the chapter discusses problems underlying this challenge: that there is almost no research on the education of mathematics teacher educators, the ill-defined nature of the field of offering education to practising mathematics teachers, and the lack of information on the practice of mathematics teacher educators working with practising teachers. The second part uses the MANOR project to illustrate how these problems might be addressed in a practical setting, introducing the construct of knowtice as a lens to capture the essence of what educators need to learn and develop. Finally, directions for future research are suggested.

There is general agreement today that the continued professional development of teachers is key to improving students' opportunities to learn mathematics (Advisory Committee on Mathematics Education, 2002; Even & Ball, in press; Loucks-Horsley, Hewson, Love, & Stiles, 1998). Yet, past experience suggests that supporting practising teachers' learning is not a trivial task, and typical learning opportunities for teachers are often believed to be inadequate (Borko, 2004; Wilson, & Berne, 1999; Zaslavsky, Chapman, & Leikin, 2003). Consequently, there is a growing interest in the international community in the education of mathematics educators that will be able to work with practising teachers on developing their teaching (Elliot, 2005; Even, 1999a, 2005; Farah & Jaworski, 2006; Sztajn, Ball, & McMahon, 2005; Zaslavsky & Leikin, 2004). This chapter centers on the challenge of educating mathematics educators to work with practising teachers. The first part of the chapter discusses problems underlying this challenge. The second part uses the MANOR project to illustrate how these problems might be addressed in a practical setting. In this part, the construct of *knowtice* is introduced as a lens to capture the essence of what educators need to learn and develop. Finally, directions for future research in the area of educating mathematics educators to work with practising teachers are suggested.

The focus of this chapter is on the education of mathematics educators to work with *practising* mathematics teachers; it is reasonable to assume that this challenge

---

[1] The major part of this chapter was written while on sabbatical leave as a Lappan-Phillips-Fitzgerald visiting scholar at the Division of Science and Mathematics Education in Michigan State University.

might be different from that which is associated with the education of mathematics educators to work with prospective teachers. Education of prospective and that of practising mathematics teachers are usually of different nature, often occurring in different settings, and not necessarily conducted by the same people. Education of prospective teachers is usually part of a formal programme in an academic institute leading to a formal certificate, whereas education of practising teachers is less structured, and is conducted in various places, such as, schools and local non-academic centers, not granting any formal diploma.

PROBLEMS UNDERLYING THE CHALLENGE TO EDUCATE MATHEMATICS EDUCATORS TO WORK WITH PRACTISING TEACHERS

A conventional chapter would normally include here a review of the literature. This chapter, instead, starts with a "review" of missing literature. It does so by presenting three problematic aspects in the current literature, which underlie the challenge to prepare mathematics educators to work with practising teachers. The first problem is that there is almost no research on the education of mathematics teacher educators, the second problem is the ill-defined nature of the field of offering education to practising mathematics teachers, and the third problem is the lack of information on the practice of mathematics teacher educators working with practising teachers.

Missing Research on the Education of Teacher Educators

The focus and nature of the education of prospective and practising mathematics teachers have received immense international attention in recent years (e.g., Advisory Committee on Mathematics Education, 2002; Even & Ball, in press; Mcnamara, Jaworski, Rowland, Hodgen, & Prestage, 2002; Reference Group on Teacher Standards, Quality and Professionalism, 2003). The past two decades have seen substantial increase in scholarship on mathematics teacher education. This is reflected, for example, in the establishment of the international *Journal of Mathematics Teacher Education* in 1998, and in the rapidly growing number of research studies presented at the International Group of Psychology of Mathematics Education (PME) meetings on this topic, marking a major distinction between current and past work of the PME Group (Hershkowitz & Breen, 2006). Whereas the first milestone PME book (Nesher & Kilpatrick, 1990) was devoted solely to cognitive research related to student learning of various mathematical topics and concepts, one of the five main research domains of current interest to the PME Group, as presented in the second milestone PME book (Gutiérrez & Boero, 2006), is the professional life of mathematics teachers and their education. Moreover, the International Commission on Mathematical Instruction (ICMI), the largest international mathematics education organization, acknowledged the need to attend to prospective and practising teachers' learning and professional development opportunities, and launched an ICMI study focused on the professional education of mathematics teachers around the world (Even & Ball, in press). ICMI also devoted plenary sessions on the development of research on mathematics teacher education and on knowledge for teaching mathematics, as part of the programmes at the tenth International Congress on Mathematics Education (ICME-10) in 2004 and the eleventh International Congress on Mathematics Education (ICME-11) in 2008 (respectively).

Although publication of peer-reviewed articles, book chapters, and books about the education of prospective and practising teachers of mathematics is on the rise, the education of mathematics teacher educators (of both prospective and practising teachers) are rarely discussed in the scholarly literature. When reviewing the focus of the papers in the *Journal of Mathematics Teacher Education* that were published between 1998-2002, Even, Robinson and Carmeli (2003) found only two papers that centered on educators of practising teachers (Even, 1999a; Halai, 1998). In her review of the literature, Elliot (2005) found only three professional opportunities for educators of practising teachers within Developing Mathematical Ideas (DMI) (Davenport & Ebby, 2000) and the QUASAR project (Stein, Smith & Silver, 1999) in the United States and the MANOR project in Israel (Even, 1999a). Elliot's (2005) own research was conducted within the Leadership Curriculum for Mathematics Professional Development (LCMPD) project in the United States. Other recent research studies related to the education and learning of educators of practising teachers report on a special M.Ed. Programme in Pakistan (Jaworski, 2001), and on growth of mathematics teacher educators through their practice in Israel (Zaslavsky & Leikin, 2004), and in Norway (Goodchild, 2007).

Taking into consideration the focus on the education of prospective and practising mathematics teachers in the last two decades, it is remarkable that the education of teacher educators has been almost neglected until now. Expecting the education of practising teachers to play a critical role in improving the quality of mathematics teaching and learning at school requires greater attention to educators of practising teachers. In a way, the recent focus on mathematics teacher education with lack of attention to the teacher educators mirrors, to some degree, the early research in mathematics education, which centered on student learning but lacked attention to teachers, teaching and teacher learning.

The Ill-Defined Nature of the Field of Educating Practising Teachers

An examination of recent literature suggests that there is no one word or phrase, which is used to refer to educators working with practising teachers. This is illustrated also in this chapter. Common terms, sometimes not distinguishing between those working with prospective and those working with practising teachers, are professional development (PD) providers, professional development teachers, professional developers, teacher developers, facilitators, teacher-leaders, teachers of teachers, teacher educators, and in-service teacher educators (Advisory Committee on Mathematics Education, 2002; Elliott, 2005; Even, 1999a, 2005; Farah & Jaworski, 2006; Higgins, 2005; Sztajn, et al., 2005; Zaslavsky & Leikin, 2004). This lacking of a common term reflects the ill-defined nature of the field; not only the scholarly field, but also the practice itself. And, indeed, one of the problems identified by Sztajn et al. (2005) as underlying the challenge of educating mathematics teacher educators was that this group is not well defined. In many countries, the group of educators who work with practising teachers includes university faculty as well as school teachers; educators whose major occupation is to work with practising teachers and those who do it only as an add-on part-time temporary activity; those who work also with prospective teachers and those who work solely with practising teachers.

Moreover, it is not only the group of educators of practising teachers that is not well-defined, but, the system of providing professional development opportunities for practising teachers is "random, sometimes voluntary, sometimes mandated, always fragmented system" (Wilson & Berne, 1999, p. 197). In addition, in many cases, the different terms used reflect different views regarding desired practice of educating practising teachers. For example, the term 'facilitator' conveys a specific meaning regarding the practice, of assisting teachers by encouraging them to find their own solutions to problems or tasks whereas 'professional development provider' sounds more business like, and infers lesser participation in the process of defining and achieving the goals of the participating teachers.

Lack of Information on the Practice of Mathematics Teacher Educators

The mathematics education literature suggests numerous ideas of how to design professional education experiences for practising teachers of mathematics so that they have an impact on mathematics teaching and learning in school, advocating teacher inquiry into, and reflection on, their practice (e.g., Ball & Cohen, 1999; Jaworski, 1998; Krainer, 1998; Loucks-Horsley et al., 1998; Mcnamara et al., 2002; Zaslavsky et al., 2003). However, the literature offers only limited empirical information about the practice of mathematics teacher educators working with practising teachers. A cursory review of the focus of the papers in the international *Journal of Mathematics Teacher Education* indicated that current empirical work tends to focus on the learning of (prospective and practising) teachers who participated in professional education activities, and not on the nature of the practice of offering professional education. Moreover, the findings of a recent survey of research in mathematics teacher education (Adler, Ball, Krainer, Lin, & Novotna, 2005) showed that almost all of the empirical research related to teacher professional education consists of self-reports of teacher educators on their own work, and thus represent only a fraction of such work – solely that conducted by teacher educators who also publish in scholarly publications (some are internationally renowned university-based researchers). The survey further showed that almost all publications on research in mathematics teacher education came from countries where English is the national language, again limiting the information about teacher professional education around the world.

A glimpse at the nature of the practice of offering professional education for practising mathematics teachers by people who are not university based, and at who they might be, is presented in Even et al. (2003). The authors describe the work of two experienced junior-high school teachers who worked with 7th grade teachers in Israel on implementing a new mathematics curriculum programme – one of the 15 specific professional development strategies or learning experiences described by Loucks-Horsley et al. (1998). Although their background and work conditions were quite different from each other the study revealed that there were similar characteristics in their work practices with teachers. The most salient ones were acting out lessons, analyzing principles of the new curriculum programme, encouraging the teachers to explicate their concerns, and asking teachers to solve concrete practical problems related to the reservations they had about specific components of the new curriculum programme. Stein et al. (1999) reported similar but also other practices of university-based educators who attempted to help teachers learn new paradigms of teaching and learning mathematics, such as, confronting teachers with conflicts between new ideas and their existing beliefs and practices. The empirical literature does not inform us whether these or other are common practices of offering professional education for teachers of mathematics. Previous rhetoric suggests that there is a need to change traditional ways of offering professional education for teachers of mathematics (Loucks-Horsley et al., 1998; Wilson & Berne, 1999), but these claims are often based on beliefs, and not on systematic empirical research on the nature of the practice of educating practising mathematics teachers.

The ill-defined nature of the field together with lack of systematic investigation of the practice of educators of practising mathematics teachers combined with the limited opportunities to learn from the experiences of others, present a genuine challenge for today's designers of programmes for the education of educators to prepare them to work with practising mathematics teachers. We move now to examine how this challenge may be met, by analyzing a case of one such programme, part of the MANOR project in Israel, which was established in 1993 and operated until 2003.

THE CASE OF MANOR

Below I first sketch briefly the context of the development of the MANOR project. Then I explicate the decisions made regarding participants, curriculum and pedagogy: decisions regarding *what* educators of practising teachers need to learn and regarding *how* they would be taught. I also introduce the construct of *knowtice* (a combination of *know*ledge and prac*tice*) as a way to capture the essence of what needs to be developed.

Context

Traditionally, opportunities for practising teacher learning in Israel resembled those described by U.S. researchers (Ball & Cohen 1999; Wilson & Berne, 1999):

> [F]ormal and informal, mandatory and voluntary, serendipitous and planned-stitched together into a fragmented and incoherent "curriculum". (Wilson & Berne, 1999, p. 174)

In 1992 the Superior Committee on Science, Mathematics and Technology Education in Israel (an ad-hoc committee set up by the Ministry of Education) published the *Tomorrow 98 Report* on reform in science, technology and mathematics education. The report included recommendations for new mathematics and science curricula, special projects, and changes, both educational and structural, to take place by 1998. The report acknowledged the important role of the teacher in improving students' opportunities to learn mathematics and science, and recommended various ways of offering professional education to practising teachers, including the establishment of regional teacher centres as sites for life-long learning of practising mathematics and science teachers.

Following the publication of the *Tomorrow 98 Report* the Ministry of Education allocated a large budget for innovative curriculum development projects, as well as for professional education of practising mathematics and science teachers. The Ministry invited university mathematics and science education researchers to help in achieving the goals of the reform. In general the atmosphere was supportive to innovative ideas. In response to the Ministry's call I proposed to develop and conduct a programme for the preparation of educators of practising secondary school mathematics teachers (MANHIM in Hebrew – which means guides), and to develop resource materials for them to use in their work. The proposal was submitted in May 1993, and in August 1993 we were asked to start the MANOR Programme in that academic year (1993/4).

Participants

One of the first decisions needed to be made was who should be invited to participate in the programme. This was not a trivial question, because, as mentioned earlier, the group of educators of practising teachers was (and still is) not well defined. The magnitude of the task of preparing enough educators for large-scale (nation-wide) teacher learning, there being about 4500 secondary school mathematics teachers in Israel, led us to aim at people who had professional contact with teachers, such as teacher mentors, mathematics coordinators in school, or staff of curriculum implementation projects. In many cases these people did not regard themselves as teachers of teachers, nor did they usually support teacher learning. Yet, we believed that they had the appropriate conditions to become authentic teacher educators for practising teachers.

Admission to the MANOR Programme was selective and took into account the applicants' background. Formal education for, and experience in, teaching secondary school mathematics were required. An important admission criterion was the candidates' potential to influence mathematics teaching at school (as indicated by the application forms, the interviews and the recommendation letters accompanying the applications). Attention was also given to accepting candidates from different parts of the country and from different sectors, so that the whole country would be "covered". Indeed, participants in the programme came from the most southern part of the country up to the northern part, from Jewish and Arab sectors, and from orthodox religious and secular sectors. Almost all of them were practising secondary school mathematics teachers when they applied to the programme. Teaching experience varied considerably, from 5 to 30 years. All held a first degree either in mathematics or in a mathematics-related field, such as a B.Ed. with a mathematics major, or a B.Sc. in Chemistry. Most held only a bachelors' degree, some had a masters' degree, and a few held a Ph.D, the latter either in mathematics education or in education. Many already had experience in work with practising secondary teachers before coming to the programme, either as providers of professional education in one way or another, or as mathematics coordinators in their schools.

The two educators of practising teachers in the Even et al. (2003) study are two examples of participants' background and work conditions. One was a teacher college graduate. She had 30 years of experience in junior-high school mathematics teaching, and was also a member of a team that developed a new curriculum programme for 7th and 8th grade heterogeneous classes. She taught a year-long bi-weekly two-hour workshop for 7th grade teachers who started to teach, as a result of the Ministry of Education's requirement, in heterogeneous classes. Twenty teachers from six schools participated in this workshop organized by the municipality. The other participant was a university graduate with a B.A. in mathematics and statistics. She had 14 years of experience in junior-high mathematics teaching. When starting the MANOR Programme, she was formally assigned by the principal to work with the mathematics teachers in the school where she was also a teacher in order to advance the teaching of this subject at the school – at the same time another teacher served as the mathematics coordinator at the school. She met with the teachers weekly for two hours during the school year.

What Should Educators of Practising Teachers Learn?

Decisions about curriculum are obviously related to aims. The MANOR Programme aimed to develop a professional group of educators whose role is to promote practising secondary school mathematics teachers' learning about mathematics teaching, and to support the development of teacher knowledge and practice. As exemplified above, this could take place in a variety of settings with different roles played by the teacher educator, and with heterogonous groups of teachers by means of education, teaching experience, socio-cultural background, motivation, etc. Examples include, working with a group of teachers from several schools in a regional teacher center on expanding their view of mathematics teaching and learning, as part of the teachers' life-long learning, or teaching short courses on a specific topic to a large group of teachers from all over the country. Another way is to work with the teachers in one's own school, expanding the role of the school mathematics coordinator to include also education of the school teachers. Yet, another setting is to work in a curriculum implementation project, focusing on teachers' professional growth, and so on. We wanted the MANOR curriculum to fit different, often unforeseen possibilities, and to enable the programme graduates to be able to develop their own ways of supporting teacher learning and professionalism according to their specific future contexts and roles, the teachers involved, and personal preferences.

Discussions about the preparation of educators tend to start with questions of knowledge: What should future educators know? (e.g., Shulman, 1986). The MANOR project was no exception. However, discussions about curriculum cannot focus exclusively on knowledge, because offering professional education is something one does, not just knows. Thus, in addition to identifying areas for professional knowledge base, the MANOR curriculum gave explicit attention to the work in which the participants would engage, by focusing on practice. Moreover, a central characteristic of MANOR was the integration of knowledge and practice (Even, 2005); integration that I term *knowtice* to signify that this integration is related to the elements that create it (*know*ledge and prac*tice*), but that the product is a new object. In other words, *knowtice* is the essence of what we wanted the MANOR participants to learn and develop. For the sake of clarity, I first focus below on the knowledge aspect and then on the practice aspect of *knowtice*.

In line with the programme's aim, the following three areas were chosen as necessary professional knowledge base:
– *Mathematics Education*: Current views of mathematics teaching, learning, and knowing.
– *Mathematics*: Secondary school mathematics and advanced mathematics.
– *Teacher education*: Current views of teaching teachers and of teacher learning.

The first two are resonant with Cooney's (1994, 2001) constructs with relation to teacher knowledge – pedagogical power and mathematical power. The third area resembles the additional type suggested by Jaworski (2001) with relation to the teacher educator's knowledge – educative power. The three areas of the professional knowledge base are also in line with the levels proposed by Jaworski (2001) (see also Zaslavsky et al., 2003).

Related programme activities included, for example, activities that aimed to develop richer, more empowering vision about good school mathematical activities, understanding about students' ways of learning, and the appreciation that inquiry into student learning of mathematics is important for teaching. Activities also aimed at developing knowledge about change initiatives in school mathematics and a disposition towards teachers as thoughtful learners and professionals. (More details about these activities can be found in Even, 1999a, 1999b, 2005; Even & Bar-Zohar, 1997.)

In addition to focusing on developing knowledge about teacher learning and education, we also taught practices related to the teacher educator's work:
– *Practices of teacher education*: Ways of educating practising mathematics teachers.

These included general practices, such as, planning, conducting and assessing activities, workshops and courses for practising teachers, and specific ones, such as, deciding on aims for professional development activities, or reflecting on, and discussing with colleagues the professional development activities they conducted.

Another practice in which teacher educators often engage is making a presentation. This could be part of a regular work with teachers, in professional conferences for teacher educators, or in meetings with policy makers (e.g., when working on making a change in school mathematics). These specific practices may seem obvious. But they were not. There was a real need to learn them, as we found out.

For example, at the beginning of the programme the participants usually could not explain the aims of the professional development activities they conducted as part of the programme requirements. They often described their aims in general vague statements (e.g., "didactic workshop"), sometimes mixing up aims and strategies (e.g., "worksheets"). Similarly, at the beginning of the programme, many of the participants were unable to explain their actions and decisions related to the professional development activities they conducted. They often felt that they were under attack, focusing on defending their actions. Many of the participants were also unable to deliver a coherent presentation and finish it on time when starting the programme. Consequently, there was a need to focus purposefully in the programme on learning these and other specific practices related to the work of teacher educators (Even & Bar-Zohar, 1997). In light of the idea of developing *knowtice*, this was done in an integrated focus on developing knowledge in the areas mentioned above, so that knowledge and practice were dealt with as one entity – *knowtice*, as discussed in the following.

How Should Educators of Practising Teachers Be Taught?

The education of educators of practising teachers in MANOR comprised several components. One was a formal two-year preparation programme (although the first cohort extended over three years). In addition to its novel focus on the preparation of educators of practising teachers, the MANOR Programme set new standards and norms for professional development of mathematics educators in Israel in non-academic positions. These include, (a) a formal multi-year programme of study, (b) requirements of high commitment and work investment outside formal meetings (e.g., experimenting with innovative learning materials, reading research articles, conducting small scale research studies, writing papers, conducting weekly professional development activities for practising teachers, initiating change in school mathematics teaching and learning, preparing annual portfolios, etc.), and (c) inclusion of an advanced academic component for which the participants received graduate credit from the Weizmann Institute (in Israel, courses for practising teachers rarely carried academic credit of any kind). This programme was designed to contribute to the learning of new knowledge and practices and the development of *knowtice* for educating practising teachers (for more details, see Even, 1999a). During the years 1993–2003 three groups – 75 graduates – have finished their studies at the MANOR Programme.

Programme graduates and other educators of practising teachers were invited to participate in a long-term monthly Forum, and in bi-annual national Conferences, as a means of continuous professional development, support for the construction of a professional community of educators of practising teachers, and the building-up of professional identity and confidence. In contrast with the design of the MANOR Programme, the Forum participants shared responsibility with the MANOR staff for the content and framework of the Forum meetings. Similarly, MANOR Programme graduates and staff organized together the Conferences. We also developed an interactive Internet site for educators of practising secondary school mathematics teachers, and provided continuous individual support and consultation either by phone or face-to-face conversation, and regular or electronic mail.

MANOR also developed resource materials, based on the assumption that teacher educators require not only adequate preparation and continuous opportunities to participate in professional activities, but also adequate materials for the purpose of designing learning experiences for practising teachers. Five resource files have so far been developed: on algebra, functions, limits, , and on teaching mathematics in heterogeneous classes. The major themes in these files are: (a) historical view on the main topic of the file, (b) selected mathematical aspects relevant to the topic, (c) students' conceptions and ways of learning and thinking, (d) aspects of mathematics lessons and teaching. The resource files contain detailed suggestions of learning activities for secondary school mathematics teachers. Several models for activities are exemplified, in order to suggest and illustrate ways of working with practising teachers, and also convey the message that there is more than one way.

Not having an adequate theoretical/conceptual framework to serve as the basis for the design of learning experiences for teacher educators, as discussed in the first part of the chapter, MANOR drew on several theoretical and conceptual orientations that focus on learning knowledge and practice in general and teacher education in particular. Below is a brief description (for more details, see Even, 2005).

One is an approach that has been promoted in recent years for *student learning of mathematics*, and was adopted in MANOR to the case of *learning to educate teachers of mathematics*. This approach reflects an amalgamation of a constructivist/cognitivist approach to learning with a socio-cultural approach (e.g., Even & Lappan, 1994; Lampert, 1990; Yackel & Cobb, 1996). It views learning as an active construction of meanings by making sense of mathematics problem situations, combined with the development of social norms and practices, such as, inquiry, reasoning, explanation, justification, argumentation, and intellectual autonomy. In line with this approach, classroom discourse is organized so that students learn to explain their ideas and solutions to mathematics problems, rather than focusing entirely on whether answers are correct. Students interact with each other: they formulate and evaluate questions, hypotheses and explanations, and propose and evaluate evidence and arguments presented by other students. Particular attention is given to those norms of discourse involving respectful attention to others' opinions and efforts to reach mutual understandings based on mathematical reasoning. We further incorporated to this approach ideas related to *teacher learning* which advocate teacher inquiry into, and reflection on, their practice (e.g., Ball & Cohen, 1999; Jaworski, 1998; Krainer, 1998; Loucks-Horsley et al., 1998; Zaslavsky et al., 2003), again adopting these ideas to the case of *learning to educate teachers of mathematics*.

Thus, the design of MANOR activities aimed at encouraging work on solving authentic problems of teaching mathematics and of educating teachers of mathematics, making what was typically assumed and taken for granted, questionable and examinable. For example, the MANOR activity 'What is a good problem in school mathematics?' (Even, 1999a) offered the opportunity to work on solving an authentic problem of teaching mathematics, and to study closely an important teaching practice. Examining, explaining reasoning to others, learning what others had experienced, and reflecting on unexpected outcomes, encouraged awareness of the need for careful consideration when choosing or designing activities for students, and attention to the different activities that may emerge from a written mathematics problem. Similarly, the 'Mini study' activity (Even, 1999b), in which participants "replicated" a study with students and with teachers, and compared their findings with the findings of the original study, offered the opportunity to work on solving a different problem of teaching mathematics, namely, students' learning processes, and on solving a problem of educating teachers of mathematics. This activity encouraged an appreciation of the idea that students construct their knowledge in ways which are not necessarily identical to the instruction. The activity challenged existing conceptions and beliefs about student learning of mathematics, and fostered the development of better understanding about what the constructivist view might mean in a mathematics classroom. For example, some realized that learning processes are complicated, no matter how "clear" the instruction; others learned that, against expectation, students were able to deal with sophisticated mathematical ideas. Moreover, the activity provided opportunities to focus on issues which commonly were not attended to by the participants, like teacher knowledge of mathematics and students.

Furthermore, via the different components of MANOR, we aimed at having the participants collaborate on solving problems of teaching mathematics and of teaching teachers of mathematics, suggest alternative solutions, explain their reasoning to their peers, examine each other's solutions, and construct meanings, new knowledge and ideas related to mathematics, to teaching mathematics, and to educating teachers of mathematics. For example, the 'Change initiatives' activity (Even, 1999a) required each participant to choose an aspect of school mathematics, and to work with a group of teachers on planning, conducting, and evaluating change initiatives related to this topic (e.g., building a mathematics room, developing a programme for students to work on projects in mathematics, teaching in heterogeneous classes, using new technologies in the teaching of mathematics, helping to prevent at-risk high school students from dropping out and not taking the matriculation exam). Those who chose the same topic worked as a team, collaborating on planning, implementing and evaluating activities, sharing and discussing ideas, difficulties and challenges.

Another theoretical orientation on which MANOR drew is related to the learning of a practice. We were inspired mainly by the situated learning approach

(Lave & Wenger, 1991), in which knowing is viewed as the practices of a community and the abilities of individuals to participate in those practices, and learning is the strengthening of those practices and participatory abilities, that takes place in a participation framework. We also embraced the framework of three types of teacher development (personal, professional and social), suggested by Bell and Gilbert (1994), which attends to issues of developing a professional identity and ways of working with others in the community. Thus, from the beginning, we offered activities that allowed the participants in the MANOR Programme, Forum and Conferences to be both members in the community of practice and learners. For example, as part of the programme each participant conducted weekly two-hour professional development activities for a group of secondary school teachers of mathematics in a school or in a regional teacher center. The support provided in the MANOR Programme for this activity emphasized thinking, examination, analysis and reflection on and about the experiences, as a means to improve *knowtice*, building on the idea of legitimate peripheral participation (Lave & Wenger, 1991), a process by which the learner becomes a full participant in the socio-cultural practices of a community.

Another opportunity to participate in the community of practice was the use of the participants' writings on the activities they conducted for practising teachers of mathematics as resources for other participants and other teacher educators. Similarly, programme graduates participated in the organization of the MANOR Conferences, where Conference participants presented their work, discussed professional issues, and shared ideas and materials with each other. Still, another example is the organization of conferences by programme participants, to which school principals, superintendents, and policy makers in the educational system were invited, to promote contacts with people who could either assist them in their future work or be assisted by them. Such MANOR activities enabled and encouraged different ways and different levels of participation in the complex practice of educating practising teachers. These activities also offered opportunities to learn ways of work with others in the community, and promoted connections with, and support from, experts, peers, and other members in the community of teacher educators of practising teachers and the larger community of mathematics educators and the educational system.

Drawing on the diverse theoretical perspectives we tried to provide a supportive and intellectually and professionally demanding environment. In all components of MANOR (Programme, Forum, Conferences, Resource Files, and individual support) we aimed at designing activities where participants needed to solve real problems of practice, combined with opportunities for reflecting on and analyzing these solutions, in the light of academic and practical knowledge (Even, 1999b; Leinhardt, McCarthy, Young, & Merriman, 1995). We focused on the development of norms of interaction that encourage the study and critique of one's own and others' practice, combined with the actual enactment of knowledge, as a means to develop knowledge, skills, dispositions and practices situated in the practice of educating practising teachers, i.e., *knowtice*.

LOOKING TO THE FUTURE

The first part of the chapter discussed three problematic aspects in the current literature which underlie the challenge to prepare mathematics educators to work with practising teachers: that there is almost no research on the education of mathematics teacher educators, the ill-defined nature of the field of educating practising mathematics teachers, and the lack of information on the practice of mathematics teacher educators working with practising teachers. The second part of the chapter examined how the MANOR project in Israel met the challenge to prepare mathematics educators to work with practising teachers, explicating decisions made regarding participants, curriculum and pedagogy. The construct of *knowtice* was introduced as the essence of what needs to be developed in a programme that prepares mathematics educators to work with practising teachers, highlighting its simultaneously objective identity and close connection to knowledge and practice – the components whose integration produces *knowtice*. The last part of the chapter revisits the problems underlying the challenge to prepare mathematics educators to work with practising teachers with an eye towards the future, suggesting directions for future research.

Research on the Education of Teacher Educators

The main reason for the limited literature on the education of mathematics teacher educators is that until recent years, there were essentially no formal programmes that prepared mathematics educators to work with teachers in general, and with practising teachers in particular (Zaslavsky et al., 2003). Tzur's (2001) personal account on how he became a mathematics teacher educator through his practice highlights this lack of institutional and professional support. With the expanding current interest in this issue in different countries, and the emergence of pioneering work in structuring the education of mathematics teacher educators as outlined at the beginning of this chapter, the timing is right for a more comprehensive research effort on the education of mathematics teacher educators for practising teachers. Questions central to such investigation should address various aspects of curriculum, pedagogy and structure of preparation programmes, as well as continuous professional development experiences. For example, What do educators of practising teachers need to learn, and when should they learn that? How should they be taught? How might the preparation be organized? What are useful learning experiences, and for what purposes? What kinds of support are helpful? How might the preparation of educators for practising teachers be distinguished from that of educators for prospective teachers? The research should also address issues of theoretical and conceptual frameworks. For example, what theoretical orientations are helpful and for what purposes? Are different theoretical orientations compatible?

The Nature of the Field of Educating Practising Teachers

There is a need to develop better understanding about the field of educating practising teachers of mathematics. For example, we need information about the heterogeneous group of people who might be regarded as educators of practising mathematics teachers. Questions central to such investigation include, who educates practising teachers of mathematics in different settings? And in different countries? Who should/could be recruited to perform this role? There is also a need to understand better the nature of the professional development system. Wilson and Berne (1999) describe the nature of professional development for practising teachers in the USA: "Some teachers pursue any opportunity to learn with passion, while others attend workshops when mandates arrive in their school mailbox" (p, 197). The apparently institutionalized education of practising teachers in elementary schools in Japan in the form of lesson study (Yoshida, 1999) suggests that there may be differences among countries. The attempt of the Israeli Ministry of Education to build a semi-formal professional development system in the 1990's suggests that political and socio-cultural aspects may play an important role. The Japanese experience suggests also that there may be differences between the nature of educational opportunities for practising teachers who teach different grade levels in the same country. Thus, research should also address questions related to characteristics of different professional development systems and to the nature of productive practice of educating practising mathematics teachers.

Information on the Practice of Mathematics Teacher Educators

Clearly we need to understand better what mathematics educators who work with practising teachers do, in order to construct an informed understanding about the nature of adequate education for educators of practising teachers. Questions central to such investigation include, what are common practices of mathematics educators who work with practising teachers? Are there practices that are common to university-based educators and others to non university-based educators? Are there practices that characterize specific professional development strategies described by Loucks-Horsley et al. (1998), such as those conducted in the context of curriculum implementation?

The community of mathematics educators recognizes the significance of educating practising mathematics teachers. Yet, the current absence of opportunities to learn how to do that means that educators of practising mathematics teachers are often ill-prepared for the complex demands of the work that they face. In this chapter I suggested that the development of *knowtice* of education of practising mathematics teachers is key, and have pointed out what is missing in current literature in the hope that these ideas will inform future dialogue, work and education of educators of practising mathematics teachers.

REFERENCES

Adler, J., Ball, D. L., Krainer, K., Lin, F. L., & Novotna, J. (2005). Mirror images of an emerging field: Researching mathematics teacher education. *Educational Studies in Mathematics, 60*, 359–381.

Advisory Committee on Mathematics Education (2002). *Continuing professional development for teachers of mathematics.* London, UK: ACME, the Royal Society.

Ball, D. L., & Cohen, D. (1999). Developing practice, developing practitioners. In L. Darling-Hammond & G. Sykes (Eds.), *Teaching as the learning profession: Handbook of policy and practice* (pp. 3–32). San Francisco, CA: Jossey Bass.

Bell, B., & Gilbert, J. (1994). Teacher development as professional, personal, and social development. *Teaching and Teacher Education, 5*, 483–497.

Borko, H. (2004). Professional development and teacher learning: Mapping the terrain. *Educational Researcher, 33*(8), 3–15.

Cooney, T. (1994). Teacher education as an exercise in adaptation. In D. Aichele & A. Coxford (Eds.), *Professional development for teachers of mathematics: 1994 Yearbook of the National Council of Teachers of Mathematics* (pp. 9–22). Reston, VA: National Council of Teachers of Mathematics.

Cooney, T. (2001). Considering the paradoxes, perils, and purposes of conceptualizing teacher development. In F. L. Lin & T. Cooney (Eds.), *Making sense of mathematics teacher education* (pp. 9–31). Dordrecht, the Netherlands : Kluwer Academic Publishers.

Davenport, L. R., & Ebby, A. (2000, April). *Teacher leadership development in mathematics education: Stories of three apprentices.* Paper presented at the annual meeting of American Educational Research Association, New Orleans, LA.

Elliott, R. L. (2005, April). *Professional development of professional developers: Using practice-based materials to foster an inquiring stance.* Paper presented at the annual meeting of the American Education Research Association, Montreal, Canada.

Even, R. (1999a). The development of teacher-leaders and in-service teacher educators. *Journal of Mathematics Teacher Education, 2*, 3–24.

Even, R. (1999b). Integrating academic and practical knowledge in a teacher leaders' development program. *Educational Studies in Mathematics, 38*, 235–252.

Even, R. (2005). Integrating knowledge and practice at MANOR in the development of providers of professional development for teachers. *Journal of Mathematics Teacher Education, 8*(4), 343–357.

Even, R., & Ball, D. L. (Eds.) (in press). *The professional education and development of teachers of mathematics: The fifteenth ICMI Study.* New York: Springer.

Even, R., & Bar-Zohar, H. (1997). *Project for the preparation of mathematics teacher-leaders: Summary and evaluation of activities 1993–1996. Technical Report.* Rehovot, Israel: Weizmann Institute (in Hebrew).

Even, R., & Lappan, G. (1994). Constructing meaningful understanding of mathematics content. In D. B. Aichele & A. F. Coxford (Eds.), *Professional development for teachers of mathematics: 1994 Yearbook of the National Council of Teachers of Mathematics* (pp. 128–143). Reston, VA: National Council of Teachers of Mathematics.

Even, R., Robinson, N., & Carmeli, M. (2003). The work of providers of professional development for teachers of mathematics: Two case studies of experienced practitioners. *International Journal of Science and Mathematics Education, 1*(2), 227–249.

Farah, I., & Jaworski, B. (Eds.) (2006). *Partnerships in educational development.* Oxford Studies in Comparative Education. Wallingford, England: Triangle Books.

Goodchild, S. (2007). Inside the outside: Seeking evidence of didacticians' learning by expansion. In B. Jaworski, A. B. Fuglestad, R. Bjuland, T. Breiteig, S. Goodchild, & B. Grevholm (Eds.), *Learning communities in mathematics* (pp. 189–203). Norway: Caspar Forlag AS

Gutiérrez, A., & Boero, P. (Eds.) (2006). *Handbook of research on the psychology of mathematics education.* Rotterdam, the Netherlands: Sense Publishers.

Halai, A. (1998). Mentor, mentee, and mathematics: A story of professional development. *Journal of Mathematics Teacher Education, 1*, 295–315.

Hershkowitz, R., & Breen, C. (2006). Forward – Expansion and dilemmas. In A. Gutiérrez & P. Boero (Eds.), *Handbook of research on the psychology of mathematics education* (pp. ix–xii). Rotterdam, the Netherlands: Sense Publishers.

Higgins, J. (2005). Pedagogy of facilitation: How do we best help teachers of mathematics with new practices? In H. L. Chick & J. L. Vincent (Eds.), *Proceedings of the 29th conference of the International Group for the Psychology of Mathematics Education* (Vol. 3, pp. 137–144). Melbourne: Psychology of Mathematics Education.

Jaworski, B. (1998). Mathematics teacher research: Process, practice, and the development of teaching. *Journal of Mathematics Teacher Education, 1*, 3–31.

Jaworski, B. (2001). Developing mathematics teaching: Teachers, teacher educators, and researchers as co-learners. In F. L. Lin & T. J. Cooney (Eds.), *Making sense of mathematics teacher education* (pp. 295 320). Dordrecht, the Netherlands: Kluwer Academic Publishers.

Krainer, K. (1998). Some considerations on problems and perspectives of inservice mathematics teacher education. In C. Alsina, J. M. Alvarez, J. Hodgson, C. Laborde, & A. Perex (Eds.), *Eighth International Congress on Mathematics Education: Selected lectures* (pp. 303–321). Seville, Spain: S.A.E.M. Thales.

Lampert, M. (1990). When the problem is not the question and the solution is not the answer: Mathematical knowing and teaching. *American Educational Research Journal, 27*, 225–282.

Lave, J., & Wenger, E. (1991). *Situated learning: Legitimate peripheral participation*. Cambridge: Cambridge University Press.

Leinhardt, G., McCarthy Young, K. M., & Merriman, J. (1995). Integrating professional knowledge: The theory of practice and the practice of theory. *Learning and Instruction, 5*(4), 401–408.

Loucks-Horsley, S., Hewson, P. W., Love, N., & Stiles, K. E. (1998). *Designing professional development for teachers of science and mathematics*. Thousand Oaks, CA: Corwin Press.

Mcnamara, O., Jaworski, B., Rowland, T., Hodgen, J., & Prestage, S. (2002). *Developing mathematics teaching and teachers: A research monograph*. Retrieved January 4, 2008, from http://www.maths-ed.org.uk/mathsteachdev/pdf/mathsdev.pdf

Nesher, P., & Kilpatrick, J. (Eds.). (1990). *Mathematics and cognition: A research synthesis by the International Group for the Psychology of Mathematics Education*. Cambridge: Cambridge University Press.

Reference Group on Teacher Standards, Quality and Professionalism. (2003). *A national statement from the Teaching Profession on teacher standards, quality and professionalism*. Retrieved November 9, 2003, from the Australian College of Educators' Website: http://www.austcolled.com.au/whatsnewpage.html

Shulman, L. S. (1986). Those who understand: Knowledge growth in teaching. *Educational Researcher, 15* (2), 4-14.

Stein, M. K., Smith, M. S., & Silver, E. A. (1999). The development of professional developers: Learning to assist teachers in new settings in new ways. *Harvard Educational Review, 69* (3), 237–269.

Superior Committee on Science Mathematics and Technology Education in Israel (1992). *Tomorrow 98: Report*. Jerusalem, Israel: Ministry of Education, Culture and Sport (English edition: 1994).

Sztajn, P., Ball, D. L., & McMahon, T. (2005, May). *And who teaches the mathematics teachers? Professional development of teacher developers*. Contributed paper for an interactive work-session at the ICMI-15 Study conference on the professional education of mathematics teachers, Aguas de Lindoia, Brazil.

Tzur, R. (2001). Becoming a mathematics teacher-educator: Conceptualizing the terrain through self-reflective analysis. *Journal of Mathematics Teacher Education, 4*, 259–283.

Wilson, S. M., & Berne, J. (1999). Teacher learning and the acquisition of professional knowledge: An examination of research on contemporary professional development. In A. Iran-Nejad & C. D. Person (Eds.), *Review of Research in Education, 24*, 173–209.

Yackel, E., & Cobb, P. (1996). Sociomathematical norms, argumentation, and autonomy in mathematics. *Journal for Research in Mathematics Education, 27*, 458–477.

Yoshida, M. (1999, April). *Lesson Study [jugyokenkyu] in elementary school mathematics in Japan: A case study*. Paper presented at the American Educational Research Association Annual Meeting, Montreal, Canada.

Zaslavsky, O., Chapman, O., & Leikin, R. (2003). Professional development in mathematics education: Trends and tasks. In A. J. Bishop, M. A. Clements, C. Keitel, J. Kilpatrick, & F. K. S. Leung (Eds.), *Second international handbook of mathematics education* (pp. 877–915). Dordrecht, the Netherlands: Kluwer Academic Publisher.

Zaslavsky, O., & Leikin, R. (2004). Professional development of mathematics teacher educators: Growth through practice. *Journal of Mathematics Teacher Education, 7*(1), 5–32.

Ruhama Even
Department of Science Teaching
Weizmann Institute of Science

MERRILYN GOOS

4. SOCIOCULTURAL PERSPECTIVES ON LEARNING TO TEACH MATHEMATICS

Within the mathematics education research community there is growing interest in theories that view teachers' learning as a form of participation in social and cultural practices rather than as an internal mental process. This chapter explores what we can learn from research that takes a sociocultural perspective on learning to teach. The first section critically analyses selected sociocultural studies that draw on discourse, situative, and community of practice perspectives. The second section presents an alternative sociocultural approach based on Valsiner's zone theory and illustrates its use in my research with prospective and practising teachers. In the final section I propose that zone theory might provide a sociocultural framework for understanding the work of mathematics teacher-educator-researchers.

This chapter considers what we can learn from research that takes a *sociocultural perspective on conceptualising "learning to teach"*. Recent reviews of research in mathematics teacher education have noted increasing attention to the social, cultural and institutional dimensions of teachers' learning as well as attempts to integrate social and individual levels of analysis (da Ponte & Chapman, 2006; Lerman, 2001; Llinares & Krainer, 2006). These developments have also proven fruitful in research investigating factors that promote or hinder teachers' learning and professional growth.

Lerman (1996) defined sociocultural approaches to mathematics teaching and learning as involving "frameworks which build on the notion that the individual's cognition originates in social interactions (Harré & Gillett, 1994) and therefore the role of culture, motives, values, and social and discursive practices are central, not secondary" (p. 4). The present chapter explores sociocultural research that focuses on mathematics teacher education. It is not the intention to provide a comprehensive review; instead, key studies have been selected as exemplars in order to address the following questions (adapted from Adler, Ball, Krainer, Lin, & Novotna (2005):

- How can sociocultural perspectives contribute to our understanding of how teachers learn, from what opportunities, and under what conditions?
- How can sociocultural perspectives inform research and professional development interventions that aim to improve teachers' opportunities to learn?

The first section of the chapter reviews selected studies of pre-service teacher education, the transition from prospective to beginning teacher, and professional development programs to illustrate what we might learn from the various sociocultural orientations employed. The second section further develops one sociocultural approach – an application of Valsiner's (1997) zone theory, illustrates its use in my own research involving prospective mathematics teachers, beginning mathematics teachers, and mathematics teachers in professional development programs, and identifies further empirical work that is needed to support and elaborate the theory. The final section of the chapter examines, from a sociocultural perspective, what it means to "learn" from research in or into teacher education and proposes that zone theory might offer a sociocultural framework for understanding the work of mathematics teacher-educator-researchers.

THE SOCIOCULTURAL LANDSCAPE IN MATHEMATICS TEACHER EDUCATION: CONCEPTS AND KEY STUDIES

Sociocultural perspectives on learning and development grew from the work of Vygotsky in the early 20th century. Vygotsky's theoretical approach refers to the social origins of higher mental functions, and the mediation of these functions by tools and signs, such as language, writing, systems for counting and calculating, algebraic symbol systems, diagrams, and so on. Vygotsky also introduced the concept of the Zone of Proximal Development (ZPD) to explain how social phenomena are transformed into psychological phenomena. He proposed that the ZPD is created when a child's interaction with an adult or more capable peer awakens mental functions that have not yet matured and thus lie in the region between actual and potential developmental levels.

Early attempts to apply Vygotsky's theory in educational research led to studies of how children learned through collaborative interaction with adults. Through the 1980s and 1990s, a second generation of research extended the emerging sociocultural framework by giving attention to the institutional context of social interactions, the importance of interpersonal relationships in teaching and learning interactions, and the idea that modes of thinking are closely linked to forms of social practice (Forman, Minick, & Stone, 1993).

Contemporary sociocultural theory proposes that learning involves increasing participation in socially organised practices. The notion of a situated learning in a community of practice composed of experts and novices (Lave & Wenger, 1991; Wenger, 1998) has been fruitfully applied to education contexts, even though this concept was not originally focused on school classrooms, nor on pedagogy.

Recent research on students' mathematics learning has employed sociocultural perspectives in a variety of ways. Forman (2003) identifies two themes within this research: (1) a *discourse* perspective focusing on the dynamics of mathematical communication in classrooms, where interest centres on the role of semiotic mediation; and (2) a *practice* perspective that links classroom activity structures with learning and identity. These themes will also serve to organise the discussion of what we can learn from key studies in mathematics teacher education that draw on sociocultural theories. To align with the two questions that guide this chapter, for each theme I have chosen studies that aim either to *understand* the processes of teacher learning or to *improve* teachers' opportunities to learn.

Discourse: Semiotic Mediation in the ZPD

Research by Blanton and colleagues has used Vygotsky's concept of language as a mediating tool in an individual's development to investigate the discourse practices of prospective teachers and also their university supervisors. One aspect of this research (Blanton, Berenson, & Norwood, 2001a) looked at a prospective teacher's classroom discourse as a mediator of her practice as her perception of teacher and student roles shifted from teacher as teller to student as mathematical participant. The changes in this prospective teacher's classroom practice did not come about by accident: they were the result of a planned intervention by Blanton as the university supervisor (Blanton, Berenson, & Norwood, 2001b). The latter study argues that practicum supervision in its conventional form is rarely educative because it merely assesses current teaching habits and focuses too much on peripheral issues such as classroom organisation. In contrast, truly educative supervision contributes to prospective teachers' learning and development by challenging their models of teaching in the context of their practice. Lesson observations provided an entry point into "teaching episodes" in which the university supervisor guided the prospective teacher through extended conversations about classroom interactions and what these revealed about how students learn mathematics. The pedagogy of supervision that emerged from this study was characterised by open ended questions as prompts for sense making, a less authoritative voice on the part of the supervisor, and a focus on mathematical pedagogy. This approach was claimed to open up a Zone of Proximal Development where the nexus between theory and practice could be explored.

More recently Blanton's research (Blanton, Westbrook, & Carter, 2005) has focused on interpreting novice teachers' ZPDs by applying Valsiner's (1997) zone theory of child development. Zone theory re-interprets and extends Vygotsky's concept of the Zone of Proximal Development to incorporate the social setting and the goals and actions of participants. Valsiner regards the ZPD as a set of possibilities for development that are in the process of becoming actualised as individuals negotiate their relationship with the learning environment and the people in it. To explain how the actual emerges from the possible, Valsiner proposes two additional zones, the Zone of Free Movement (ZFM) and the Zone of Promoted Action (ZPA). The ZFM structures an individual's access to different areas of the environment, the availability of different objects within an accessible area, and the ways the individual is permitted or enabled to act with accessible objects in accessible areas. The ZPA comprises activities, objects, or areas in the environment in respect of which the person's actions are promoted. Valsiner explained that the ZFM and ZPA are dynamic and inter-related, and are constantly being re-organised by the adult in learning interactions with the child. Applying Valsiner's ideas to classrooms, the teacher's instructional choices about what to promote and what to allow establish a ZFM/ZPA complex that defines the learning opportunities experienced by students. Blanton et al. compared the ZFM/ZPA complexes organised by three mathematics and science teachers in their respective classrooms as a means of revealing these teachers' understanding of student-centred inquiry and hence establishing the potential for development within their own ZPDs. The researchers observed that two of the teachers created the appearance of promoting discussion and reasoning when their teaching actions did not actually allow students to experience these, and they explained this apparent contradiction by theorising the existence of an illusionary Zone of Promoted Action (IZ). For one of these teachers, the IZ appeared to signal a transitory state as she eventually changed her practice to both promote and allow student interaction. Thus existence of an IZ may indicate that inquiry based teaching practices are within the teacher's ZPD but have not yet been enacted in the way intended or perceived by that individual. Nevertheless, Blanton et al. note there is no guarantee that this transition will occur and they acknowledge the need for further research on external factors, such as the role of teacher educators, that contribute to teachers' development.

Practice: Situative Perspectives and Communities of Practice

Situative perspectives combine a psychological focus on individual behaviour and cognition with a social science focus on interaction and communication between individuals (Greeno, 2003). Learning mathematics is viewed as participation in mathematical practices, and situative research examines how students acquire and use knowledge through their participation in the activities of a classroom community. Peressini, Borko, Romagnano, Knuth, and Willis (2004) adapted a situative perspective on learning to develop a conceptual framework for learning to teach secondary mathematics, focusing particularly on teacher learning within multiple contexts such as university mathematics and teacher education courses, practicum experiences, and schools of employment. Their longitudinal study followed six secondary mathematics teachers from the teacher education program into the first two years of teaching. Analysis of two cases focused on apparent inconsistencies in the mathematical understanding and pedagogical approaches displayed by teachers across different contexts. For example, one teacher developed two different conceptions of proof as she progressed through the university program: in her mathematics content classes she adhered to formal notions of proof, while in mathematics methods classes and during practice teaching she viewed proof as an act of informal sense-making or a way to explain why a piece of mathematics works. A second teacher implemented reform-based pedagogical approaches during practice teaching but more traditional approaches in his first year of employment in response to the different affordances and constraints of these two settings. The situative perspective helped the researchers understand how context makes a difference to the development of mathematics teachers and their professional identities: they argued that teachers' knowledge and beliefs interact with the contexts in which they are embedded "to create the situations in which learning to teach occurs" (p. 68), and hence knowledge-in-practice varies with participation in different contexts.

While situative perspectives attempt to coordinate cognitive and social views of learning, some researchers have embraced a fully social theory of learning as participation in a community of practice (Lave & Wenger, 1991). The notion of community resonates with current ways of understanding teachers' learning through professional collaboration (e.g., Graven, 2004; McGraw, Arbaugh, Lynch & Brown, 2003). Bohl and Van Zoest (2003) investigated the effects of participation in different communities on the development of a beginning teacher. According to Wenger (1998), a community's modes of participation comprise mutual engagement, negotiation of a joint enterprise, and development of a shared repertoire of resources for creating meaning. A community is also characterised by regimes of accountability that regulate participation of its members. Bohl and Van Zoest applied these concepts to analyse discontinuities between the beginning teacher's facility in talking about reform based mathematics teaching and her difficulty in translating her knowledge and beliefs into classroom practice. They noted that during her teacher education program she participated successfully in communities that required only conversation about and reflection on teaching, and thus imposed undemanding regimes of accountability. When she began teaching full-time, these modes of participation continued in extra-classroom communities involving her department and school, parents, and the broader professional community. Yet she lacked the repertoire of pedagogical resources for establishing a reform oriented classroom, and she was not held accountable for doing so by her school administration. Although she described her practice in terms of the language of reform, she and her students had settled into a comfortable and undemanding regime of accountability that created a traditional teacher-centred classroom.

What Can We Learn from Research on Understanding and Improving Mathematics Teachers' Learning?

One of the most significant challenges for teacher education research is to *understand* how prospective and beginning teachers learn from their experiences in different contexts, especially when these experiences can produce conflicting images of mathematics teaching. This challenge is sometimes associated with the perceived gap between knowledge provided by university-based teacher education courses and the practical realities of classroom teaching. Studies of teacher socialisation from a functionalist perspective typically identify influences such as the beliefs that prospective teachers bring to the course from their own schooling and the classroom practices they observe and experience in the early stage of their careers (Brown & Borko, 1992). Such approaches view teachers as being passively moulded by external forces to fit the existing culture of schools – thus producing the common explanation for why many beginning teachers give up their innovative ideas in the struggle to survive and conform to institutional norms of traditional practices. However, a sociocultural perspective proposes that any examination of teachers' learning and socialisation needs to consider the "person-in-practice-in-person" (Lerman, 2001, p. 28), a unit of analysis that allows us to shift our analytical focus between the individual and the social. As demonstrated by the small sample of studies summarised above, socioculturally oriented research can explain the apparent contradictions we observe when we examine teachers' learning across multiple contexts such as the university course, the practicum, the school of employment, and other professional settings outside the school.

Sociocultural perspectives have perhaps been used less effectively to inform research on *improving* teachers' opportunities to learn, thus leaving the role of the teacher educator largely untheorised. A more elaborated sociocultural theory of teaching is therefore needed to complement sociocultural language and concepts used to describe learning in a community of practice or in the ZPD. Lave and Wenger (1991) developed the concept of community of practice from studying informal learning in apprenticeship contexts where teaching is incidental rather than deliberate, as in teacher education, and Wenger (1998) provided no discussion of teaching when he later developed the concept into a social theory of learning. Valsiner's zone theory may offer a more promising pathway as it accounts for learning (ZPD), teaching (ZPA), and context (ZFM). For example, Blanton, Berenson, and Norwood's (2001b) pedagogy for supervision could be re-interpreted as a Zone of Promoted Action for teacher education, while the multiple contexts for teacher learning investigated by Peressini et al. (2004) and Bohl and Van Zoest (2003) might represent different Zones of Free Movement. This possibility is explored in the next part of the chapter.

USING VALSINER'S ZONE THEORY TO UNDERSTAND AND IMPROVE MATHEMATICS TEACHERS' LEARNING

Over several years I have developed a research program that applies Valsiner's (1997) zone theory to teacher learning and development (Galbraith & Goos, 2003; Goos, 2005a; Goos, Evans, & Galbraith, 1994). My approach differs from that of Blanton, Westbrook, and Carter (2005) described earlier in that all zones are defined from the perspective of the teacher as learner. When I consider how teachers learn, I view the teacher's ZPD as a set of possibilities for their development that are influenced by their knowledge and beliefs, including their disciplinary knowledge, pedagogical content knowledge, and beliefs about their discipline and how it is best taught and learned. The ZFM can then be interpreted as constraints within the teacher's professional context such as students (behaviour, socio-economic background, motivation, perceived abilities), access to resources and teaching materials, curriculum and assessment requirements, and organisational structures and cultures. While the ZFM suggests which teaching actions are *allowed*, the ZPA represents teaching approaches that might be specifically *promoted* by pre-service teacher education, formal professional development activities, or informal interaction with colleagues in the school setting. For learning to occur, the ZPA must engage with the individual's possibilities for development (ZPD) and must promote actions that the individual believes to be feasible within a given ZFM. It is significant that prospective teachers develop under the influence of two ZPAs, one provided by the university program and the other by the supervising teacher(s) in the practicum school, which do not necessarily coincide.

The following sections illustrate how I have applied Valsiner's (1997) zone theory to mathematics teacher learning in two studies involving prospective and practising teachers. These particular studies are chosen to illuminate the main theoretical questions addressed by the chapter. The first study developed the adaptation of zone theory outlined above to investigate the process of teacher learning in the context of transition from pre-service to beginning teaching of secondary school mathematics. The second study used zone theory to evaluate the effectiveness of a professional development intervention. Both studies aimed to identify factors that influence teachers' use of digital technologies in secondary school mathematics.

Previous research on technology use by mathematics teachers has already identified a range of factors influencing uptake and implementation, including: skill and previous experience in using technology; time and opportunities to learn; access to hardware, software, and teaching materials; technical support; organisational culture; knowledge of how to integrate technology into mathematics teaching; and beliefs about mathematics and how it is learned (Fine & Fleener, 1994; Manoucherhri, 1999; Simonsen & Dick, 1997; Walen, Williams & Garner, 2003; Wallace, 2004; Windschitl & Sahl 2002). In terms of the theoretical framework outlined above, these different types of knowledge and experience represent elements of a teacher's ZPD, ZFM and ZPA, as shown in Table 1. A limitation of previous research is that, in simply listing these factors, it does not consider possible relationships between the teacher's setting, actions, and beliefs, and how these relationships might change over time or across different classroom or school contexts. Zone theory provides a framework for analysing these dynamic relationships.

Table 1. Factors affecting technology usage

Valsiner's Zones	Elements of the Zones
Zone of Proximal Development	Skill/experience in working with technology
	Pedagogical content knowledge (technology integration)
	General pedagogical beliefs
Zone of Free Movement	Access to hardware, software, teaching materials
	Technical support
	Organisational culture
	Curriculum & assessment requirements
	Students (perceived abilities, motivation, behaviour)
Zone of Promoted Action	Pre-service education (university program)
	Practicum and beginning teaching experience
	Formal professional development activities
	Informal professional learning with colleagues

Technology Integration: From Prospective to Beginning Teaching

In this three year longitudinal study I followed successive cohorts of my teacher education students into their early years of teaching (Goos, 2005a, 2005b, 2005c). Participants comprised prospective teachers enrolled in a secondary mathematics methods course at a university in a large Australian city. I designed and taught the course so that students experienced regular and intensive use of graphics calculators, computer software, and Internet applications. Thus the course offered a teaching repertoire, or ZPA, that emphasised technology as a pedagogical resource.

Case studies of selected participants captured developmental snapshots of experience during the final practice teaching session, towards the end of the first year of full-time teaching, and in some cases later years of teaching. Participants were selected to sample practicum school settings that differed in terms of the Zone of Free Movement (professional context) and Zone of Promoted Action (Supervising Teacher approaches) they offered. I visited them in their practicum schools and schools of employment for lesson observations, collection of teaching materials and audio taped interviews (see Goos, 2005a for details).

Data sourced from lesson observations, surveys, questionnaires, and interviews were categorised as representing elements of participants' ZPDs, ZFMs, and ZPAs. As the zones themselves are abstractions, this analytical process focused on the particular circumstances under which zones were "filled in" with new people, actions, places and meanings. This approach enabled me to explore how personal, contextual, and instructional factors came together to shape prospective and beginning teachers' pedagogical identities as users of technology.

From Pre-Service to Beginning Teaching: The Case of Adam

The school where Adam completed his practice teaching sessions had recently bought resources such as graphics calculators, data logging equipment, and software. Every mathematics classroom was equipped with computers connected to the Internet, a data projector, and a TV monitor for projecting graphics calculator screen output. A hire scheme provided calculators to all students in the final two years of secondary school, and there were also sufficient class sets of calculators for use by younger classes. Some of these changes had been made in response to new mathematics syllabuses that mandated the use of computers or graphics calculators in teaching and assessment programs. Thus the school and curriculum environment offered a Zone of Free Movement that seemed to afford the integration of technology into mathematics teaching.

Adam had previously worked as a software designer and was confident in using computers and the Internet. Although he had not used a graphics calculator before starting the teacher education course, he quickly became familiar with its capabilities and with the support of his Supervising Teacher began to incorporate this and other technologies into his mathematics lessons. In theoretical terms, the Zone of Promoted Action organised by the Supervising Teacher was consistent with the ZPA I offered in my university course and also with the ZPD that defined Adam's potential for development.

After graduation Adam was employed by the same school where he had completed his practicum. One might expect Adam to experience a seamless transition from prospective to beginning teacher; yet I found this was not the case when I visited him near the end of his first year of teaching. By this time he had discovered that many of the other mathematics teachers were unenthusiastic about using technology and favoured teaching approaches that he claimed were based on their faulty belief that learning is linear and teacher-directed rather than richly connected and student-led. He described these beliefs and teaching approaches as follows:

> You do an example from a textbook, start at Question 1(a) and then off you go. And if you didn't get it – it's because you're dumb, it's not because I didn't explain it in a way that reached you.

Because he disagreed with this approach, Adam deliberately ignored the worksheet provided for the lesson by the teacher who coordinated this subject. The worksheet led students through a sequence of exercises where they were to construct tables of values, plot graphs by hand, and answer questions about the effects of each constant in turn. Only then was it suggested that students might use their graphics calculators to check their work. Conflicting pedagogical beliefs were a source of friction in the staffroom, and this was often played out in arguments where the teacher in question accused Adam of not teaching in the "right" way. Compared with his earlier experience as a prospective teacher, Adam now found himself in a more complex situation that required him to defend his instructional decisions while negotiating a harmonious relationship with several colleagues who did not share his beliefs about learning. Adam explained:

> [Now I'm willing] to stand up and say "This is how I am comfortable teaching". I just walk away now because we've had it over and over and the kids are responding to the way I'm teaching them. So I'm going to keep going that way.

In terms of Valsiner's zone framework, Adam became aware of conflicts between his technology-rich ZFM, a ZPA that promoted, at best, fairly mundane uses of technology in his teaching, and his personal ZPD. He responded by paying attention only to those aspects of the Mathematics Department's ZPA that were consistent with his own beliefs and goals (his ZPD) and also with the ZPA offered by the university teacher education course. This, it seemed to me, was how he was able to reconcile his pedagogical beliefs (a part of his ZPD) with the ZFM/ZPA complex within his teaching environment.

Analysis of Adam's case and the experiences of other prospective and beginning teachers who participated in this study has given me a better understanding of the scope and limitations of my role as a mathematics teacher educator as I ponder the question of how I can help novice teachers implement the technology enhanced approaches I promote in my teacher education course. For many years I dealt with this question by addressing separately some of the key factors (listed in Table 1) known to influence technology integration. For example, I had my students carry out an annual technology audit of their practicum schools so that on their return to the university they could report on and debate the significance of *access to resources and technical support* and the effect of *curriculum and assessment requirements* on technology usage. In these post-practicum sessions I also structured small group discussion tasks in which students compared their own *pedagogical beliefs* about the role of technology in mathematics education with the technology-related practices demonstrated (or not) by their *supervising teachers*. These coursework activities have not changed in their classroom enactment. What has changed is the way I now integrate these and other elements of my course into a single zone-theoretical framework that suggests to me how and where I might intervene in the development of prospective and beginning teachers' identities as users of technology. The postscript to Adam's story illustrates this approach.

In his second year of teaching Adam was transferred to a school where there was limited access to computer laboratories and only one class set of graphics calculators. Adam complained that none of the mathematics teachers were interested in using technology, and they preferred the same kind of teacher-centred, textbook-oriented teaching approaches as his colleagues in his previous school. I could see that the teachers in this school, through their lack of interest in technology, promoted approaches (ZPA) that were consistent with the technology-poor environment (ZFM) but not with Adam's own beliefs and aspirations (his ZPD). I decided that the possibilities for intervention were to work with Adam to change the ZFM/ZPA complex to bring these zones into alignment with his ZPD. Clearly, neither Adam nor I had the power to change the material aspects of the teaching environment that were causing him frustration. Instead, I encouraged him to view the single class set of graphics calculators as an opportunity he could exploit, simply because he was the only teacher who wanted to use them. I also supported him in increasing his involvement in the local mathematics teacher professional association where I hoped he would find a ZPA external to the school that would nurture his potential for further development. Through these quite modest interventions I aimed to help Adam change the way he interpreted his circumstances and gain a sense of agency in his own development.

Technology Integration: A Professional Development Intervention

Participants in this study were a group of ten secondary mathematics teachers who volunteered for a training program that prepared them to deliver professional development workshops on the use of graphics calculators. The training program, conducted intensively over a single week-end early in the school year, emphasised teaching roles that treated technology as a means of developing students' understanding of mathematical concepts rather than simply a tool that replaces pen and paper calculations. These sessions engaged participants as learners in technology-rich activities that could be used in secondary school classrooms, and in discussion of associated teaching and learning issues. I was a member of the team of technology-experienced teachers and teacher educators who designed and delivered the sessions.

With a colleague I then followed the progress of three of the teachers who subsequently delivered professional development workshops at conferences or in their own schools, and interviewed them on how their views about technology had been affected by the training program (Galbraith & Goos, 2003). At the end of the year we observed several sessions of a two-day technology workshop organised and presented by one of these teachers in her role as Mathematics Head of Department for the benefit of mathematics teaching staff in her school. We drew on our workshop observations and interviews with this teacher to interpret her professional learning experiences using Valsiner's zone concepts, in the manner described previously.

A Professional Development Intervention: The Case of Lisa

By comparison with Adam, Lisa was an experienced teacher but a relative novice in the use of technology when she participated in the graphics calculator training program. When reflecting on her initial, unhelpful, professional development experiences in this field, she told us that she "got lost in the first ten seconds, and was really turned off so didn't touch them again for a while". After several more workshops she felt confident enough to use graphics calculators in her teaching, but only as a replacement for pen and paper.

The training program offered as part of this research study proved to be a turning point for Lisa as it emphasised pedagogy rather than "pushing buttons":

> It was out of that week-end that I really understood the impact that [graphics calculators] had on the pedagogy. Up to then I saw it as a tool to draw graphs and analyse statistics. But at that workshop ... they explained to us how kids start trying to help. So when we were doing that we were grabbing somebody else's calculator and sharing our data, so it made the group work thing a whole lot better. And I really valued the part where we, as groups, we went out and used the overhead projector and we presented our information back to the group. I started to see different ways of using it that I hadn't thought of before. We weren't just *doing*, we were *understanding* at a higher level. I found that really powerful. Because I had thought that all they do is save you that boring part of maths.

Constraints of the professional context (ZFM) seemed to play little part in Lisa's learning, possibly because as Head of her school's Mathematics Department she had considerable autonomy in obtaining desired resources and in managing curriculum and assessment programs. Instead, her learning can be interpreted in terms of the changing relationship between her goals and interests (ZPD) and the ZPAs offered by the professional development and training she experienced. The early workshops she attended were described as "off-putting", because the emphasis was on procedural aspects of operating the calculators and the mathematics presented was too difficult for participants to engage meaningfully with the technology. This contrasted with the approach taken in the week-end workshops offered as part of the research project. Lisa seemed to find in these professional development sessions a ZPA that matched her need to understand pedagogical, rather than procedural, aspects of using technology.

The case of Lisa illuminated for me some of the issues facing experienced teachers who are unfamiliar with new technologies such as graphics calculators. While her ZFM presented few constraints, she had to search for professional development (ZPA) that would extend and challenge, rather than accommodate, her existing ideas about teaching with technology (her ZPD). As a result of this work with Lisa and other teachers I began to use Valsiner's zone theory to design professional development interventions that give careful attention to discovering the participating teachers' epistemological and pedagogical beliefs, understanding their institutional contexts, and identifying how all these factors interact to potentially influence their learning and development (see Goos, Dole, & Makar, in press). Only by analysing this initial zone configuration do I feel able to engage with their possibilities for development (ZPD) and promote actions they believe to be feasible in their school environments (ZFM).

Valsiner's zone theory has framed my own learning as a mathematics teacher educator arising from the experience of working with teachers like Adam and Lisa. Weaving this theory into my own practice has led me to believe that there are three sets of factors influential in sustaining or limiting teachers' professional growth: teacher knowledge and beliefs (ZPD), the professional context (ZFM), and the sources of assistance available to teachers (ZPA). It is when these factors are mutually supporting that the possibilities for professional growth are maximised.

What Can We Learn from Teacher Education Research Using Valsiner's Zone Theory?

The elaboration of Valsiner's zone theory outlined above is helpful for analysing relationships between teachers' pedagogical knowledge and beliefs and the teaching repertoire offered by courses for prospective teachers, practicum and initial professional experiences, and professional development programs in order to understand how they learn in multiple contexts. One such configuration is represented in Figure 1; others can be imagined if we allow the overlap between zones to change. This representation implies that learning takes place at the intersection of the three zones.

In this chapter I have used zone theory to explain specific cases, using an inductive approach in moving from data to theory as the means of creating evidence to support an argument about teacher learning. However, the representation shown in Figure 1 also suggests that a more deductive approach is possible by imagining different ways in which the zones might overlap and asking what such cases would look like. The theory can thus be used to make projections about anticipated evidence that will test its usefulness and perhaps lead to modifications.

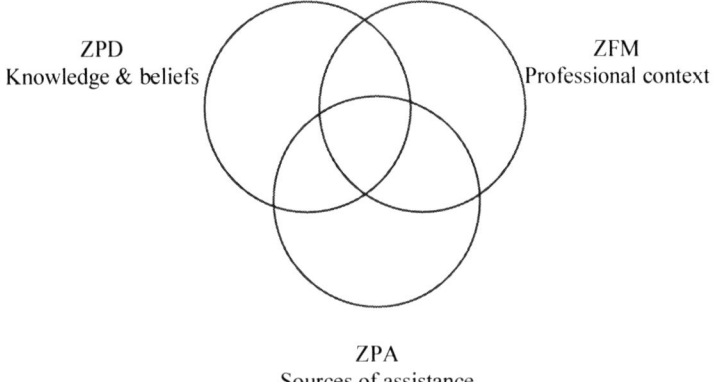

Figure 1. Representation of relationships between ZPD, ZFM, ZPA.

From a practical perspective, the theory could be used by teacher educators to improve teachers' opportunities to learn at three stages of development:
- Pre-service education: helping prospective teachers to analyse their practicum experiences (ZFM), the pedagogical models these offer (school ZPA), and how these experiences align with or contradict the knowledge gained in the university-based program (university ZPA);
- Transition to the early years of teaching: creating induction and mentoring programs that promote a sense of individual agency within the boundaries of the school environment (ZPD within ZFM);
- Professional development: designing professional learning programs for more experienced teachers (ZPA to stretch ZPD).

WHAT DOES IT MEAN TO LEARN FROM RESEARCH ON OR INTO MATHEMATICS TEACHER EDUCATION?

In reflecting on the emerging field of mathematics teacher education Krainer noted that teacher educators have the dual roles of "intervening and investigating ... of improving and understanding" (Adler et al., 2005, p. 371). Later in the same discussion Adler pointed out that this personal investment in teaching makes it difficult for us to take a critical stance towards the research we do with prospective and practising teachers, and she suggested that we need to develop effective theoretical languages to distance ourselves from what we are looking at. In our dual role as teacher-educator-researchers it seems to me that we can learn from research in two ways: by developing strong theories of teacher education that are related to broader theories of learning, and by being explicit about how we use these theories to inform our work with teachers.

Mathematics teacher-education-researchers working with sociocultural perspectives are attempting to extend existing theoretical frameworks for studying learning – Vygotskian, neo-Vygotskian, situative, community of practice – to the study of learning to teach, and this is potentially beneficial in bringing some coherence to a field marked by a proliferation of constructs and frameworks. However it is rare to find in research reports a theorised discussion of our own teaching or how our research contributes to our learning as mathematics teacher educators. This chapter has shown how Valsiner's zone theory brings teaching, learning, and context into the same discussion, and how the theory can be applied in two connected layers – to the teacher-as-teacher orchestrating classroom ZFM/ZPAs for students (Blanton, Westbrook, & Carter, 2005) as well as the teacher-as-learner negotiating the ZFM/ZPAs offered by the professional environment (Goos, 2005a). At the latter layer the teacher-educator-as-teacher comes on the scene, providing the ZPA. What if we imagine a third layer, with teacher-educator-as-learner? How does our professional context constrain our actions in culturally expected ways (ZFM), and what are our opportunities to learn (ZPA)? Could we describe a set of possibilities for our own development in the near future (ZPD)? In other words, how might zone theory help us analyse our own roles as mathematics teacher educators conducting research with prospective and practising teachers?

Let me sketch out what such an analysis might look like by applying zone theory to my own practice in the dual roles of researcher and teacher educator. As a researcher, my Zone of Free Movement is constrained by academic structures and cultures within and beyond my university. These include:
- guidelines for career development, identifying activities that are formally recognised and rewarded;
- mechanisms for managing academic workloads that seek to balance teaching and research;
- government programs for assessing the quality and impact of university research;
- competitive research grant schemes;
- the process of peer review of articles submitted for publication in scholarly journals.

Closely inter-related with these elements of my professional context is the Zone of Promoted Action represented by my initial research training (doctoral studies, early experiences as a research assistant), participation in research conferences and other activities of educational research associations, and formal or informal mentoring by more experienced colleagues. This ZFM/ZPA complex helps shape possibilities for my development as a researcher (ZPD) by defining what is allowed and what is promoted. The learning opportunities that arise in this way are well charted and form part of the enculturation of novice researchers into academic life.

As a mathematics teacher educator, I must negotiate a different zone configuration. Here, my practice is constrained by a Zone of Free Movement comprising the following elements:

- student characteristics, such as their mathematical knowledge and their beliefs about mathematics teaching and learning;
- curriculum and assessment requirements that are increasingly governed by external teacher registration authorities as well as university course accreditation processes;
- limited access to technology resources in the university;
- reduction of the hours allocated to teaching methods courses in the pre-service teacher education program;
- difficulties in finding suitable practicum placements for prospective teachers;
- the perception amongst academic colleagues that teacher education is low status work.

My ZPA as a teacher educator is less clearly defined in that it is difficult to identify people or activities that explicitly promote my development in this role, and thus difficult to describe the ZFM/ZPA complex that shapes my teacher education practice. Llinares and Krainer (2006) point out that the growth of mathematics teacher educators as learners is a new field of study, and research in this area has so far drawn on notions of reflective practice rather than sociocultural theories that take into account the settings in which practice develops. From a sociocultural perspective, I could say that my own research in teacher education acts as a ZPA that informs my practice as a mathematics teacher educator. For example, I use the diagrammatic representation of zone theory shown in Figure 1 to invite practising teachers to talk about their knowledge and beliefs (ZPD), contexts (ZFM), and professional learning experiences (ZPA) in relation to technology. Each zone, represented by a circle with its elements as listed in Table 1, is printed on separate overhead transparencies that teachers can superimpose to show the degree of overlap that matches their own circumstances and to imagine ways in which this configuration might change. My research using zone theory has also influenced how I work with prospective teachers – my own teacher education students – to help them analyse tensions between the learning experiences offered by the university course and the practicum (Goos, Stillman, & Vale, 2007). While this approach helps give coherence to my dual roles as researcher and teacher educator, further elaboration of Valsiner's zone theory is necessary to create a conceptual framework that better explains how mathematics teacher educators learn from research into teacher education.

REFERENCES

Adler, J., Ball, D., Krainer, K., Lin, F-L., & Novotna, J. (2005). Reflections on an emerging field: Researching mathematics teacher education. *Educational Studies in Mathematics, 60*, 359–381.

Blanton, M., Berenson, S., & Norwood, K. (2001a). Using classroom discourse to understand a prospective mathematics teacher's developing practice. *Teaching and Teacher Education, 17*, 227–242.

Blanton, M., Berenson, S., & Norwood, K. (2001b). Exploring a pedagogy for the supervision of prospective mathematics teachers, *Journal of Mathematics Teacher Education, 4*, 177–204.

Blanton, M., Westbrook, S., & Carter, G. (2005). Using Valsiner's zone theory to interpret teaching practices in mathematics and science classrooms. *Journal of Mathematics Teacher Education, 8*, 5–33.

Bohl, J., & Van Zoest, L. (2003). The value of Wenger's concepts of modes of participation and regimes of accountability in understanding teacher learning. In N. A. Pateman, B. J. Dougherty, & J. T. Zilliox (Eds.), *Proceedings of the 27th annual conference of the International Group for the Psychology of Mathematics Education* (Vol. 4, pp. 339–346). Honolulu, HI: PME.

Brown, C. A., & Borko, H. (1992). Becoming a mathematics teacher. In D. A. Grouws (Ed.), *Handbook of research on mathematics teaching and learning* (pp. 209–239). New York: Macmillan.

Da Ponte, P., & Chapman, O. (2006). Mathematics teachers' knowledge and practices. In A. Gutierrez & P. Boero (Eds.), *Handbook of research on the psychology of mathematics education: Past, present and future* (pp. 461–494). Rotterdam: Sense Publishers.

Fine, A. E., & Fleener, M. J. (1994). Calculators as instructional tools: Perceptions of three preservice teachers. *Journal of Computers in Mathematics and Science Teaching, 13*(1), 83–100.

Forman, E. A. (2003). A sociocultural approach to mathematics reform: Speaking, inscribing, and doing mathematics within communities of practice. In J. Kilpatrick, W. G. Martin, & D. Schifter (Eds.), *A research companion to principles and standards for school mathematics* (pp. 333–352). Reston, VA: National Council of Teachers of Mathematics.

Forman, E. A., Minick, N., & Stone, C. A. (1993). *Contexts for learning: Sociocultural dynamics in children's development*. New York: Oxford University Press.

Galbraith, P., & Goos, M. (2003). From description to analysis in technology aided teaching and learning: A contribution from zone theory. In L. Bragg, C. Campbell, G. Herbert & J. Mousley (Eds.), *Mathematics education research: Innovation, networking, opportunity* (Proceedings of the 26th annual conference of the Mathematics Education Research Group of Australasia, pp. 364–371). Sydney: MERGA.

Goos, M. (2005a). A sociocultural analysis of the development of pre-service and beginning teachers' pedagogical identities as users of technology. *Journal of Mathematics Teacher Education, 8*(1), 35–59.

Goos, M. (2005b). *Theorising the role of experience in learning to teach secondary school mathematics*. Paper presented at the Conference of the 15th ICMI Study: The Professional Education and Development of Teachers of Mathematics, Aguas de Lindoia, Brazil, 15–21 May 2005. Available http://stwww.weizmann.ac.il/G-math/ICMI/Goos_Merrilyn_ICMI15_prop.doc

Goos, M. (2005c) A sociocultural analysis of learning to teach. In H. Chick & J. Vincent (Eds.), *Proceedings of the 29th conference of the International Group for the Psychology of Mathematics Education* (Vol. 3, pp. 49-56). Melbourne: PME.

Goos, M., Dole, S., & Makar, K. (in press). Designing professional development to support teachers' learning in complex environments. *Mathematics Teacher Education and Development*.

Goos, M., Evans, G., & Galbraith, P. (1994). *Reflection on teaching: Factors affecting changes in the cognitions and practice of student teachers*. Proceedings of the 1994 annual conference of the Australian Association for Research in Education, Newcastle. Retrieved 7 October 2007 from http://www.aare.edu.au/94pap/goosm94341.txt

Goos, M., Stillman, G., & Vale, C. (2007). *Teaching secondary school mathematics: Research and practice for the 21st century*. Sydney: Allen & Unwin.

Graven, M. (2004). Investigating mathematics teacher learning within an in-service community of practice: The centrality of confidence. *Educational Studies in Mathematics, 57*, 177–211.

Greeno, J. (2003). Situative research relevant to standards for school mathematics. In J. Kilpatrick, W. G. Martin, & D. Schifter (Eds.), *A research companion to principles and standards for school mathematics* (pp. 304–332). Reston, VA: National Council of Teachers of Mathematics.

Harré, R., & Gillett, G. (1994). *The discursive mind*. London, UK: Sage.

Lave, J., & Wenger, E. (1991). *Situated learning: Legitimate peripheral participation*. Cambridge: Cambridge University Press.

Lerman, S. (1996). Socio-cultural approaches to mathematics teaching and learning. *Educational Studies in Mathematics, 31*(1–2), 1–9.

Lerman, S. (2001). A review of research perspectives on mathematics teacher education. In F.-L. Lin & T. Cooney (Eds.), *Making sense of mathematics teacher education: Past, present and future* (pp. 33–52). Dordrecht: Kluwer Academic Publishers.

Llinares, S., & Krainer, K. (2006). Mathematics (student) teachers and teacher educators as learners. In A. Gutierrez & P. Boero (Eds.), *Handbook of research on the psychology of mathematics education: Past, present and future* (pp. 429–459). Rotterdam: Sense Publishers.

Manoucherhri, A. (1999). Computers and school mathematics reform: Implications for mathematics teacher education. *Journal of Computers in Mathematics and Science Teaching, 18*(1), 31–48.

McGraw, R., Arbaugh, F., Lynch, K., & Brown, C. A. (2003). Mathematics teacher professional development as the development of communities of practice. In N. A. Pateman, B. J. Dougherty, & J. T. Zilliox (Eds.), *Proceedings of the 27th annual conference of the International Group for the Psychology of Mathematics Education* (Vol. 3, pp. 269–276). Honolulu, HI: PME.

Peressini, D., Borko, H., Romagnano, L., Knuth, E., & Willis, C. (2004). A conceptual framework for learning to teach secondary mathematics: A situative perspective. *Educational Studies in Mathematics, 56*, 67–96.

Simonsen, L. M., & Dick, T. P. (1997). Teachers' perceptions of the impact of graphing calculators in the mathematics classroom. *Journal of Computers in Mathematics and Science Teaching, 16*(2/3), 239–268.

Valsiner, J. (1997). *Culture and the development of children's action: A theory of human development.* (2nd ed.) New York: John Wiley & Sons.

Walen, S., Williams, S., & Garner, B. (2003). Pre-service teachers learning mathematics using calculators: A failure to connect current and future practice. *Teaching and Teacher Education, 19*(4), 445–462.

Wallace, R. (2004). A framework for understanding teaching with the Internet. *American Educational Research Journal, 41*, 447-488.

Wenger, E. (1998). *Communities of practice: Learning, meaning and identity.* Cambridge, MA: Cambridge University Press.

Windschitl, M., & Sahl, K. (2002). Tracing teachers' use of technology in a laptop computer school: The interplay of teacher beliefs, social dynamics, and institutional culture. *American Educational Research Journal, 39*, 165–205.

Merrilyn Goos
School of Education,
The University of Queensland

ORIT ZASLAVSKY

5. MEETING THE CHALLENGES OF MATHEMATICS TEACHER EDUCATION THROUGH DESIGN AND USE OF TASKS THAT FACILITATE TEACHER LEARNING

This chapter presents seven unifying themes of tasks used in mathematics teacher education. These themes reflect broad goals for mathematics teacher education, and are closely related to the knowledge required for teaching mathematics. The challenges that teacher-educators face in having to design and implement productive tasks for teachers are discussed and illustrated through a specific task. The iterative process of design of this task is attributed, on the one hand, to the teacher-educator's reflection in and on the task implementation; on the other hand, to her professional growth that is expressed in her increased appreciation of and attention to the above central themes. In addition, the interplay between teacher-educators' roles as: researchers, facilitators of teacher learning, and designers of tasks for teacher education are addressed. The chapter concludes with a framework for examining the demands on and expectations of teacher-educators as facilitators of teacher learning and the role of tasks in enhancing teacher learning.

TEACHER-EDUCATORS' CHALLENGES

Conceptual Framework

Teacher education seeks to transform prospective and/or practicing teachers from neophyte possibly uncritical perspectives on teaching and learning to more knowledgeable, adaptable, analytic, insightful, observant, resourceful, reflective and confident professionals ready to address whatever challenges teaching mathematics presents. This view of teacher education presents a great challenge for teacher-educators, who are in charge of facilitating teacher learning towards these goals. In this chapter I focus specifically on the teacher-educator role as task designer.[1]

There is a consensus among mathematics educators that learning occurs through engagement in tasks (Krainer, 1993; Simon & Tzur, 2004; Zaslavsky, 2005). Hiebert & Wearne (1993) assert that what people learn "is largely defined by the tasks they are given" (p. 395). Along these lines, Kilpatrick, Swafford and Findell

---

[1] Although there are many other roles that can be considered that do not necessarily involve explicit tasks (e.g., general professional mentoring and support).

(2001) claim that the quality of instruction depends to a large extent "on whether teachers select cognitively demanding tasks, plan the lesson by elaborating the mathematics that the students are to learn through those tasks, and allocate sufficient time for the students to engage in and spend time on the tasks" (ibid, p. 9). Sierpinska (2004) considers the design, analysis and empirical testing of tasks one of the major responsibilities of mathematics education.

Transforming this idea to teacher education, the underlying assumption is that tasks play a critical role in teacher learning, similar to their role in students learning. The above transformation occurs optimally through constructive engagement in tasks that foster knowledge for teaching mathematics. Ideally such tasks provide a bridge between theory and practice, and serve to challenge, surprise, disturb, confront, extend, or provoke examination of alternatives, drawn from the context of teaching. In this context, tasks are seen as problems or activities that, having been developed, evaluated and refined over time, are posed to teacher education participants. Such participants are expected to engage in these tasks collaboratively, energetically, and intellectually with an open mind and an orientation to future practice. These tasks might be similar to those used by classroom teachers or idiosyncratic to teacher education.

The demands on teacher-educators, in terms of knowledge and qualities, are enormous and multifaceted. An overarching demand is for teacher-educators to be reflective practitioners (Schön, 1987). They need to constantly reflect on-action and in-action, in all phases of their work (e.g., planning, interacting with teachers, observing teachers' work). Based on Steinbring (1998), Figure 1 suggests a general recursive mechanism of mutual construction of knowledge of both facilitator and learners. Although Steinbring referred in his model to teacher-student learning, this model can be extended to teacher-educators as facilitators and teachers as learners; it explains how teacher-educators learn through their practice. Thus, it provides insight into the role of teacher-educators[2] as designers and orchestrators of tasks that foster teacher learning, and at the same time highlights the dynamic nature of teacher-educators' practice and development.

Mostly, teacher-educators are 'self made'. They make their own transitions from their experiences as mathematics teachers and/or as researchers of mathematics learning and teaching. There are relatively few explicit curricula for mathematics teacher education, particularly for programs for practicing teachers (with some exceptions described by Even, this volume), nor agreed upon content that needs to be "covered". Consequently, teacher-educators are free to make choices with minimal constraints, yet they need to be clear about the goals they attempt to address in their work with teachers. Examining some of the "big ideas" for teacher education and looking at the goals that are put forth for approaches to teacher education points to a very broad and sound knowledge base required of mathematics teacher-educators.

---

[2] In this chapter the term 'teacher-educator' refers to mathematics teacher-educators that are responsible for the learning of prospective and/or practicing mathematics teachers.

TEACHER-EDUCATORS AS TASK DESIGNERS

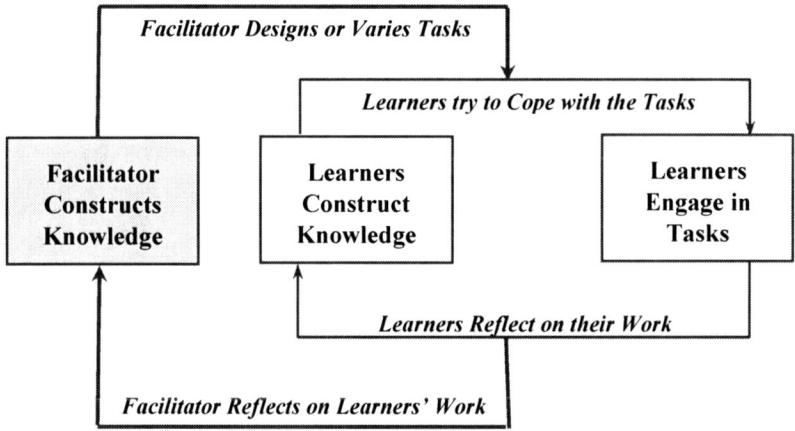

Figure 1. Facilitator-Learner mechanism of construction of knowledge through dealing with tasks (inspired by Steinbring, 1998)

In order to understand the role and responsibility of teacher-educators in orchestrating mathematics teacher learning and development, and discuss the knowledge base teacher-educators draw on and the kind of practice and qualities they need to develop, I propose to examine a number of unifying themes that reflect goals for mathematics teacher education.[3] These interrelated and partly overlapping themes are not based on the conventional content topics of teacher education (e.g., teaching decimals, grouping practices); they concern qualities and kinds of competence and knowledge that mathematics teacher education seeks to promote in prospective and practising teachers in a broad sense.
The themes are:
1. Developing adaptability
2. Fostering awareness to similarities and differences
3. Coping with conflicts, dilemmas and problem situations
4. Learning from the study of practice
5. Selecting and using (appropriate) tools and resources for teaching
6. Identifying and overcoming barriers to students' learning
7. Sharing and revealing self, peer, and student dispositions

I turn to a succinct description of the above themes, and the challenges with which they present teacher-educators. It should be noted that each theme can be regarded from both mathematical and pedagogical perspectives. Following the description of

---
[3] The central themes around which this chapter is built were developed jointly with Peter Sullivan in the course of structuring a forthcoming edited book, by Springer: Zaslavsky, O. & Sullivan, P. (Eds.), *Constructing knowledge for teaching secondary mathematics: Tasks to enhance prospective and practicing teacher learning*.

these themes, I illustrate how a teacher-educator may meet these challenges through carefully designed tasks.

Theme 1: Developing adaptability: A unifying theme in many aspects of teacher education is the development in teachers of an orientation to being adaptable, to considering variations to questions, tasks, and intended curriculum, to searching for alternatives to unsuccessful approaches, and to adapting existing resources to intended goals. This kind of orientation can be considered as *adaptability* and concurs with Cooney's (1994) ideas of *adaptation*, as well as notion of *contingency* discussed by Rowland, Huckstep, & Thwaites (2005). Especially in a teaching and learning environment that encourages active learning by students, there is a need for teachers to be prepared to make active responses, and these cannot be planned in advance. Thus, adaptability is inter-related to flexibility (Leikin & Dinur, 2007). Indeed, it is not only a desirable orientation, but also a desirable personal quality. Teacher adaptability can be useful in diverse situations. Often teacher adaptability and flexibility are tightly connected to knowing to act in the moment (Mason, 1999).

In order to help teachers become adaptable a teacher-educator first must be adaptable and exhibit this quality in the course of working with teachers. Thus, it is important that the teacher-educator orchestrates situations in which he or she can model flexibility and the ability to vary and consider alternatives; moreover, it is also critical to provide experiences for teachers to engage in activities that require in the moment decisions that encourage flexibility and adaptability to unexpected situations.

Theme 2: Fostering awareness to similarities and differences. Noticing similarities and differences, in the broad sense, is at the heart of learning and teaching (Mason, 1998). It is well known that the gradual process of associating concepts with categories is a critical aspect of learning. Classification of different objects according to various criteria may enhance awareness of ways in which they are related to each other (Silver, 1979). This process requires the identification of similarities and differences between objects along several dimensions; in the course of classifying objects, objects that initially appear different, may be considered the same from a certain criterion. For example, when classifying numbers according to their parity, 2 and 3 are different; however, when using the criterion of "primeness", 2 and 3 are the same. Awareness to similarities and differences between mathematical objects is considered fundamental to mathematical thinking; comparing and contrasting is needed also in order to identify patterns and make connections between and across topics, contexts, types of problems and even between teaching approaches. In the latter, pedagogical considerations are involved.

In order to help teachers develop a tendency to notice and an ability to identify similarities and differences as a state of mind, and particularly in classroom situations, the teacher-educator must engage teachers in activities that require this kind of noticing, and in addition exhibit such awareness, not only in the planning stage, but also in in-the-moment decisions and interactions with teachers.

Theme 3: Coping with conflicts, dilemmas and problem situations. Teachers constantly face dilemmas and need to make decisions and choices under conflicting constraints, and deal with uncertainty and complexity (Sullivan & Mousley, 2001; Sullivan, 2006); they need to be problem solvers in the broad sense of this term, and enhance their students' ability to solve mathematical problems and cope with impasses and cognitive conflict. Thus, it is imperative that teachers are prepared for dealing with this complex terrain, both as teachers – that is, as designers and orchestrators of such situations for their students, and as learners – that is, as active problem solvers. Sánchez & García, this volume, describe and analyse the dilemmas teacher-educators face in their decision making processes, which to some extent mirror teachers' dilemmas.

Examination of what it takes for a mathematics teacher-educator to deal with the challenges of this theme points to a very broad knowledge base and several personal qualities that are required. In addition to being a competent problem solver and familiar with the relevant content and pedagogy, a teacher-educator is expected to be confident enough to engage teachers in open ended problem situations to which the possible solutions and new questions that may arise are not necessarily known to him/her in advance. A teacher-educator must be open minded and willing to accept and explore in real time unexpected approaches and ideas that teachers may suggest. This is similar to the demands on teachers to exhibit the same approaches with their students.

Theme 4: Learning from the study of practice. Many teacher education programs are seeking ways to enhance the practical relevance of their curriculum, while allowing prospective teachers opportunities to review key theoretical perspectives, and ultimately to develop a career long orientation to learning from the study of their own teaching or the teaching of others. There have been several approaches to learning from the study of practice. These include the realistic simulations offered by videotaped study of exemplary lessons (Clarke, 2000); interactive study of recorded exemplars (e.g., Merseth & Lacey, 1993); case methods of teaching dilemmas that problematise aspects of teaching (e.g., Stein, Smith, Henningsen, & Silver, 2000); and Lesson Study that engages teachers in thinking about their long-term goals for students, developing a shared teaching-learning plan, encountering tasks that are intended for the students, and finally observing a lesson and jointly discussing and reflecting on it (e.g., Lewis, Perry, & Hurd, 2004; Fernandez & Yoshida, 2004). Each of these requires appropriate prompts to critical analysis to be effective. In each case, the teacher learning is through the opportunity to view and review exemplars, to discuss with peers interpretations of the exemplars, to engage in critical dialog on the experience, and to hear informed analysis of both the practice and the experience of critique.

Teacher-educators who design and implement these experiences are presented with great challenges. They need to be able to capture problematic and insightful classroom situations, and translate them into challenging cases for teachers to ponder. Fostering critical discussions regarding such cases requires high level metacognitive and mentoring skills.

Theme 5: Selecting and using (appropriate) tools and resources for teaching. Selecting appropriate tools for mathematics teaching and using them effectively is a major challenge for teachers. Tools can be text books, additional readings, manipulatives, construction and measuring devices, transparencies, graphical calculators, and other technological environments. Making educated choices regarding what tools to use for certain purposes and how to use them requires familiarity with a wide range of tools from both a learner's and a teacher's perspective. It also requires awareness of the potential and limitations of each tool, for various purposes and contexts, and confidence in using it for teaching.

Tools can be seen in their broadest sense, to include many different kinds of resources, including human and cultural resources, such as language and time. Adler (2000) argues that increasing attention should be given to resources in mathematics teacher education from two aspects:

> First, mathematics teacher education programmes need to work with teachers to extend common-sense notions of resources beyond material objects and include human and cultural resources such as language and time as pivotal in school mathematical practice. Second, attention in professional development activities needs to shift from broadening a view of *what* such resources are to *how* resources function as an extension of the mathematics teacher in the teaching-learning process. (ibid, p. 207).

From a mathematics teacher-educator perspective, enhancing teachers' competence in selecting and effectively using tools for teaching requires a familiarity with a wide range of available tools and appreciations of their potential for accomplishing various goals; it also requires great sensitivity to teachers' reluctance to incorporate unfamiliar innovative tools in their teaching.

Theme 6: Identifying and overcoming barriers to students' learning. Education and schooling strive to overcome some real, and in some cases substantial, barriers and create opportunities for all students, especially those who would not otherwise have those opportunities. One of the challenges for teacher education is to educate prospective and practising teachers on the existence and sources of barriers, and of strategies that can be effective in assisting students to overcome those barriers (Sullivan, Zevenbergen, & Mousley, 2003). The barriers are associated with diversity, and might be due to: epistemological aspects of mathematics (e.g., informal vs. formal approaches, modes of representation; missed prior learning opportunities; learning styles); cultural factors including community expectations, gender, school/home aspirational mismatches; language barriers and usage;

physical and other disabilities; socio-economic factors, and others (Trentacosta & Kenney, 1997).

A teacher-educator, who intends to address this theme, must be aware of such barriers not just for students learning but also for teacher learning. One way to enhance teachers' awareness and appreciation of barriers to student learning is to experience overcoming of barriers to their own learning. To do this, a teacher-educator must understand the nature and causes of such barriers (e.g., mathematical, representational, communicational), and be familiar with possible productive interventions. He or she needs to be able to address any prejudices or knowledge mismatches within the prospective or practising teachers, and design experiences that can assist teachers in intervening effectively to overcome barriers for their own learning as well as for their students. These experiences should help teachers develop ways to engage all students, especially those who may sometimes feel alienated from mathematics and schooling, in meaningful mathematical thinking and learning.

Theme 7: Sharing and revealing self, peer, and student dispositions. In the multidimensional endeavour of teaching and learning mathematics, and learning to teach mathematics, a key dimension is the disposition of the (prospective and practicing) teacher as a learner, the teacher as a teacher, and the pupil as a learner.

The dimension of disposition is itself multifaceted. It can include the following overlapping categories:
- Beliefs about: the nature of mathematics; the utility of mathematics; the way mathematics is learned; one's own ability to learn mathematics;
- Self-regulatory behaviours such as: persistence; self-efficacy; motivation; resilience;
- Attitudes such as: liking for mathematics; enjoyment of mathematics; mathematics anxiety;

Indeed almost all aspects of teacher education have an attitudinal or dispositional dimension that should be considered. There is an agreement that helping teachers become aware of their own beliefs, as well as their students' dispositions, is "a significant step toward improving students' opportunities to learn mathematics" (Mewborn & Cross, 2007, p. 262).

It follows that a teacher-educator needs to know about the multifaceted dimension of beliefs and dispositions and their effects on various aspects of learning and teaching mathematics. Moreover, it is important for a mathematics teacher-educator to exhibit positive dispositions and enthusiasm towards mathematics and learning mathematics.

The above themes convey some of the challenges of mathematics teacher education. As mentioned earlier, tasks play a critical role in (teacher) learning, thus, in meeting these challenges. From a teacher-educator's perspective, designing and using worthwhile tasks for prospective or practicing teachers is a non-trivial task. The following sections deal with the special role of teacher-educators as designers of tasks that foster mathematics teacher learning.

ORIT ZASLAVSKY

MEETING TEACHER-EDUCATORS' CHALLENGES THROUGH WELL DESIGNED TASKS

Where Do Tasks Come from?

One of the main differences between the role of a teacher-educator and of a teacher in selecting and designing appropriate tasks for enhancing learning is the availability of appropriate resources. While a teacher has many accessible textbooks, teacher guides, and enrichment material that are easily accessible, a teacher-educator has very few resources to draw on directly. First, many teacher-educators programs have broad goals (such as the ones described above) with no specific curriculum. It is often left to the teacher-educator how to address these goals, in terms of content and coverage. In addition, there are hardly any textbooks for mathematics teacher learning. Thus, the teacher-educator has much freedom and flexibility in the choice he or she makes, and at the same time has very little material to build on. The main sources that are specific for teacher-educators (that is, in addition to the resources available for teachers) are professional journals and books, research papers and other accounts of practitioners, and personal experience, all related rather indirectly to tasks for teacher education. Experienced teacher-educators have a state of mind of constantly searching for ideas that can be turned into productive tasks for teachers. Such ideas are often encountered through interactions with colleagues at professional conferences and other occasions.

Another source of ideas for tasks for teacher education is related to the interplay between research and practice. There have been numerous calls for connecting research to teacher practice (e.g., Jaworski & Gellert, 2003; Heid et al., 2006). Connecting research and the practice of teacher-educators lends itself well, since most teacher-educators are also active researchers, who tend to conduct long term research programmes, or at least integrate action research in their work (e.g., Jaworski, 2001, 2003). An example of such connection is the insightful account García, Sánchez, & Escudero (2006) provide, highlighting the interrelations between their roles as researchers and teacher-educators.

Often, research informs and enhances teacher-educators' teaching and vice versa through tasks. Zaslavsky (2005) points to the dual purpose tasks serve for teacher-educators. On the one hand, tasks are the means and content by which teacher-educators may enhance teacher learning. On the other hand, through a reflective process of designing, implementing, and modifying tasks, they turn into means of the teacher-educator's learning. This kind of learning can be seen as learning through a form of action research of the teacher-educator, who constantly researches his or own practice. However, research may provide a rich source of ideas for tasks for teachers in many other ways as well. Zaslavsky and Zodik (2007) developed tasks for teachers based on their study of teachers' use of examples. Their classroom observations led them to realize some features of example choice and use of which they were not aware. They used their findings for eliciting teachers' discussions surrounding problematic aspects of exemplification, based on real practice. Similarly, the work of Stein et al. (2000) has emerged from authentic classroom observations, which were used as the basis for designing cases in the form of tasks for teacher education.

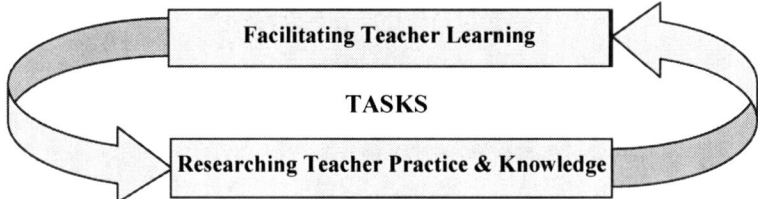

Figure 2. The role of tasks in the interplay between teacher-educators' teaching and research practices

Thus, there is a fruitful interplay between teacher-educator's roles as: researchers, facilitators of teacher learning, and designers of tasks for teacher education (Figure 2). This interplay is a driving force for effective learning of the entire community of mathematics educators, including mathematics teachers, teacher-educators, and educators of teacher-educators (see Goodchild, this volume).

In order to get a glimpse at what is involved in designing a productive task for mathematics teacher education, I turn to an example of a task that has evolved over years of experience, through an iterative reflective process, until reaching a stage where unexpected responses from participants hardly occur. My experience as teacher-educator shows that at the initial stages, tasks tend to undergo substantial changes as a result of experiencing them with teachers. However, after several iterations, with many different groups of teachers, the process stabilizes, and in most cases no further modifications seem to be needed. Through the following example I draw connections between the design of the task and the seven themes discussed above, and show how these themes are reflected in the process of task selection, design and modification. This evolution is influenced by ongoing reflection on the task implementation (as described in the model in Figure 1), but also by a gradual professional growth of the teacher-educator in which appreciation of the above themes develops beyond the scope of a specific task.

The Spiral Task

The following task (Figure 3, Figure 4) was inspired by Burke (1993). It is very similar to the original (ibid) and can be seen as a well structured and directed investigation.

As a teacher-educator searching for ideas for challenging mathematical tasks for prospective mathematics teachers, over a decade ago I came across this activity that dealt with a perplexing and surprising problem of the existence of a pair of triangles that are not congruent yet have 5 congruent parts/elements (that is a combination of angles and sides).

5-CON TRIANGLES

The purpose of this activity is to investigate the properties of the following shape, and to continue it by adding on triangles on each side, satisfying the same pattern. You have the same shape on a yellow sheet of thick paper. You may cut it into 4 triangles, and use them for measuring and comparing angles and sides. Before cutting the triangles, label the interior of each angle with the letter of its corresponding vertex.

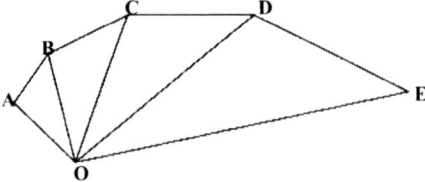

Find as many congruent parts for each pair of adjacent triangles, by comparing the cut-out triangles one pair at a time, and summarize your findings in the following table (note the example in the 1st column).

△OAB and △OBC	△OBC and △OCD	△OCD and △ODE
∡AOB=∡BOC		
∡OAB=∡OBC		
∡OBA=∡OCB		
OA=BC		
OB=OB		
AB≠OC		

You have found pairs of non-congruent triangles with 5 congruent parts. We term such triangles "5-con pair of triangles". What else can you say about each pair?

Now you can extend the spiral, by adding triangles on both sides of the spiral (one connected to △OAB, and the other connected to △ODE).

Figure 3. The initial version of the "Spiral Task": Structured investigation - part 1 (based on Burke, 1993)

This task addresses a counter-intuitive case, since students and teachers tend to think, based on their experiences with the triangles congruency theorems, that if two triangles have more than 3 congruent elements, (particularly if they have 5), then the triangles are necessarily congruent. It seemed appropriate for prospective and practising mathematics teachers because it dealt with challenging mathematics, for both students and teachers; it lends itself to making connections to other areas of mathematics and to the real world. Indeed, the teachers were highly engaged in

> **NECESSARY AND SUFFICIENT CONDITIONS FOR CONSTRUCTING OTHER SPIRALS CONSISTING OF 5-CON TRIANGLE PAIRS**
>
> 1. For each triplet of numbers, try to construct a triangle with side lengths corresponding to the given triplets:
>   (a) 6, 18, 54; (b) 18, 54, 162; (c) 16, 24, 36; (d) 24, 36, 54. Which cases are possible and which not? What properties does each triplet have?
>
> 2. Construct the ("possible") triangles corresponding to triplets (c) and (d), cut them out, and compare their angles. What can you infer about these triangles? Are they 5-con triangles? Can you place them one next to the other so they form part of a spiral?
>
> 3. As you saw, all the above triplets form an increasing geometric sequence. For (a) and (b) no corresponding triangles exist, while for (c) and (d) there are corresponding triangles that exist. Moreover, the two corresponding triangles (one with side lengths of 16, 24, 36 and the other with side lengths 24, 36, 54) are 5-con triangle pairs and placed accordingly, can form part of a spiral. If we denote the elements of the sequences as: $a, aq, aq^2, aq^3, ...$, what is the difference between the triplets for which a corresponding triangle exists and those for which no corresponding triangle exists?
>
> 4. From your above investigation, it follows that a necessary condition for a geometric sequence of 3 elements to have a corresponding spiral is that: $q>1$, $a+aq>aq^2$, that is: $1+q>q^2$. What conditions does this imply for q? Are these sufficient and/or necessary conditions for a spiral of 5-con triangle pairs?
>
> 5. What do you know about the number: $\frac{1+\sqrt{5}}{2}$. How does it relate to this investigation?

Figure 4. The initial version of the "Spiral Task": Structured investigation - part 2 (based on Burke, 1993)

this task, and most of them were genuinely surprised and felt they learned a lot from it, mostly mathematically. They also appreciated the structure of the task, that is, the way one question leads to another. In addition to the mathematical knowledge and insight that this task promoted, the task offered teachers a mathematical investigation that could be offered to their students in the same way. As mentioned previously, not all tasks for teachers are necessarily similar to tasks for students. Through tasks that are (almost) equally as suitable for teachers as for students, teachers may gain a better appreciation of what is entailed in the task by actually coping with the activity as learners.

After a rather long period of time using this task successfully with prospective as well as practicing teachers, and based on my personal development as teacher-educator, which included an increasing appreciation for open ended tasks, I decided to modify the activity turning it into a more open task (in the spirit of Zaslavsky, 1995). I also was concerned with teachers' reluctance to engage in and facilitate cooperative learning. This led to the second version of the Spiral Task (Figure 5).

TRIANGLE-SPIRALS

In the following drawing there are 4 triangles arranged in a spiral shape. You have the identical drawing on a transparency sheet.

You are asked to investigate the special properties of this spiral, find connections between the triangles that constitute it, and finally continue the spiral by adding two triangles to it.

Try to use the transparency sheet for comparing measures of the different elements of the triangles.

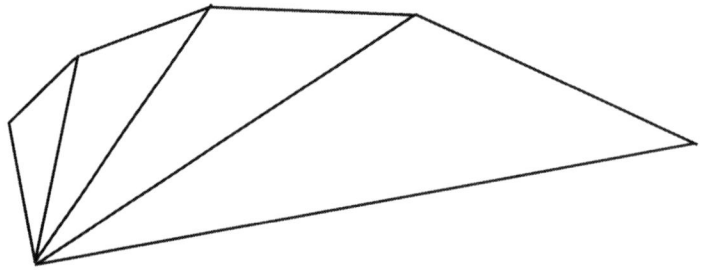

Look at your peers and compare: Did you continue the spiral in the same way?

Consider the limitation of the accuracy of the drawing and your measuring tools: How might they affect your conjectures and conclusions?

In your investigation you probably found that each pair of adjacent triangles has exactly 5 congruent elements, yet they are not congruent. Such triangles are called 5-con triangles. Can you construct other pairs of 5-con triangles that are not part of the given spiral?

Figure 5. The second version of the "Spiral Task"

In the second version, there is no example disclosing the target connections, thus, there was a "risk" that not all of these connections would be identified. Instead, there is an invitation to peer interactions and comparisons. The intention was to reinforce the participants' awareness of the social aspect of learning in problem solving situations. Generally, the second version of the Spiral Task led to more expressions of surprise, excitement, and enthusiasm when used with teachers. It evoked genuine collaborative work, with full participation of all group members.

Note that there was also a change in the measuring tools that were suggested: instead of cutting out triangles and using them to compare angles and sides, in the second version there is a use of a transparency sheet. That is, attention was also given to *tools*. Clearly, there is a difference if one needs to measure actual lengths of sides and angle measures, or whether one needs to compare the measures of pairs of sides or angles, just to detect equal or non-equal measurements. For the latter, transparencies can be very useful. Since teachers' use of transparencies up to then was mostly for presentations, I thought this would be an opportunity to draw teachers' attention to other possible uses of transparencies. (As a matter of fact, at that time I designed other activities for teachers with the use of transparencies, which turned insightful not only in terms of the mathematics but also in terms of expanding their familiarity and appreciation of tools. These activities included explorations of the properties of inverse functions, as well as a qualitative approach to constructing the graph of the derivative function directly from the graph of a function).

Figure 6 illustrates the outcome of a group of teachers who rotated counter-clockwise the spiral (congruent to the given one) that was drawn on the transparency. This rotation made the comparison between adjacent triangles transparent and the connections obvious. Not only had the congruence of angles and sides become clear but also the similarity between each two adjacent triangles. To most teachers, this was a new way of using transparencies. It demonstrated a qualitative way of identifying patterns, without "numbers". Since some teachers used other tools (e.g., a protractor and ruler) there was a natural opportunity to discuss the affordances and limitations of each tool.

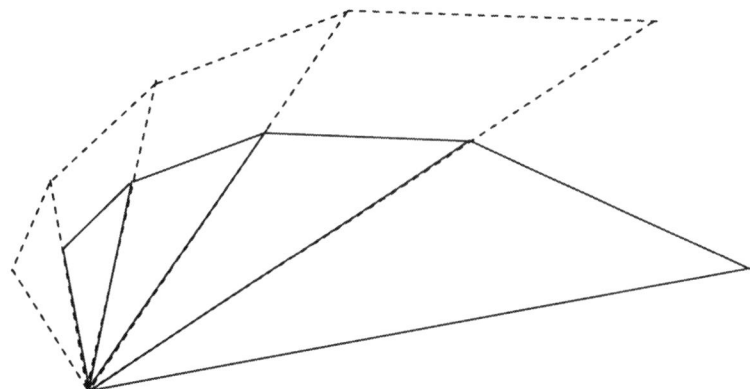

Figure 6. Comparing measurements with a transparency sheet in the second version of the "Spiral Task"

Following the above experiences with teachers, I realized that this task could also serve for enhancing teachers' use and appreciation of advanced technological tools, and could involve more uncertainty with respect to the existence of such

Triangle Spiral (of the kinds described by Zaslavsky, 2005). Hence, I tried to create a situation which would also allow teachers to experience some degree of frustration, followed by ways to eliminate their frustration, with the hope that this would be a springboard for gaining appreciation to some possible barriers to students' learning. Thus, the third modification was done with these goals in mind (Figure 7).

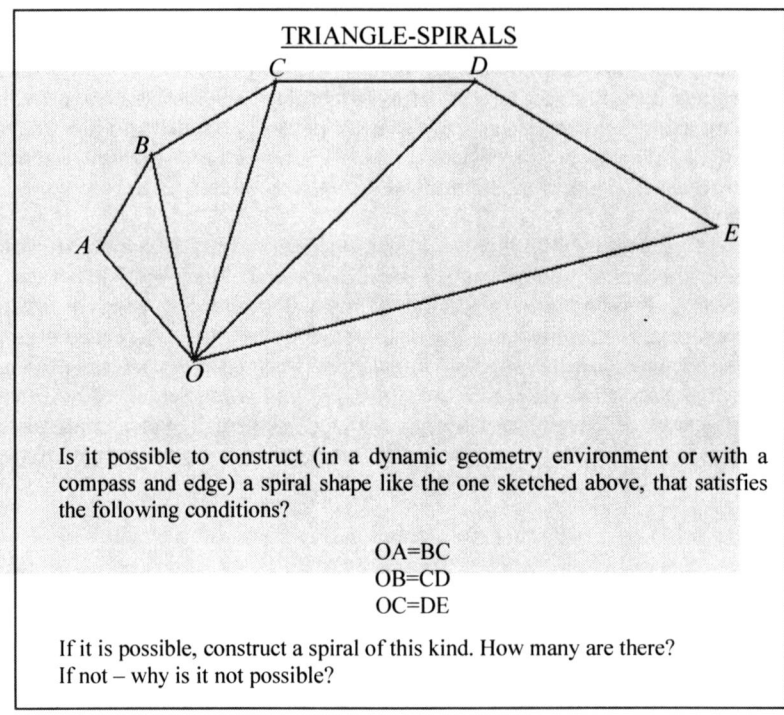

Figure 7. The third version of the "Spiral Task"

In the third version the questioned connections between the sides and angles of each pair of adjacent triangles are given, and the task is to find out if these can co-exist, and if so – under what conditions. Through the attempts to construct such a spiral, without knowing for sure whether or not it is at all possible, teachers were engaging in authentic exploration that arose from the need to answer the "big" question. Some used a DGE and others turned to a compass and edge. Similarly as in the second version, this created a natural opportunity to compare the advantages and limitations of each tool for this particular investigation.

Most teachers begin their attempts to construct the requested spiral with no prior analysis. Very soon they notice that there is more to this than meets the eye. Once they try constructing the spiral, they face sort of an impasse due to unexpected results on the screen. Figure 8 depicts two types of impasse that usually occur in such cases: when trying to construct the second triangle, the two sides do not intersect at the required point (C in Figure 7). This impasse is similar to what Hadas, Hershkowitz, and Schwarz (2000) discuss. It leads some teachers to switch to the conjecture that it is actually impossible to construct a spiral satisfying the given conditions, and others, to turn to a deep mathematical analysis of the implications of the given constraints. Thus, some begin analysing what the given connections imply in terms of the initial triangle (OAB). This analysis is triggered by an inner need, that is, by an impasse they face.

The given properties of the spiral imply that the initial triangle (and actually every triangle in the spiral) has a special property: the lengths of its sides form a geometric sequence, within certain restrictions. It is very unlikely that a triangle chosen at random will satisfy these conditions, thus it is very unlikely that someone will construct such a spiral without conducting this analysis. At the end, through social interactions, most groups manage to resolve the state of uncertainty, through which they encounter many surprising connections and insights (similar to those suggested by Movshovitz-Hadar, 1988). A film documenting the underlying processes of teachers and students dealing with the Spiral Task highlights the similarities and differences between the two populations with respect to the uncertainty they experienced and the understandings they reached (I.F.A., 1997). This film serves for teacher workshops in which they relate to possible implementation of the task in their classrooms.

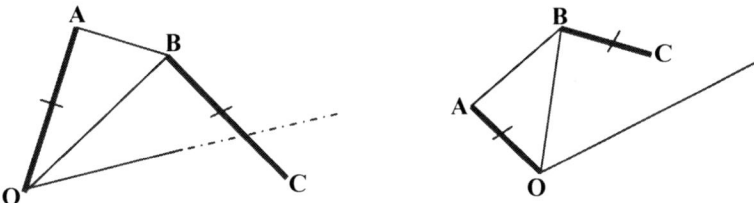

Figure 8. Two typical types of impasse teachers face with the third version of the "Spiral Task"

As mentioned above, this version evoked much uncertainty. It was implemented with many groups of prospective and practicing teachers who in turn tried it out with their students. To the vast majority of teachers and students the Spiral Task looked rather straightforward at first and the construction seemed feasible.

It should be noted that in all three versions, before handing out the task, teachers were prompted for their notions about the maximum number of pairs of congruent elements two non-congruent triangles could have. My experience is that in most groups there is a unanimous agreement that 3 is the maximum number, and that not just any 3 such pairs can exist in two non-congruent triangles. Few think that 4 could be, but all are convinced that if there are 5 pairs of congruent elements in two triangles, the triangles are necessarily congruent. By this prompt, the surprise later on becomes more dramatic. Not only that they find out at the end that non-congruent triangles with 5 congruent elements exist, but they realize that there are an infinite number of such cases, and they even learn how to construct many different cases. This approach aims at addressing some beliefs teachers and students hold with respect to mathematics; it demonstrates that mathematics is stimulating and can be exciting.

Reflection on the processes teachers undergo while dealing with this version of the task leads to the distinction between the "machine's" feedback and that provided by the facilitator. In this case, realizing that an arbitrary choice of sides and angles may not form a triangle, through observation of the outcome on the screen, is a manifestation that something "went wrong", and this in itself motivates a further investigation or change of strategy, with no interference of the facilitator. This situation is an incentive for the participants to turn to a mathematical analysis of the given conditions. Peer discussions and debate evolve naturally, involving sound reasoning including questions of inferences, such as, what can be inferred if it appears not to work. Does it mean that such a spiral does not exist or does it imply that other approaches need to be applied before any conclusion can be reached?

Having gained experience with the three versions of the Spiral Task, I became more aware of the significance of varying the task, and began using all three with various groups of teachers. This was an opportunity to use the task for enhancing adaptability and dealing with task variation. After coping with one version (different participants begin with different versions), teachers moved to the other two and compared the three along several dimensions, such as challenge, openness, motivation and disposition, mathematical insights, learning opportunities and student diversity (with respect to the suitability of the different versions to different students). Teachers were also asked to suggest other versions for their own classroom. One offered to try out a lesson based on the Spiral Task with her 9th grade students, and invited other teachers to observe. Her lesson was documented and used for discussions focusing on the way tasks unfold in the classroom, and how similar or different the task unfolded in a teacher workshop compared to her classroom.

What Is the Spiral Task an Example of?

I use the Spiral Task to illustrate several issues. First, it conveys the challenges involved in designing worthwhile learning situations for prospective and practising mathematics teachers. These kinds of activities are not readily available for teacher-educators to offer. It is awareness of the teacher-educator to all the earlier discussed themes and goals for mathematics teacher education that may lead him or her to an ongoing process of addressing as many aspects of those themes in the actual work with teachers.

The Spiral Task also illustrates how a task may evolve in a way that addresses to some extent all seven of the themes presented earlier. It involved reflection and adaptability on the part of the teacher-educator, in addition to other personal traits, that were modelled and discussed. It provided opportunities for teachers to engage in challenging problem solving and exploration, within school mathematics, coping with uncertainty and conflict, and required making comparisons and noticing connections. Each version of the task dealt to some extent with tools and their affordances. In the course of working with the third version teachers experienced as learners a certain degree of frustration, followed by ways of overcoming it, and reflected on the implications of these experiences for teaching. The task as a whole became a means for developing motivation and positive dispositions toward mathematical explorations.

Table 1 gives an overall picture of some of the ways in which each version addresses the seven themes, and conveys the added value of each version in terms of these themes. Movement from left to right in most rows indicates an increase in the amount of ways in which the themes are addressed. Moreover, the specific aspects addressed become more subtle and sophisticated.

Table 1. Some affordances of each version of the Spiral Task in terms of the seven themes for teacher education

Versions / Themes	*Initial Version*: Individual guided Investigation of the properties of a given spiral	*Second Version*: Collaborative open investigation of the properties of different spirals	*Third Version*: Collaborative open investigation of the conditions for the existence of a spiral with given properties	Comments
Adaptability & flexibility	The task ended with an analysis of its merits and limitations, and suggested ways to adapt it for students.	The task ended with an analysis of its merits and limitations, and suggested ways to adapt it for students.	The task ended with an analysis of its merits and limitations, and suggested ways to adapt it for students.	Teachers were given various versions, and considered the possible use of the alternative approaches with their students.
Awareness to similarities, differences (incl. making connections & identifying patterns)	Comparison between different pairs of adjacent triangles; Connection to similar triangles, Fibonacci numbers, golden section, real world phenomena.	In addition to what the initial version triggered: Comparison between different spirals – each group explored a different one;	Search for necessary and sufficient conditions for this kind of spiral; Classification of triangles according to their possible inclusion in such spirals;	The three task versions provided opportunities for comparing them along several dimensions, including pedagogical & epistemological.

109

Versions \ Themes	Initial Version: Individual guided Investigation of the properties of a given spiral	Second Version: Collaborative open investigation of the properties of different spirals	Third Version: Collaborative open investigation of the conditions for the existence of a spiral with given properties	Comments
Coping with conflicts, dilemmas, & problem situations	A counter-intuitive outcome; Challenging mathematics; Surprising results	Overcoming the conflict between intuition and evidence; Awareness to social aspects of open-ended problem solving situations.	Coping with uncertainty; Subtle logical inferences; Need for careful analysis in order to deal with the impasse.	Teachers' notions were elicited in advance, regarding the maximum number of pairs of congruent elements of two non-congruent triangles.
Learning from practice	The nature of the task and its structure allowed readily use with students.	One teacher invited her peers to learn from observing her use this version in her classroom.	This version appealed to teachers who felt comfortable with DGE, and tried it out with their students.	Teachers tried out the different versions of the task with their students and shared their experiences with their peers.
Judicious use of tools	Possible use of ruler and protractor; Use of cut-out triangles for comparing side lengths and angle measurements.	Use of transparency for comparing side lengths and angle measurements; Appreciation of the limitations of measuring tools.	Use of ruler and compass, vs. DGE for constructions and comparisons; Non-judgmental feedback provided by a technological tool – by gap between expected vs. actual outcome.	Experiencing all three versions and reflecting on the different tools employed, created a powerful opportunity to developing appreciation of tools.
Overcoming barriers		Overcoming uneasiness regarding open-ended investigation; Appreciation of inner motivation.	Overcoming impasses and dealing with frustration; Sense of joy in overcoming difficulties.	Reflection on personal and group encounters; Discussing barriers, motivation, and inclusion, associated with learners.

Versions Themes	Initial Version: Individual guided Investigation of the properties of a given spiral	Second Version: Collaborative open investigation of the properties of different spirals	Third Version: Collaborative open investigation of the conditions for the existence of a spiral with given properties	Comments
Sharing and revealing dispositions	Sharing personal dispositions regarding the task and the mathematics involved in it.	Sharing feelings of perplexity, enthusiasm and excitement, elicited by the social setting.	Sharing frustration in facing impasse, elicited by the social setting; Sharing enthusiasm and satisfaction.	During and after each task time was devoted to reflection on personal feelings and dispositions.

For example, looking at the theme *Judicious use of tools*, the initial version is based on simple measuring tools (ruler and protractor); the second version offers a special use of transparency for comparing measurements and making connections transparent; the third version lends itself to technology – in which although measurements can be instantly displayed, the formulation of the task leads to better use of technology, by conveying the idea that the tool cannot replace a mathematical analysis. Instead, it can offer accurate drawings which form powerful feedback, particularly when there is a mismatch between what is expected and what actually happens.

It should be noted that not surprisingly, this one task, with its three versions, does not address all aspects of each theme. It would take many more carefully designed tasks, to capture the multifaceted goals and themes described earlier. However, as illustrated by the Spiral Task, keeping these multifaceted goals and themes in mind, could lead to gradual increase in the affordances of tasks with respect to meeting these goals and challenges.

THE DEMANDS ON TEACHER-EDUCATORS AS FACILITATORS OF TEACHER LEARNING

As reflected in the previous sections, the demands on mathematics teacher-educators are heavy, in terms of their knowledge-base, personal traits, and responsibilities. There are several personal traits that are considered desirable and even imperative for mathematics educators (teachers, teacher-educators and educators of teacher-educators). For example, a mathematics educator needs to be reflective, adaptive, flexible, open minded, risk taker, sensitive, confident, and enthusiastic about what he or she teaches. Some of these traits may be naturally developed over years of experience; however, tasks for teacher education can be seen as springboards for enhancing such personal traits as well. A common way of doing so is by modelling these traits in the course of using tasks. Yet, this is not enough. The implicit exhibition of such traits should be made explicit. That is, if a teacher-educator exhibits flexibility and is open to an unexpected idea, it is important to come back to that at another stage and reflect on the kind of flexibility that was exhibited and what teachers gained from it as learners and what they can take from it as teachers.

Zaslavsky (2007) conceptualizes the enormous demands on teacher-educators, based on a collection of articles dealing with the nature and role of tasks in mathematics teacher education (Zaslavsky, Watson & Mason, 2007). Accordingly, the knowledge base of teacher-educators consists of the knowledge for teaching mathematics (that is, mathematics teachers' knowledge base) as well as knowledge of how to enhance teacher learning. This could include the model offered by Zaslavsky & Leikin (2004) for discussing mathematics teacher-educators' knowledge base, but also includes theoretical and practical knowledge of the seven central themes discussed earlier. In addition, teacher-educators are expected to hold and exhibit several personal traits that are needed for facilitating learning. Based on their knowledge and personal traits they are expected to engage teachers in productive tasks, during which they may model certain behaviours and traits, and later reflect on them and make them an explicit focal topic for discussion. The ultimate goal of this aspect of mathematics teacher-educators' practice is to help teachers construct knowledge for teaching mathematics and develop personal traits needed for becoming a competent and effective teacher.

When discussing tasks, in general, and tasks for teacher education, in particular, it is hard to make a clear distinction between the actual task and how it unfolds in the learning setting. Tasks for teachers have several layers. As described above, and unlike tasks for students, a mathematical task for teachers rarely deals with just the mathematics. It can be seen as an opportunity to generalize from it to a large class of tasks, and to deal with many other aspects of teaching mathematics as well. Moreover, as Zsalvsky (2007) points out, "tasks evolve over years of reflective practice" of mathematics teacher-educators.

This chapter highlights the complexity and challenges of one (although central) aspect of mathematics teacher-educators' roles and responsibilities. It conveys the high demands, and raises the question of how to prepare teacher-educators for these roles. There are relatively few programmes for developing mathematics teacher-educators (e.g., Even, 1999), and none address all the above themes in a substantial way. Learning from their own practice is certainly a way to develop (e.g., Zaslavsky & Leikin, 2004; Tzur, 2001), yet – it would be valuable to begin articulating the goals for mathematics teacher-educators' education, and ways of achieving them.

REFERENCES

Adler, J. (2000) Conceptualising resources as a theme for mathematics teacher education. *Journal of Mathematics Teacher Education, 3*(3), 205–224.

Burke, M. (1993). 5-Con Triangles. In E. M. Maletsky (Ed.), *Teaching with student math notes*, Vol. 2 (pp. 107–112). Reston, VA: National Council of Teachers of Mathematics.

Clarke, D. J. & Hollingsworth, H. (2000). Seeing is understanding: Examining the merits of video and narrative cases. *Journal of Staff Development 21*(4), 40–43.

Cooney, T. J. (1994). Teacher education as an exercise in adaptation. In D. B. Aichele & A. F. Coxford (Eds.), *Professional development for teachers of mathematics: 1994 Yearbook* (pp. 9–22). Reston, VA: National Council of Teachers of Mathematics.

Even, R. (1999). Integrating academic and practical knowledge in a teacher leaders' development program. *Educational Studies in Mathematics, 38,* 235–252.

Fernandez, C. & Yoshida, M. (2004). *Lesson Study: A Japanese approach to improving mathematics teaching and learning.* Mahwah, NJ: Lawrence Erlbaum Associates, Inc., Publishers.

García, M., Sánchez, V., & Escudero, I. (2006). Learning through reflection in mathematics teacher education. *Educational Studies in Mathematics, 64,* 1–17.

Hadas N., Hershkowitz R., & Schwarz, B. (2000). The role of contradiction and uncertainty in promoting the need to prove in dynamic geometry environments. *Educational Studies in Mathematics, 44,* 127–150.

Heid, M. K., Middleton, J. A., Larson, M., Gutstein, E., Fey, J. T., King, K., Strutchens, M. E., & Tunis, H. (2006). The challenge of linking research and practice. *Journal for Research in Mathematics Education, 37*(2), 76–86.

Hiebert, J., & Wearne, D. (1993). Instructional tasks, classroom discourse, and student learning in second grade. *American Educational Research Journal, 30,* 393–425.

I.F.A. (Israel Film Association) (1997). Teaching and Learning Mathematics: Spirals, Triangles and Similarity. A documentary film, produced within the framework of the "Tomorrow 98 in the Upper Galilee" Project (in Hebrew).

Jaworski, B. (2001). Developing mathematics teaching: Teachers, teacher educators, and researchers as co-learners. In F.-L. Lin & T. J. Cooney (Eds.), *Making sense of mathematics teacher education* (pp. 295–320). Dordrecht, the Netherlands: Kluwer Academic Publishes.

Jaworski, B. (2003). Research practice into /influencing mathematics teaching and learning development: Towards a theoretical framework based on co-learning partnerships. *Educational Studies in Mathematics, 54*(2–3), 249–282.

Jaworski, B. and Gellert, U. (2003). Educating new mathematics teachers: Integrating theory and practice, and the roles of practising teachers. In A. J. Bishop, M. A. Clements, C. Keitel, J. Kilpatrick & F.K.S. Leung (Eds.), *Second international handbook of mathematics education* (pp. 823–876). Dordrecht, the Netherlands: Kluwer Academic Publishes.

Kilpatrick, J., Swafford, J., & Findell, B. (Eds.) (2001). *Adding it Up: Helping children learn mathematics.* Washington, DC: National Academy Press.

Krainer, K. (1993). Powerful tasks: A contribution to a high level of acting and reflecting in mathematics instruction. *Educational Studies in Mathematics, 24,* 65–93.

Leikin, R. & Dinur, S. (2007). Teacher flexibility in mathematical discussion. *Journal of Mathematical Behavior, 18*(3), 328–247.

Lewis, C., Perry, R. & Hurd, J. (2004). A deeper look at lesson study. *Educational Leadership, 61*(5), 18–23.

Mason, J. (1998). Enabling teachers to be real teachers: Necessary levels of awareness and structure of attention. *Journal of Mathematics Teacher Education, 1*(3), 243–267.

Mason, J. & Spence, M. (1999). Beyond mere knowledge of mathematics: The importance of knowing-to act in the moment. *Educational Studies in Mathematics, 38,* 163–187.

Merseth, K. K. & Lacey, C. A. (1993). Weaving stronger fabric: The pedagogical promise of hypermedia and case methods in teacher education. *Teaching and Teacher education, 9*(3), 283–299.

Mewborn, D. S. & Cross, D. I. (2007). Mathematics teachers' beliefs about mathematics and links to students' learning. In W. G. Martin, M. E. Strutchens, & P. C. Elliot (Eds.), *The learning of mathematics: Sixty-ninth yearbook* (pp. 259–269). Reston, VA: National Council of Teachers of Mathematics.

Movshovitz-Hadar, N. (1988). School mathematics theorems – An endless source of surprise. *For the Learning of Mathematics, 8*(3), 34–40.

Rowland, T., Huckstep, P., & Thwaites, A. (2005). Elementary teachers' mathematics subject knowledge: The knowledge quartet and the case of Naomi. *Journal of Mathematics Teacher Education, 8*(3), 255–281.
Schön, D. A. (1987). *Educating the reflective practitioner.* Oxford, UK: Jossey-Bass.
Sierpinska, A. (2004). Research in mathematics education through a keyhole: Task problematization. *For the Learning of Mathematics, 24*(2), 7–15.
Silver, E. A. (1979). Student perceptions of relatedness among mathematical verbal problems. *Journal for Research in Mathematics Education, 10,* 195–210.
Simon, M. A., & Tzur, R. (2004). Explicating the role of mathematical tasks in conceptual learning: An elaboration of the Hypothetical Learning Trajectory, *Mathematical Thinking and Learning, 6*(2), 91–104.
Stein, M. K., Smith, M. S., Henningsen, M. A., & Silver, E. A. (2000). *Implementing standards-based mathematics instruction: A casebook for professional development.* New York: Teachers College Press.
Steinbring, H. (1998). Elements of epistemological knowledge for mathematics teachers. *Journal of Mathematics Teacher Education, 1*(2), 157–189.
Sullivan, P. (2002).Using the study of practice as a learning strategy within mathematics teacher education programs. *Journal of Mathematics Teacher Education, 5*(4), 289–292.
Sullivan, P. (2006). Dichotomies, dilemmas, and ambiguity: Coping with complexity. *Journal of Mathematics Teacher Education, 9*(4), 307–311.
Sullivan, P. & Mousley, J. (2001).Thinking teaching: Seeing mathematics teachers as active decision makers. In F-L. Lin and T. Cooney (Eds.) *Making sense of mathematics teacher education* (pp. 147–164). Dordrecht/Boston : Kluwer Academic Publishers.
Sullivan, P., Zevenbergen, R., & Mousley, J. (2003). The context of mathematics tasks and the context of the classroom: Are we including all students? *Mathematics Education Research Journal, 15*(2), 107–121.
Trentacosta, J. & Kenney, M. J. (1997) (Eds.). *Multicultural and gender equity in the mathematics classroom: 1997 yearbook.* Reston, VA: National Council of Teachers of Mathematics.
Tzur, R. (2001). Becoming a mathematics teacher-educator: Conceptualizing the terrain through self-reflective analysis. *Journal of Mathematics Teacher Education, 4*(4), 259–283.
Zaslavsky, O. (1995). Open-ended tasks as a trigger for mathematics teachers' professional development. *For the Learning of Mathematics, 15*(3), 15–20.
Zaslavsky, O. (2005). Seizing the opportunity to create uncertainty in learning mathematics. *Educational Studies in Mathematics, 60,* 297–321.
Zaslavsky, O. (2007). Mathematics-related tasks, teacher education and teacher educators. *Journal of Mathematics Teacher Education, 10,* 433–440.
Zaslavsky, O., & Leikin, R. (2004). Professional development of mathematics teacher educators: Growth through practice. *Journal of Mathematics Teacher Education, 7*(1), 5–32.
Zaslavsky, O., Watson, A., & Mason, J. (Eds.) (2007). The nature and role of tasks in mathematics teacher education. Special issue of the *Journal of Mathematics Teacher Education, 10,* 201–440.
Zaslavsky, O. & Zodik, I. (2007). Mathematics teachers' choices of examples that potentially support or impede learning. *Research in Mathematics Education, 9,* 143–155.

Orit Zaslavsky
Department of Education in Technology & Science
Technion – Israel Institute of Technology

OLIVE CHAPMAN

6. MATHEMATICS TEACHER EDUCATORS' LEARNING FROM RESEARCH ON THEIR INSTRUCTIONAL PRACTICES

A Cognitive Perspective

This chapter discusses the learning of mathematics teacher educators based on research they conducted on their instructional practices used with prospective teachers. The focus is on three themes: characteristics of the instructional approaches, characteristics of the learning outcomes for the instructional approaches, and characteristics of the teacher educators' learning. A review of current, related studies forms the basis for discussing these themes. These studies embodied a cognitive perspective as a basis of the instructional practices and what the educators can learn from them.

INTRODUCTION

In this chapter, I discuss the learning of mathematics teacher educators based on research they conducted on their instructional practices. Such research has the potential for providing teacher educators with professional development and for allowing them to know something important about their practice that they did not know before. It can also inform others of what mathematics teacher educators can learn or have learned, and how they learn, from conducting these studies. Thus, focusing on this category of studies by teacher educators could provide a representative basis of their learning from their own research.

Current research in the mathematics education literature includes studies conducted by mathematics teacher educators focusing on instructional practices they developed for use with prospective teachers. These studies seem to be in response to current reform movements, in particular, reform recommendations for mathematics education in North America (e.g., National Council of Teachers of Mathematics [NCTM], 1991, 1989) and the resulting interest in educating prospective and practicing teachers accordingly. The shift in perspective of mathematics education towards the "reform approach" made it necessary to prepare prospective teachers for using a type of school curriculum different from what they experienced as students. As Lampert, Heaton, and Ball (1994) suggested, a new pedagogy of teacher education is required if teacher education is to prepare prospective teachers to be responsive to visions of mathematics education advocated by the reform. They argued that conventional teacher education programmes, which present prospective teachers with ideal methods and techniques derived from a synthesis and interpretation of educational theory and research, do not represent the complexity and uncertainty of teaching found in reform-based instruction. Thus, the nature of alternative approaches for preparing prospective teachers to become teachers envisioned by the reform emerged as an important consideration for mathematics teacher educators.

The growth in research of mathematics teacher educators' instructional approaches can also be associated with the influence of cognitive psychology that resulted in many studies on teacher thinking. These studies raised concerns about the nature and role of prospective teachers' initial beliefs, conceptions and knowledge of mathematics and mathematics teaching during teacher education programmes (Ponte & Chapman, 2006). This knowledge, acquired during many years of being a student, can be a powerful influence on the way the prospective teachers approach teacher education and what they learn from it (Britzman, 1991; Calderhead & Robson, 1991). This influence is likely to be negative if the goal is for them to develop teaching practices that are different from those they experienced as learners of mathematics. It also is based on their taken-for-granted preconceptions or tacit personal theories that are not readily accessible to make explicit (Polanyi, 1958). Such factors have contributed to the challenge of preparing future mathematics teachers. Thus, a critical issue for mathematics teacher educators continues to be how to help prospective teachers develop a vision of mathematics, learners and instruction that departs from their past experiences as learners of mathematics. Teacher educators are faced with the question of what types of experiences prospective teachers would need in order to become effective teachers of mathematics. An ongoing issue in mathematics education, then, is how to design instruction that can influence the nature and quality of prospective teachers' thinking and practice. Mathematics teacher educators have addressed this challenge by creating university courses in which prospective teachers can learn mathematics and its teaching and learning in ways consistent with current reform views. Central to the instructional approaches of these courses is the assumption that experiencing mathematics differently as learners will allow prospective teachers to reconstruct their beliefs, assumptions, and ultimately their practice (Ponte & Chapman, in press).

Mathematics teacher educators, then, seem to have both external and internal motivation for researching their instructional approaches as a basis for learning from them. In this chapter, I consider examples of these studies involving the mathematics teacher educators' practice, in particular, the instructional approaches they developed, implemented and formally investigated. The settings for these studies are mathematics-for-teachers courses, mathematics education courses focusing on both content and pedagogy (e.g., 'methods' courses), and field experiences (e.g., student teaching). I focus only on studies involving prospective teachers and on current published work over the last 10 years, most of which are based in North America. My intent is not to provide a historical overview of studies mathematics teacher educators have engaged in that relate to their practice, or to document all possible examples of these studies, but to use the focus on current published work involving prospective teachers as a basis to discuss the teacher educators' learning from researching their own practice. Based on my review of these studies, three themes emerged in relation to the teacher educators' learning: characteristics of the instructional approaches, characteristics of the learning outcomes for the instructional approaches, and characteristics of the teacher educators' learning. These themes are used to organize and form the focus of the chapter.

CHARACTERISTICS OF INSTRUCTIONAL PRACTICES STUDIED

Although it may not be explicitly stated in the research surveyed, it can be inferred that the instructional practices themselves can be a basis of mathematics teacher educators' learning. However, as discussed later, for the most part, this learning was presented as what other teacher educators could learn about the nature of these approaches to instruction rather than the actual learning of the persons conducting the research. This section, then, discusses mathematics teacher educators' learning from this perspective with a focus on the influences, theoretical orientations and specific features that are characteristic of the instructional approaches they used with prospective teachers.

In keeping with the influence of the NCTM standards on most of the studies discussed in this chapter, collectively they covered learning situations involving problem solving, investigations, reasoning, connections, communication, and technology – all key components of the recommended reform curriculum (NCTM, 1989). These studies also suggest an accepted trend that a pedagogy of teacher education parallels the mathematical pedagogy of reform efforts (e.g., NCTM, 1991). A few of the studies referred to NCTM explicitly as the basis for the content and goals of the approach. Artzt (1999), for example, explained that her approach included a focus on the current goals of secondary mathematics instruction as outlined in the NCTM Standards as a basis for the prospective teachers' reflective activities. Cramer (2004) used an approach for facilitating prospective teachers' growth in content knowledge consisting of a pedagogical model based on the NCTM Standards. Lee (2005) explained that two aspects of reform in mathematics education were of interest to her study: the importance of problem solving, and the increased availability and use of a variety of learning tools, especially technology. Fernández (2005) aimed at helping the prospective teachers' development of reform-oriented teaching. Other studies (e.g., Goodell. 2006; Szydlik, Szydlik, & Benson, 2003) referred to the importance of enabling the prospective teachers to fulfill the roles prescribed for them by the NCTM Standards.

From a theoretical perspective, a cognitive orientation was implied in the instructional practices studied. In particular, although not always explicitly stated, these practices generally involved notions of constructivism, consistent with the influence of the NCTM Standards. Thus learner-centeredness in terms of inquiry-based learning activities and student-centered interactions to facilitate the prospective teachers' construction of knowledge about mathematics and mathematics pedagogy were common features of the approaches. Implied in some of the approaches were influences of the perspective associated with the work of the Cognitively Guided Instruction group (Carpenter et al., 1999) that focused on the development of teachers' understanding about children's learning. In a broader cognitive context, the approaches also contained an underlying orientation towards viewing the prospective teachers as having personal experiences, knowledge, beliefs, intentions, and ways of thinking that are integral to how and what they learn. Thus, reflection was a central theoretical construct of many of the approaches as a basis of the future teachers' learning. Some of the researchers explicitly emphasized specific aspects of these theoretical orientations and only this subset of the studies reviewed are offered as examples of the mathematics teacher educators' considerations for framing their instructional approaches.

Some studies focused on the importance of self or one's thinking in defining one's behavior with the goal of understanding or changing such thinking. In these studies, notions of beliefs and ways of accessing or changing beliefs formed the dominant basis of framing the instructional approaches. For example, Artzt (1999) explained that her approach was based on the premise that cognitive components of teaching play a critical role in shaping a teacher's instructional practice. In particular, teachers' knowledge, beliefs, and goals directly impact their instructional practice. They affect the way teachers design a lesson, the way they monitor and regulate their instruction during the teaching process, and the way they analyze the lesson after it has been concluded. Ambrose's (2004) approach was based on views of beliefs, the process of belief change and the ways that teacher educators can promote belief change in prospective teachers. She described her theoretical framework as including several aspects of beliefs: their origins, their effects on one's interpretations of experiences, the ways separate beliefs combine to create belief systems, and the ways beliefs change within this framework. Masingila and Doerr (2002), in considering knowledge of competent professionals to be tacit and implicit in their actions, drew on the work of Schön (e.g., 1987) as a basis for their approach, which involved a multimedia case study of an experienced teacher designed so that prospective teachers could have access to her thinking. As a final example, Goodell (2006) drew on notions of 'critical incidents' as a basis for her prospective teachers to access their thinking about their own instruction. She explained that a critical incident can be thought of as an everyday event encountered by a teacher in his or her practice that makes the teacher question the decisions that were made.

A few of the studies explicitly identified specific notions of constructivism as guiding or playing an important role in the instructional approach. For example, Ebby (2000) described her approach as positioning prospective teachers as creators rather than receivers of knowledge about teaching and encouraged them to take a critical stance towards teaching and schooling. Goodell (2006) explained that she wanted her course to be not only about constructivism as a theory of learning, but also constructivist in nature itself to take account of the suggestion that there should be constructivist conditions of interventions for educating teachers. McDuffie (2004) also explained that she followed constructivist frameworks for student-teaching supervision that emphasized the developing of self-directed, reflective practitioners. Lee (2005) described the perspective of her approach as involving individuals' constructive process of resolving perturbations through reflecting on their actions. The perturbations occurred for prospective teachers as they observed and interacted with students and reflected on their own and students' interactions with each other, the mathematics, and the available tools.

The preceding examples of theoretical orientations used by these teacher educators to frame the instructional approaches they studied suggest what they valued and sought to understand or validate, directly or indirectly, as a basis of their learning. This also highlights the personal orientation of the approaches that could limit their transferability to other contexts and use as a basis of other educators' learning. However, specific characteristics of the approaches could be informative in a broader sense. Thus, in the remainder of this section, I summarize six key characteristics that are representative of instructional practices and cognitive perspectives used in the aforementioned and other studies reviewed for this chapter. I inferred from the studies that these were central features of the instructional approaches for prospective teachers' learning from this category of research. Only brief abstracts of relevant aspects of the studies are provided to indicate the nature of these characteristics as they relate to the instructional approaches. Not all characteristics apply to all of the studies, so it is only when they are explicit in the description of the approaches that the studies are included in what follows.

1. Child-Study

Some of the instructional approaches incorporated "child-study" to various degrees. In child-study, the prospective teacher works with a child to observe, interview, and document information about this child as a learner of mathematics. This process can be structured or guided in a variety of ways depending on the goal of the instructional approach as illustrated in the following examples. Ebby's (2000) approach to facilitate the prospective teachers' development of a conception of their teaching role included an investigation of children's learning of mathematical concepts by observing and analyzing a child as a mathematical learner. This approach seems to be open-ended in terms of the aspect of mathematics the prospective teachers were required to investigate in the child-study. Other studies, however, indicated a specific focus for the child-study, often centered on the child performing a specific mathematical task. For example, Ambrose's (2004) approach to initiate belief change and help prospective teachers to understand the importance of subject matter knowledge in the teaching of mathematics included having them work with children in an elementary school. The goal was for them to explore whole and rational numbers and analyze children's invented approaches to problems in order to make sense of non-standard methods commonly created by students, the reasoning behind the methods, and how the structure of number is used in these calculations. Pairs of prospective teachers worked with a child using specific tasks and activities designed to elicit the child's thinking. One in each pair led the problem-solving session, and the other took notes. Each partner had a chance to perform each role several times with different children. In Chapman (2005), the prospective teachers were required to select a non-algorithmic problem appropriate for a secondary school student and to use it to observe a pupil solving it while thinking aloud. The goal was for them to focus on the students' thinking, consider when and how they wanted to intervene in it, compare it to their own thinking, and collaborate with peers to develop a model of problem solving based on the students' thinking.

In some approaches, the prospective teachers first solved mathematics problems then used the same problems in the child-study. For example, Lee (2005) provided her prospective teachers with experiences with children to help prepare them for the pedagogical challenges of effectively engaging students in problem solving. Individually, the prospective teachers solved a mathematics problem using a java applet, planned a hypothetical learning trajectory for students based on Simon (1995), and then interacted with two students (individually) as they solved the same problem with a java applet. This was followed by a revision of the hypothetical learning trajectory and interaction with two different students as they solved the same problem with java applet. Nicol (1999) provided opportunities for her prospective teachers to work with children to investigate their teaching. In one of their weekly class sessions, they worked together on a mathematics problem and thought about how they might pose such a problem to students. In the second weekly class, working in pairs, they posed their extended or adapted problems to sixth and seventh grade (11 – 12 year old) students in the school.

Other approaches emphasized the interview process in conducting the child-study. In particular, Schorr's (2001) approach focused on the use of clinical interviews as a way to help prospective teachers develop a deeper understanding of the ways in which children learn mathematics and thus to devise better ways of teaching. The prospective teachers learned the clinical interview method, interviewed children, and reflected on the interviews in class. During weekly classes, they investigated a particular mathematics idea by solving problems related to the idea. When appropriate, they watched a videotaped interview involving a child or series of children grappling with the same or similar mathematical ideas and shared reflections about the questions posed and the interview techniques used. They also discussed the children's mathematical thinking and pedagogical implications of teaching and learning the mathematical ideas. They were encouraged to interview a child about the same ideas and share the results in class and were required to interview a child every other lesson. In Mewborn's (2000) approach, involving how to communicate with children about mathematics, the prospective teachers had opportunities to listen to, and make sense of, the mathematics that children generate and to consider how children's mathematical thinking might impact on effective instruction. The approach included the prospective teachers observing a classroom, watching videotapes of children solving mathematics problems and then conducting task based-interviews with four children.

2. Self-Study

Most of the instructional approaches in the studies reviewed involved some form of "self-study" on the part of the prospective teachers. Self-study involves inquiring into one's thinking, learning and instructional practices. It is influenced by the notion that what prospective teachers take away from their coursework has much to do with the beliefs, dispositions, and experiences that they bring with them. How the prospective teachers teach is a reflection of the knowledge and beliefs they have about the mathematical content, the particular students they teach, and the way students learn. Thus understanding, deepening, and/or restructuring their own thinking form a central theme in the instructional approaches particularly in terms of the use of reflective processes that allows them to inquire into their selves as in the following examples.

A few studies focused explicitly on the prospective teachers' self-study of their thinking independent of their actual practice. In Ebby's (2000) approach, this involved having the prospective teachers reflect on their own beliefs about teaching and learning mathematics. In Chapman's (2005) study of problem solving, the prospective teachers were required to respond to prompts that included: Choose a grade and make a mathematics problem that would be a problem for those students. What did you think of to make the problem? Why is it a problem? Is it a 'good' mathematics problem? Why? What process do you go through when you solve a problem? These prompts were intended to capture their initial thinking about problems and problem solving. Bolte (1999) discussed the use of concept maps and the writing of interpretive essays in enhancing and assessing prospective teachers' integration and expression of mathematical knowledge. She explained that in the process, the prospective teachers actively participated in the worthwhile task of developing connections between related mathematics concepts, reflecting on their thinking, and engaging in mathematical discourse about their thinking.

Many of the approaches focused on prospective teachers' thinking and behaviors in relation to practice. Artzt's (1999) approach consisted of a detailed model to guide the self-study. It was used to facilitate the prospective teachers' reflection as a basis of their learning during practicum teaching. It consisted of pre-lesson and post-lesson reflective activities involving the prospective teachers' analysis of their cognition and instructional practice before, during, and after their lessons. The prospective teachers wrote lesson plans and a paper of their pre-lesson thoughts and concerns; engaged in a post-lesson conference with the supervisor and cooperating teacher that included a self-assessment of their teaching, then wrote a paper describing their post-lesson thoughts; and documented their thoughts and experiences through weekly entries in their journals. Goodell (2006) had her prospective teachers select a 'critical incident' that happened either to them or to their cooperating teachers. After discussing the incidents in small groups, which they then reported to the whole class, each prospective teacher submitted a written report describing the incident, why it happened that way, how he or she might handle the situation differently, and what the implications for her or his practice might be. Goodell explained that critical incident reflections required more than just restating what happened. A reconstruction of the event was a central part of the critical incident protocol. In the approach used by Fernández (2005), self-study was implied in the process of helping the prospective teachers' development of reform-oriented teaching through micro-teaching lesson study. The prospective teachers were involved in three cycles of planning, teaching, analyzing and revising a mathematics lesson. Written assignment for each cycle mirrored the phases of lesson study and included pre-lesson thoughts, analysis of the lesson based on a video recording of it, and revisions to the lesson plan.

Other studies with a self-study component include: Lee (2005) where the prospective teachers reflected on their role in facilitating students' problem solving with technology and their understanding of what the students understood about the problem; Ebby (2000) where they taught and critiqued a lesson; Nicol (1999) where they were required to reflect critically on aspects of their practice, to problematize their teaching, and to consider what teaching could be or ought to be; Masingila and Doerr (2002) where they engaged in a case study of an experienced teacher's practice to support their analysis and reflection on their own emerging practices; and Chapman (1999) and Lloyd (2006) where the prospective teachers wrote stories of actual teaching and/or fictional stories as a basis of their self-reflection. In addition, McDuffie (2004) suggested a framework of teaching as a deliberate practice that could make prospective teachers aware of their thinking.

3. Case Study of Experienced Teachers

Another characteristic of the instructional approaches studied is the use of case studies of experienced teachers. While cases can be in written text form, in some studies, they were video-based. They were viewed as virtual field experiences for the prospective teachers to engage in analysis of practice. In the Masingila and Doerr's (2002) study, the case seemed to represent exemplary teaching. The authors explained that the case lessons were designed to engage the students (in the case) in actively learning mathematics, generating and critiquing mathematical arguments, and expressing their mathematical ideas about problematic situations. The case captured the records of practice in a classroom in a way that would reflect the complexities of classroom interactions, teacher decisions, and students' mathematical thinking. The goal for the prospective teachers was to make sense of the case-study teacher's practice and thinking and their own emerging practices. In contrast, the cases in the Santagata, Zannoni, and Stigler's (2007) study did not necessarily represent exemplary teaching, did not reflect local school context but focused on analysis of the lessons. These authors explained that the three videotaped, eighth-grade mathematics lessons used in the course were selected not because they were necessarily examples of good teaching but because they illustrated common teaching situations that would provide a context to develop analysis skills. The authors also explained that lessons from other countries were chosen because the exposure to alternative practices helped observers to become aware of their own cultural routines. The prospective teachers were guided in viewing the lessons by a framework designed to direct their analysis to the connections between learning goals, teaching strategies and student learning. They watched each lesson three times, each time focusing on a different aspect of the lesson respectively: parts of the lesson and learning goals; students' thinking and learning; and alternative teaching strategies.

4. Mathematical Tasks

Some of the instructional approaches for the education (as opposed to mathematics) courses included, or were centered on, a component that consisted of mathematical tasks framed in a context of inquiry or investigation. The prospective teachers' understanding of problems and problem solving was the goal of some of these studies. For example, in Chapman (2005), the prospective teachers were provided with a list of problems to compare and contrast without solving them in order to investigate the nature and role of problems in the school curriculum. The list consisted of a non-verbal, algebraic exercise; a simple translation algebraic word problem; a complex translation algebraic word problem; a process (non-routine) word problem; an applied (open) problem, and a puzzle problem. These labels, taken from Charles and Lester (1982), were not included with the problems. Szydlik, Szydlik, and Benson (2003) described their tasks as authentic mathematical experiences centered on solving and discussing non-routine mathematics problems in order to foster autonomous mathematical behaviors in relation to problem solving. Prospective teachers also produced written reports that focused on their mathematical thinking by describing the problem, discussing the strategies they used to work on the problem, providing a solution, and arguing that their solution was complete and valid. In Roddick, Becker, and Pence's (2000) study, some tasks involved problem solving, problem posing, and modeling. The prospective teachers spent significant time on topics such as: what is a problem; finite differences; examination of problem solving in traditional and innovative curricula; and assessment of problem solving. They also reflected on their problem solving.

In other studies, the focus of the tasks was the development of content and pedagogical content knowledge. The number strand of the school curriculum was most represented. In Chapman (2007), prospective teachers investigated a variety of word problem situations in order to understand the meaning of the arithmetic operations with whole numbers. Tirosh's (2000) approach focused on operations with fractions with a specific goal to promote development of prospective teachers' content knowledge as well as their awareness of the nature and the likely sources of related common misconceptions held by children. Ilany, Keret, and Ben-Chaim (2004) also dealt with rational numbers, but for the concepts of ratio and proportion. They described their core tasks as authentic investigative activities dealing with ratio, rate, scaling, and indirect proportion. Heaton and Michelson (2002) created opportunities for prospective teachers to learn the process of statistical investigation in the context of inquiry about elementary mathematics teaching.

Technology also played a role in some approaches involving investigative or problem-solving tasks. For example, in Ponte, Oliveira, and Varandas (2002), the prospective teachers explored the use of Geometer's Sketchpad from an investigative perspective, starting with simple mathematical questions about properties of triangles and quadrilaterals, then moving on to features of conics and, finally, invariant properties of certain geometrical transformations. Zbiek's (1998) prospective teachers used computing tools to develop and validate functions as mathematical models of real-world situations. Activities for exploring these functions involved using a variety of technology, such as graphing tools, symbolic manipulators, and spreadsheets. In Lee's (2005) approach, the prospective teachers individually solved a mathematics problem using a java applet.

5. Student-Centered Interactions

All of the studies reviewed included some form of student-centered interaction in the instructional approaches. Depending on the activity, the prospective teachers worked in pairs, groups of three to six, and whole-class settings facilitated by the teacher educator. Szydlik, Szydlik, and Benson (2003) explained the nature of the teacher educator's facilitation in their approach after the prospective teachers worked on a problem in small groups, then convened in a large semicircle for a discussion of their findings, strategies, solutions and arguments. In these discussions, the teacher educator saw her primary roles as that of a motivator, scribe, challenger and guide. In these roles, she often engaged consciously in the following behaviors: intent listening, feigned (and sometimes real) confusion, skepticism and silence. The goal was for the essential mathematics and underlying structure of each problem to be revealed, but by the prospective teachers, not the teacher educator. When one of the prospective teachers made a conjecture that the teacher educator could not immediately evaluate, she acted as one of the community, joining in the attempt to argue or create counter-examples.

In organizing the prospective teachers to participate in these interactions, the teacher educators, collectively, used various combinations of individual, small-groups and whole-class work as part of the instructional approaches. For example, in Schorr's (2001) approach, the prospective teachers reflected on interviews individually and collectively in class. In Chapman's (2005) approach, they first worked individually, then shared and compared their individual reflections and their findings from the inquiry activities in small groups, and finally engaged in whole-class sharing of the small-groups' findings. Other studies involved group work followed by whole-class discussions. Tirosh's (2000) prospective teachers formed groups, discussed the questions and prepared a class presentation of the process they went through with their final conclusions. This was followed by class discussions. Similarly, Ilany, Keret, and Ben-Chaim's (2004) prospective teachers worked in groups followed by whole-class discussions of results. Roddick, Becker, and Pence's (2000) approach included substantial in-class time working in groups on problems and giving presentations and justifications to the class. In Ambrose's (2004) child-study activity, the prospective teachers worked in pairs, helped each other, exchanged ideas about which problem to present next and what question to ask, and discussed the experience afterward considering issues that arose during the session.

6. Connecting Course and Field Experiences

A final characteristic of some of the instructional approaches is the connection of course work and some form of field (practicum) experience. This occurred in approaches that involved child-study activities, self-study of actual teaching and classroom observations tied to course assignments. The prospective teachers were involved in their regular field experiences that occurred during the same period as their methods courses (e.g., Ebby, 2000; McDuffie, 2004) or field experiences specially organized by the instructor (e.g., Mewborn, 2000). These experiences provided a context through which prospective teachers could investigate, interpret, and discuss the situations encountered within the methods course. As Ebby (2000) explained, coursework helped each of the prospective teachers to think about the children in their fieldwork classroom in new ways, and observing children's learning in the fieldwork classroom helped them clarify their thinking about what they were learning in coursework. McDuffie (2004) also noted that much of her prospective teachers' learning from the critical incidents was made possible because their methods course was linked to an extended practicum experience. Through this linkage, the prospective teachers had the opportunity to put into practice the methods they were learning in class and concurrently to reflect on their experiences, so that they could create their own.

The preceding six characteristics of the instructional approaches studied by teacher educators illustrate examples of what they knew and what other mathematics teacher educators could learn about the nature of these approaches from a cognitive perspective that could inform their practice with their students. All of the studies offered conclusions to suggest that instructional approaches with these characteristics could make an effective contribution to the prospective teachers' development. They enabled the prospective teachers to engage in critical examination of self, teaching and learning and to delve more deeply into issues revealing the complexities of teaching. This led to the prospective teachers' thinking being challenged and to their rethinking, reconstruction, and new constructions of what they thought they knew towards what the researchers considered to be more desirable or useful for understanding and teaching mathematics. The next section discusses these learning outcomes further.

CHARACTERISTICS OF LEARNING OUTCOMES FOR THE INSTRUCTIONAL APPROACHES

Although it was not explicitly stated in the research surveyed, it can be inferred that, in addition to the characteristics of the instructional approaches being studied, the learning outcomes (e.g., the prospective teachers' thinking or knowledge) resulting from these approaches were also being offered by the researchers as another basis of mathematics teacher educators' learning. However, similar to the preceding section on characteristics of the instructional approaches, for the most part, this learning was presented from the perspective of what other teacher educators could learn about the prospective teachers' learning from the approaches and not about the actual learning of the teacher educators conducting the research. Thus, as in the preceding section, this section also discusses teacher educators' learning from this perspective. In it, I summarize three factors that are representative of the learning outcomes reported in the findings of the studies: change, reflection, and guidelines for instruction. I deduced from the studies that these factors are central features of the findings that could form a basis of mathematics teacher educators' learning from this category of research. Again, only samples of the studies where these features are explicit in the description of the approaches that were studied are included in what follows.

1. Change

The first factor that is associated with the findings is *change*. All of the studies showed that the instructional approaches resulted in changes in the prospective teachers' content or pedagogical knowledge, depending on the goals of the approaches. Studies that focused on content knowledge reported change in the way the prospective teachers viewed mathematics and/or understood concepts or procedures they studied. For example, Blomm (2004) reported that the barriers that kept content areas compartmentalized started to break down and participants began to see mathematics as a continuum of interrelated concepts, rather than a set of isolated skills and formulas. Ilany, Keret, and Ben-Chaim (2004) found significant changes in the prospective teachers' understanding of ratio and proportion. They exhibited greater success in solving problems, used different strategies and were better at oral and written explanations. Roddick, Becker, and Pence's (2000) prospective teacher experienced considerable growth in her views of problem solving and its role in instruction while Chapman's (2005) prospective teachers showed deeper understanding of problems and problem solving.

Most of the studies focused on changes in pedagogical knowledge in terms of the prospective teachers' thinking or understanding of teaching, learning and learners. Ambrose's (2004) participants began to recognize that teaching requires more than simply presenting information to students. They grew to appreciate the importance of multiple solution strategies in mathematics. They began to consider the importance of providing children time to think when solving mathematical problems. Ebby (2000) found that the prospective teachers all moved towards a conception of teaching as facilitating student learning and of students as active meaning makers rather than passive receivers of knowledge, while Fernandez (2005) found growth in the prospective teachers' understanding and implementation of reform-oriented teaching. In Schorr's (2001) study, the prospective teachers revised their ideas about the teaching and learning of mathematics, and developed a deeper understanding of the ways in which children build mathematical ideas. They noticed that children often invent their own strategies when solving problems and that getting the right answer does not necessarily mean that the student understands the mathematics. Tirosh's (2000) prospective teachers became familiar with various sources of students' errors in division of fractions. They were aware of students' tendencies to attribute properties of division of natural numbers to division of fractions, of the constraints that the partitive model imposed on the operation of division, and of intuition-based mistakes. Masingila and Doerr (2002) found that the prospective teachers began to understand the issue of supporting student ideas and thereby keeping participation at a high level. For Santagata, Zannoni, and Stigler (2007), the prospective teachers improved their analyses of teaching by moving from simple descriptions of what they observed to analyses focused on the effects of teacher actions on student learning.

Some studies suggested that change resulted when the instructional approaches allowed the prospective teachers to struggle or be challenged. Such experiences led to the necessity for them to become aware of alternative ways of thinking and patterns of behavior as well as to make informed choices. As Lee (2005) found in her approach, the planning–experience–reflection cycle provided opportunities for the prospective teachers to begin to struggle with issues of facilitating students' problem solving. They were able to make their struggle an open and reflective activity and used it as an opportunity to improve their practice. They struggled to pose questions and critically assess students' problem solving while using a technology tool. Nicol's (1999) prospective teachers often experienced tensions with the kinds of questions posed and the reasons for posing them, with what they were listening for, and with how they responded to students' thinking and ideas. They began to pose questions to their students, listen to their students, and respond to their students differently as the course progressed. In Heaton and Mickelson's (2002) study, the approach challenged prospective teachers' conceptions and images of the roles and responsibilities of teachers and learners and allowed them to construct their own understandings of the culture and environment of elementary classrooms and to see a different role for learners in their teaching.

2. Reflection

The findings of the studies supported the importance of reflection in the instructional approaches by revealing what the prospective teachers were able to accomplish through the reflective activities and how they experienced the reflective processes. For example, Bolte's (1999) participants felt that the construction of the concept maps and the writing of the corresponding interpretive essays encouraged them to reflect on their knowledge and enhanced their ability to make mathematical connections. Artzt (1999), whose approach was centered on reflection on practice, explained that when prospective teachers are made more aware of the monitoring they do of their students, they become more conscious of the need to change their instruction accordingly. When they reflect on their lessons in a structured way, they can attend to the most critical facets of classroom instruction: tasks, learning environment, and discourse. Artzt concluded that, for prospective teachers, the structure for reflection appears to be a powerful tool for facilitating their continual professional development. They are encouraged to be more analytical about their teaching and to examine and attend to the underlying assumptions and beliefs that drive their practice. They are also encouraged to think about why they make the decisions they do in light of their goals for students and how their knowledge and beliefs regarding the content, their students, and methods of teaching impact the design of their lessons.

Mewborn (2000) offered another perspective on what influences reflection. She explained that when the locus of authority was internal to the prospective teachers, they engaged in reflective thinking and generated their own problems and solutions. Three elements of the ecology of the field experience seemed particularly crucial to promoting the shift in locus of authority that precipitated reflective thinking among the prospective teachers, an inquiry approach, the cohort group, and school-university collaboration. Lee (2005) also suggested that engaging in an iterative planning–experience–reflection cycle allows prospective teachers to reflect critically upon and thereby improve their practice. In McDuffie's (2004) study, she examined the reflective practices of prospective teachers during their practicum teaching and concluded that the long-term reflection exhibited by the prospective teachers seemed to be an important part of their reflective practice for future teaching. Through long-term reflection, prospective teachers can put together the pieces of individual reflective episodes to realize a pattern that informed their practices.

3. Guidelines for Instruction

Many of the researchers explicitly offered suggestions or general guidelines that teacher educators should consider or be aware of in their practice with prospective teachers. One suggestion involves providing the prospective teachers with multiple experiences of the same approach. For example, Ambrose (2004) explained that while providing prospective teachers with intense experiences that involve them intimately with children poses a promising avenue for belief change, the change described in her study was incremental rather than monumental; suggesting that building upon prospective teachers' existing beliefs will be a gradual process. Thus, she suggested that teacher educators might consider creating several such experiences throughout the teacher preparation programme, especially while prospective teachers are doing their subject-matter preparation, to ensure that the beliefs become well connected. Another suggestion for instruction relates to helping prospective teachers develop useful dispositions for future learning from practice. Ebby (2000) recommended viewing methods courses not as being about developing new knowledge and beliefs that we can only hope will be implemented in practice, but as being about developing habits of mind to learn from their classrooms. Her prospective teachers developed habits of mind that have been identified as being transformative for practicing teachers, such as learning to make sense of children's understanding and learning to take a reflective stance towards one's own teaching. But, for this to occur it is not enough for coursework and fieldwork to be simultaneous experiences; methods courses need to be explicitly oriented towards learning from fieldwork.

Other suggestions offered for instructional approaches deal with how prospective teachers can be actively involved. Artzt (1999) claimed that prospective teachers must continually be called upon to share, reexamine, and question their knowledge, their beliefs, and their goals for students. They must be made aware of the need to monitor student understanding during the lesson as a means for reconstructing their own meanings about what is going on in their classrooms. Lee (2005) explained that if we wish prospective teachers to become comfortable with methods and solutions that are different from their own habits, we need to provide opportunities for them to analyze and discuss students' mathematical problem solving. She suggested that teacher educators should engage prospective teachers in exploring mathematics with various tools to acquire a solid understanding of the affordances and constraints of using these tools in their own learning. In other studies, such as Szydlik, Szydlik, and Benson (2003), problem-solving behavior was emphasized. They explained that focusing on problem solving using a variety of strategies, reflection on the process of problem solving, and engagement in the process of exploration, conjecture, and argument can help prospective teachers develop mathematical beliefs that are consistent with autonomous behavior. Specifically, a classroom culture in which they solve challenging problems without external help and where mathematical conviction is determined based on logic and consistency can alter their beliefs about mathematics.

A variety of other suggestions for instruction includes the following: Masingila and Doerr (2002) considered case study as a site to investigate, analyze, and reflect on another teacher's practice could enable the prospective teachers to begin to understand the complexities of practice, in particular, the use of their ideas when teaching, and to reflect upon their own teaching practice through the case study teacher's practice. For Fernandez (2005), a "micro-teaching lesson study" approach can help prospective teachers understand and begin implementing teaching practices that are consistent with reform-oriented teaching. Santagata, Zannoni, and Stigler (2007) noted that the initial responses prospective teachers gave to the analysis task confirmed the need for a framework to guide their observations of the videotaped lessons in the case studies. Initial comments prior to learning the framework were merely descriptive; when evaluations were included, they were mostly based on preconceptions and did not reflect a close analysis of the lesson. Finally, Beckmann et al. (2004) discussed the importance of integrating K-12 mathematics materials in college-level mathematics courses for prospective teachers. They concluded that inclusion of grades K-12 mathematics activities enabled the prospective teachers to acquire a deeper and more connected understanding of the mathematical content.

The preceding three factors from the findings of the studies by these mathematics teacher educators illustrate what they and others could learn about the usefulness of the instructional approaches that could inform their practice with prospective teachers. These factors provide evidence that engaging prospective

129

teachers in learning opportunities with the six characteristics, previously discussed, can be a promising avenue to promote their development as "reform-oriented" mathematics teachers. However, it is necessary to be cautious regarding the effectiveness of the approaches reported in the studies because of the small number of participants (one or two in many studies) on which the findings are based. While this makes it difficult to draw conclusions about the broader effectiveness of the particular approach involved, the key characteristics discussed could still be viewed as having potential to be effective. Nevertheless, there is also the issue that there is rarely sufficient detail offered about characteristics such as the classroom environment, the ways of interacting with the prospective teachers, instructor expectations and the influence of school practices to make it possible to replicate the approaches in any meaningful sense.

CHARACTERISTICS OF MATHEMATICS TEACHER EDUCATORS' LEARNING

Based on the studies reviewed, for the most part, it was easier to deduce what teacher educators in general could possibly learn from such research than what the teacher-educator researchers themselves actually learned and how they learned from the research. Most of the studies were reported as if the teacher-educator researchers were 'outsiders' and the information was being offered to educate others about the instructional approaches investigated. It seemed that most of the teacher educators engaged in the research of their practice more for broader purposes of the production, enhanced understanding and advancement of knowledge about mathematics teacher education practices than for purposes of their own personal-professional development, i.e., ongoing improvement of their own pedagogical practice. This perspective determined what has formed the basis for what has been highlighted from the studies reviewed in this chapter, i.e., two broad categories of what teacher educators could learn about the nature of instructional practices and their impact on the prospective teachers' learning. This, however, does not address the issue of how teacher educators' development is, or can be, facilitated by engaging in research on their instructional practices in order to know something important about their practice that they did not know before. The dilemma is that the teacher educators do not explicitly discuss subsequent impact on their practice or thinking or on what the research provided as a basis for their learning. Thus it is not clear that the teacher educators' understandings of prospective teacher education were enhanced at all, or in what ways their learning was enhanced, through the research, except in a few cases.

In these few studies, there was some indication of what the teacher educators learned that actually informed their practice. Goodell (2006) explained that she learned how to foster reflection in her prospective teachers, and how to ensure that they were learning from their reflections. She also learned that merely providing the opportunities for them to reflect does not ensure that they are learning anything. It is important to monitor the class discussions, and keep them focused on mathematics teaching as much as possible. Her own reflections about this process were essential in helping her to recognize that the reports had no value unless the analysis was complete. She now takes an objective stance towards the prospective teachers' concerns and insists that the incident reports are completed. She concluded that without this study, she may never have made that realization. Artzt (1999) referred to her transformation from often finding herself frustrated by her inability to help prospective teachers develop as she wished. She had often sat in the back of their classrooms bewildered by the things she saw. When interacting with them she focused on their instructional practice and addressed the issues that she felt needed attention. She tried to hide her frustrations to protect their fragile egos and often left the conference with them disturbed. Since she used the approach in her study, much has changed. By getting into the minds of the prospective teachers in a structured way, she is better able to make sense of what they do and is therefore better able to help them progress. Heaton and Mickelson (2002) learned about the difficulties and complexities of teaching and learning statistics using statistical investigations. They continued to believe that it is important that their prospective teachers get experiences with statistical investigation first as learners and then as teachers. They learned about the contextual and situational issues affecting statistics as an element of the elementary curriculum, the knowledge of elementary teachers, and a teacher education programme. They also learned about both processes of inquiry and the contribution of inquiry to learning at a multiplicity of levels, as well as the shortcomings of inquiry in terms of curricular needs and the establishment of traditional knowledge. McDuffie (2004) provided a glimpse of her learning when, for example, she realized that by having the pre-lesson conference with her participant prior to the day of the lesson, she could better support the prospective teacher in understanding the complexities and issues in teaching this lesson. The participant was not able to act on the ideas McDuffie suggested for her to consider just 15 minutes ahead of teaching. Finally, in my experience with my studies, I have learned about the importance of my role in the approaches, for example, helping the prospective teachers to notice what is interesting to pursue mathematically; engaging them so they continue to think about it; having them explain the sense they are making; and trying to understand their ideas. I have learned about the importance of involving the prospective teachers in a cycle of initial reflection, inquiry, and post-reflection which now frames most of my teaching. I have also developed and show more understanding for the prospective teachers as learners – instead of being judgmental about and impatient for them to develop clear and adequate ideas, I recognize that putting ideas in relation to each other can be confusing and try to use alternative ways to empower them to build the breadth and depth that will give significance to their knowledge.

What, then, are the conditions that enabled this learning to happen? Again, this is not clear from the description of the studies. However, based on my experience and what is implied from these studies that included some information of the teacher educators' learning, the following seems to be important to the process: the teacher educator (1) suspends judgment about the instructional approach; (2) gets into the minds of the prospective teachers as they engage in the approach; (3) experiences conflicts between intent or initial thinking and prospective teachers' behaviours; and (4) examines critically the prospective teachers' learning and his or her role in promoting it. In addition to these factors of the teacher educator in the role as researcher, the design of the research on his or her instructional practice is also important to facilitate the professional development. In particular, the following seems to be necessary, the research: (1) should be based on an approach that is, or is intended to be, part of the teacher educators' regular teaching; (2) should engage prospective teachers in a way that exposes their thinking during the process and as a central part of the instructional approach – e.g., their thinking regarding what they are learning and doing, whether they are aware of the pedagogical strategies as possibilities for themselves, and what the teacher educator-instructor is doing; and (3) should include self-reflection of the teacher educator – before, during and after the research process, i.e., some form of self-study in which the teacher educator is being thoughtful about his or her work.

CONCLUSION

To conclude, an explicit goal of mathematics teacher educators' research of their practice should be self-understanding and professional development, which most of the studies reviewed here did not articulate. One can research one's teaching without this being a goal and thus development leading to change in thinking and practice may not occur. If professional development is a goal of the studies, then research reports need to include how the teacher-educator-researchers reflected, what practical knowledge they acquired, and how this knowledge impacted or is likely to impact their future behavior in working with their students. This will allow such research to contribute to greater theoretical understanding about mathematics teacher education and mathematics teacher educator learning and ultimately to the improvement of practice.

REFERENCES

Ambrose, R. (2004). Initiating change in prospective elementary school teachers' orientations to mathematics teaching by building on beliefs. *Journal of Mathematics Teacher Education, 7*, 91–119.

Artzt, A. (1999). A structure to enable preservice teachers of mathematics to reflect on their teaching. *Journal of Mathematics Teacher Education, 2*, 143–166.

Beckmann, C. E., Wells, P. J., Gabrosek, J., Billings, E. M. H., Aboufadel, E. F., Curtiss, P., et al., (2004). Enhancing the mathematical understanding of prospective teachers: Using Standards-based, grades K-12 activities. In R. R. Rubenstein & G. W. Bright (Eds.), *Perspectives on the teaching of mathematics* (pp. 151–163). Reston, VA: National Council of Teachers of Mathematics.

Blomm, I. (2004). Promoting content connections in prospective secondary school teachers. In R. R. Rubenstein & G. W. Bright (Eds.), *Perspectives on the teaching of mathematics* (pp. 164–279). Reston, VA: National Council of Teachers of Mathematics.

Bolte, L. (1999). Enhancing and assessing preservice teachers' integration and expression of mathematical knowledge. *Journal of Mathematics Teacher Education, 2*, 167–185.

Britzman, D. (1991). *Practice makes practice: A critical study of learning to teach.* Albany, NY: State University of New York Press.

Calderhead, J., & Robson, M. (1991). Images of teaching: Student teachers' early conceptions of classroom practice. *Teaching and Teacher Education, 7*, 1–8.

Carpenter, T., Fennema, E., Franke, M., Levi, L., & Empson, S. (1999). *Children's mathematics: Cognitively Guided Instruction.* Portsmouth, NH: Heinemann.

Chapman, O. (1999) Reflection in mathematics teacher education: The storying approach. In N. Ellerton (Ed.), *Mathematics teacher development: International perspectives* (pp. 32–57). West Perth, Australia: Meridan Press.

Chapman, O. (2005). Constructing pedagogical knowledge of problems solving: Preservice mathematics teachers. In H. L. Chick & J. L. Vincent (Eds.), *Proceedings of the 29th Conference of the International Group for the Psychology of Mathematics Education* (Vol. 2, pp. 225–232). Melbourne, Australia: University of Melbourne.

Chapman, O. (2007, April,). *Contextual tasks for developing mathematical knowledge for teaching arithmetic operations.* Paper presented at the annual meeting of the American Educational Research Association, Chicago, IL.

Charles, R., & Lester, F. (1982). *Teaching problem solving: What why & how.* Palo Alto, CA: Dale Seymour Publications.

Cramer, K. (2004). Facilitating teachers' growth in content knowledge. In R. R. Rubenstein & G. W. Bright (Eds.), *Perspectives on the teaching of mathematics* (pp. 180–194). Reston, VA: National Council of Teachers of Mathematics.

Ebby, C. B. (2000). Learning to teach mathematics differently: The interaction between coursework and fieldwork for preservice teachers. *Journal of Mathematics Teacher Education, 3,* 69–97.

Fernández, M. L. (2005). Exploring "lesson study" in teacher preparation. In H. L. Chick & J. L. Vincent (Eds.), *Proceedings of the 29th Conference of the International Group for the Psychology of Mathematics Education* (Vol. 2, pp. 305–310). Melbourne, Australia: University of Melbourne.

Goodell, J. E. (2006). Using critical incident reflections: A self-study as a mathematics teacher educator. *Journal of Mathematics Teacher Education, 9,* 221–248.

Heaton, R. M., & Mickelson, W. T. (2002). The learning and teaching of statistical investigation in teaching and teacher education. *Journal of Mathematics Teacher Education, 5,* 35–59.

Ilany, B.-S., Keret, Y., & Ben-Chaim, D. (2004). Implementation of a model using authentic investigative activities for teaching ratio and proportion in pre-service teacher education. In M. J. Høines & A. B. Fuglestad (Eds.), *Proceedings of the 28th Conference of the International Group for the Psychology of Mathematics Education* (Vol. 3, pp. 81). Bergen University College, Norway.

Lampert, M., Heaton, R, & Ball, D. L. (1994). Using technology to support a new pedagogy of mathematics teacher education. *Journal of Special Education Technology, 12(3),* 276–289.

Lee, H. S. (2005). Facilitating students' problem solving in a technological context: Prospective teachers' learning trajectory. *Journal of Mathematics Teacher Education, 8,* 223–254.

Lloyd, G. (2006). Preservice teachers' stories of mathematics classrooms: Explorations of practice through fictional accounts. *Educational Studies in Mathematics, 63,* 57–87.

Masingila, J. O., & Doerr, H. M. (2002). Understanding pre-service teachers' emerging practices through their analysis of a multimedia case study of practice. *Journal of Mathematics Teacher Education, 5,* 235–263.

McDuffie, A. R. (2004). Mathematics teaching as a deliberate practice: An investigation of elementary pre-service teachers' reflective thinking during student teaching *Journal of Mathematics Teacher Education, 7,* 33–61.

Mewborn, D. S. (2000). Learning to teach mathematics: Ecological elements of a field experience. *Journal of Mathematics Teacher Education, 3,* 27–46.

National Council of Teachers of Mathematics. (1989). *Curriculum and evaluation standards for school mathematics.* Reston, VA: Author.

National Council of Teachers of Mathematics. (1991). *Professional standards for teaching mathematics.* Reston, VA: Author.

Nicol, C. (1999). Learning to teach mathematics: Questioning, listening, and responding. *Educational Studies in Mathematics, 37,* 45-66.

Polanyi, M. (1958). *Personal knowledge.* Chicago: University of Chicago Press.

Ponte, J. P., & Chapman, O. (in press). Preservice mathematics teachers' knowledge and development. In L. English (Ed.), *Handbook of international research in mathematics education* (2nd ed.). Mahwah, NJ: Lawrence Erlbaum Associates.

Ponte, J. P., & Chapman, O. (2006). Mathematics teachers' knowledge and practices. In A. Gutierrez & P. Boero (Eds.), *Handbook of research on the psychology of mathematics education: Past, present and future* (pp. 461–494). Rotterdam, the Netherlands: Sense Publishers.

Ponte, J. P., Oliveira, H., & Varandas, J. M. (2002). Development of pre-service mathematics teachers' professional knowledge and identity in working with information and communication technology. *Journal of Mathematics Teacher Education, 5*, 93–115.

Roddick, C., Becker, J. R., & Pence, B. J. (2000). Capstone courses in problem solving for prospective secondary teachers: Effects of beliefs and teaching practices, *Proceedings of the 28th Conference of the International Group for the Psychology of Mathematics Education* (Vol. 4; pp. 97–104). Bergen University College, Norway.

Santagata, R., Zannoni, C., & Stigler, W. J. (2007). The role of lesson analysis in preservice teacher education: An empirical investigation of teacher learning from a virtual video-based field experience. *Journal of Mathematics Teacher Education. 10*, 123–140.

Schön, D. (1987). *Educating the reflective practitioner*. San Francisco: Jossey Bass.

Schorr, R. (2001). A study of the use of clinical interviews with prospective teachers, *Proceedings of the 25th Conference of the International Group for the Psychology of Mathematics Education* (Vol. 4, pp. 153–160). Utrecht University, the Netherlands: Psychology of Mathematics Education.

Simon, M.A. (1995). Reconstructing mathematics pedagogy from a constructivist perspective. *Journal for Research in Mathematics Education, 26*, 114–145.

Szydlik, J. E., Szydlik, S. D., & Benson, S. R. (2003). Exploring changes in preservice elementary teachers' mathematics beliefs. *Journal of Mathematics Teacher Education, 6*, 253–279.

Tirosh, D. (2000). Enhancing prospective teachers' knowledge of children's conceptions: The case of division of fractions. *Journal for Research in Mathematics Education, 31*, 5–25.

Zbiek, R. M. (1998). Prospective teachers' use of computing tools to develop and validate functions as mathematical models. *Journal for Research in Mathematics Education, 29*, 184–201.

Olive Chapman
Faculty of Education
University of Calgary

SECTION 2

REFLECTION ON DEVELOPING AS A MATHEMATICS TEACHER EDUCATOR

RON TZUR

7. PROFOUND AWARENESS OF THE LEARNING PARADOX (PALP)

A Journey towards Epistemologically Regulated Pedagogy in Mathematics Teaching and Teacher Education

> Most teachers have had the experience of asking students a question and getting an apparently totally nonsensical reply. They have also had the experience of trying to guide a student who apparently does not see something obvious such as a common factor, a rearrangement, or a simplification, and finding it very hard to believe that the student does not see it. These are examples of students not seeing what the teacher sees. Although not seeing is an obvious and common phenomenon, it is often obscured behind apparently taken-as-shared discourse. (Mason, 1998, p. 246)

In this chapter I address a twofold problem: (a) what might constitute an epistemological foundation for mathematics teaching and hence mathematics teacher education and (b) how might such a foundation evolve. To this end, following Ma's (1999) notion of profound understanding of fundamental mathematics (PUFM), I introduce a new construct, profound awareness of the learning paradox (PALP). I use this construct to explicate epistemological roots common to traditional and reform-oriented pedagogical approaches, and contrast both with a constructivist, conception-based perspective. Using my own journey toward a PALP as a case, the paper unfolds in a flashback manner. First, I present my current thinking about mathematics teaching, which draws on a constructivist theory of learning. This includes a discussion of why a pedagogy rooted in PALP is likely to promote intended conceptual understanding where reform and traditional approaches so frequently fail. Then, I articulate the process by which I arrived at my current position, through reflection on fragments of my experience as a mathematics teacher educator. This process revolved around becoming gradually aware of teacher disbeliefs about lessons we co-planned and implemented, and of plausible reasons why these lessons seemed counter-intuitive to the teachers. I conclude with proposing implied capacities to be fostered in mathematics teacher educators and examples of tasks that, drawing on my journey, can help promoting emergence of those capacities.

For you, Shiri.

INTRODUCTION

Drawing on Mason's (1998, this volume) core construct of awareness, in this chapter I discuss a profound awareness that has been evolving for me through mathematics teaching and mathematics teacher education endeavours. This awareness is clearly indicated by the quotation above: quite often teachers struggle and fail to 'show' students mathematical entities and/or relationships that are painfully obvious to the teachers. To a person who adheres to the key constructivist notion of assimilation (Piaget, 1971, 1985), such common situations can frequently be explained in terms of the learning paradox (LP) (Bereiter, 1985; Pascual-Leone, 1976). In order to assimilate and complete tasks/activities in which a teacher engages her or his students to promote their learning of a new concept – the students need to have already established that concept prior to learning it. To address the LP, I argue in accord with Steinbring (1998) that a shift in epistemological stances is indispensable. This shift should relate to teaching of mathematics, education of mathematics teachers, and mentoring of mathematics teacher educators.

Realising and characterising this shift evolved particularly through my reflections on mutually empowering, long-term collaborations with three dedicated, first-rate teachers – Chris, Edna, and Nevil (all names are pseudonyms). These teachers generously shared with me their classrooms, practices, accomplishments, and struggles. Most importantly, time and again they revealed their disbeliefs regarding lesson plans I shared with them. Those lessons were created on the basis of a conceptual framework that addresses the LP. Yet, the teachers simply could not 'see' what, for me, seemed obvious: Why would these lessons bring about students' learning of the intended concepts? Then, after having observed and/or implemented the lessons, they continued pondering why, contrary to their intuitive expectations, the lessons were so successful in terms of both student engagement and outcomes.

Initially, I termed the deviating pedagogical stances as intuitive teaching (teachers) and counter-intuitive teaching. Feedback from colleagues and further contemplation convinced me that such terminology is problematic. One hopes that, eventually, mathematics teachers (MTs) and mathematics teacher educators (MTEs) would develop a profound awareness of the learning paradox (PALP) and intuitively use a pertinent, epistemologically regulated pedagogy. Thus, this chapter focuses on development of MTEs' awarenesses regarding teacher progress toward a conception of teaching that is rooted in a profound awareness of the learning paradox (PALP) and an implied capacity to address the LP.

DISTINGUISHING AN EPISTEMOLOGICALLY REGULATED PEDAGOGY

In this section I distinguish a perspective and approach to mathematics teaching that is rooted in an epistemological stance toward mathematics learning. Initially, most MTs and evolving MTEs may view such a perspective and approach as counter-intuitive. I will point out (a) why such a view is expected, (b) shifts in awarenesses needed for more powerful (PALP) intuitions to arise, and (c) why the latter are likely to succeed where, too often, the former fail.

Traditional and Reform Pedagogical Approaches Overlook the LP

Two fundamental questions that any MT and MTE is constantly faced with are *what* should or could her/his students learn next, and *how*. Let us begin with the "How?" question. In the last three decades, we have witnessed substantial reforms around the world (Australian Council of Deans of Education, 2001; National Council of Teachers of Mathematics, 2000; The Royal Ministry of Education, 1999). These reforms may be rooted in socio-cultural theoretical stances (Lave & Wenger, 1991; Lerman, 2006; Wenger, 1998; Wertsch & Toma, 1995), cognitive stances (Steffe, 1990a, 1992; von Glasersfeld, 1995a, 1995b), or both (Bauersfeld, 1988; Cobb & Bauersfeld, 1995; Cobb, Wood, Yackel, & McNeal, 1992; Dewey & Bentley, 1949). Common to these reforms is a renouncement of traditional pedagogies, which are characterised by teachers first showing-and-telling to their rather passive students the wished-for mathematics, followed by students drilling/memorising similar examples – predominantly facts and procedures. Instead, reform-oriented (or child-centred) teaching-learning discourses promote student engagement in actively solving realistic problems while using a variety of tools/manipulatives (including computers and calculators), communicating and reasoning about these solutions (small groups, whole-class), reflecting on their solutions (e.g., writing individual journals, discussing whether solutions are mathematically different), and frequently practicing procedures/facts for mastery as conceptual understanding is being achieved. In short, reform-oriented, child-centred pedagogical approaches revolve around the psychological premise of 'learning is an inter- and intra-active social endeavour.'

Proceeding to the "What?" question, consider a teacher who deeply understands the mathematical terrain – scope and sequence – to be learned by students. She or he may also have a pretty good grasp of research-based findings regarding developmental order of mathematical concepts (e.g., counting-all develops before counting-on). The teacher triangulates several data sources and assesses that a group/class of students evidently (a) did not understand or (b) understood concept "X." In the former case she or he naturally re-teaches concept "X" and in the latter case teaches the next-in-sequence "Y" concept. Simply put, in *both* cases one intuitively teaches what students do not yet know.

Such intuitions seem to be rooted in teachers' lengthy past experiences as learners in mathematics classrooms. In those classrooms, logical reasoning of those who already knew the mathematics dictated a staircase-like outlook on concepts and the sequence for introducing them to students. The staircase metaphor seemed to entail treating the training of behaviour and understanding as nearly identical. Accordingly, and in line with a desire to be most effective, a teacher attempts to direct learners' attention to what is novel for them.

Remarkably, this intuitive tendency is prevalent among *both* traditional and reform-oriented pedagogies for, in part, two interrelated reasons. First, this tendency is rooted in an entrenched presumption that underlies human communication. When interacting, people customarily do not question the sense others make of the words one utters or writes (von Glasersfeld, 1995b). At times, people may bump into experiences marked by lack of compatible meanings. In such instances they engage in clarification of meanings to restore mutual understanding. However, most people rarely question the presumption itself.

Second, once people have formed mathematical conceptions that afford 'seeing' the world in a certain way, they cannot 'return' to not seeing it like this (von Glasersfeld, 1995b). In line with the aforementioned presumption, they also naturally attribute to fellow human beings the capacity/potential for such 'seeing.' Thus, although using markedly different instructional tools/methods, both traditional and reform-oriented teachers strive to create experiences for students to have an equivalent (to the teacher's) 'seeing.' Combined, these two reasons underlie what, in learning-is-active reform orientations, my colleagues and I characterised as a perception-based perspective (Simon, Tzur, Heinz, Kinzel, & Smith, 2000; Tzur, Simon, Heinz, & Kinzel, 2001). We contrasted this perspective with an epistemologically incongruent perspective rooted in a profound awareness of the learning paradox (PALP), which we termed 'conception-based.' Below, I briefly present a conception-based theoretical account of learning and plausible roles for teaching that depict, in the cognitive realm, the current situation of my journey.

Pedagogy Rooted in PALP: Epistemology Takes Centre-Stage

The theoretical account I use to explain the meaning of 'having a mathematical conception' and 'constructing a new (to the learner) mathematical conception' revolves around three core constructivist notions: assimilation, anticipation, and reflection (Dewey, 1933; Piaget, 1985). It evolved from my empirical studies on children's learning of fractions (Tzur, 1996, 1999, 2000, 2004, 2007a), and has been further reorganised in three recent theoretical papers (Simon & Tzur, 2004; Simon, Tzur, Heinz, & Kinzel, 2004; Tzur & Simon, 2004). This reorganisation comprises a mechanism for forming a new mental entity (mathematical conception), two stages in this constructive process, and a cyclic process of teaching through mathematical tasks. Due to space limitations, here I briefly present only the latter.

Mathematical Tasks within a Cyclic Process of Teaching. For teaching mathematics students must be engaged in tasks that serve three principal functions: (a) fostering assimilation of tasks into their available, relevant conceptions, (b) fostering orientation of their focus of attention so that they notice effects of their work on the task intended by the teacher, and (c) fostering students' reflection on and distinction/formation of the new, intended conception. Teachers whose intuitions are rooted in traditional or reform perspectives grossly overlook the first function. Accordingly, they design and implement instructional tasks and problem situations for teaching, straight away, the next-in-sequence mathematical idea.

When such teachers contemplate a lesson plan that begins with (assimilation into) what students do know they typically cannot foresee how it can foster the intended conception.

In contrast, in a PALP-rooted, conception-based pedagogy tasks are designed and implemented relative to *both* the mathematical sequence of ideas and to the developmental sequence of children's mathematical conceptions. That is, tasks are dynamically created, evaluated, and adjusted to fit within the ever changing window-of-opportunity that proceeds from students' extant conceptions to the intended mathematical understandings (Noss & Hoyles, 1996). Using Mason's (1998, this volume) notions, teachers use tasks to promote *re*-structuring of learners' awarenesses through orienting the learners' attention. The italicised term (*re*) stresses the explicit awareness of the first function of tasks (assimilation) that MTs and MTEs need to bring into play.

The account of learning and the threefold function of tasks entail the following cyclic process of seven principal activities that constitute an epistemologically (PALP) regulated pedagogy. This cycle further extends Simon's (1995; Simon, 2006b) notion of hypothetical learning trajectory. Although a cyclic process, the starting point is not arbitrary. Rather, to account for and take stock of assimilation, teaching must proceed from and draw heavily on articulating student conceptions. All seven activities should be considered in both the planning and the implementation phases of teaching. However, in planning more focus is put on the first three, in implementing more focus is on the last three, and activity #4 is equally important in both.

1) *Specifying students' current conceptions.* A teacher needs to inquire into mental processes that might underlie students' work on previous tasks in terms of goals *they* plausibly set, activities *they* plausibly executed, and effects *they* plausibly noticed and linked to the activities. The critical awareness to accompany this activity is twofold. First, it is geared toward providing the teacher with an image of the mathematical world that students may 'see' *while still lacking* the conception to be learned. Second, such analyses represent, at best, an observer's (teacher) successful organisation of her or his experiential reality of interactions with students, not an ontological positing of what is actually going on in students' minds (Steffe, 1990a, 1990b). 'Successful' is used in the sense of rather consistently effecting/explaining those teacher-student inter-actions.

2) *Specifying the intended conception.* A teacher needs to decompose the conceptions students are expected to learn into their components – activities, effects, and relationships among them, that is, to articulate the conceptual advance intended. This principal activity is consistent with Dewey's (1902) contention that subject matter "must be restored to the experience from which it was abstracted. It needs to be psychologised; turned over, translated into the immediate and individual experiencing within which it has its origin and significance" (p. 29). Simon and Tzur (2004) asserted that this is not a trivial undertaking, because it is insufficient merely to specify what a student will be able to do (traditional)

or understand (reform). To promote the intended learning effectively the teacher also needs to specify the differences between and transformations (shifts in awareness) needed from a current state to an intended state.

3) *Identifying an activity sequence.* The teacher needs to postulate a *mental* activity sequence that, when 'triggered' in students, is likely to generate the intended effects and relationship. At times, it may be useful to promote students' development of such a sequence through 'tailoring' available activities anew (e.g., by learning to play a new game). The crucial point is that an activity sequence must be identified relative not only to the intended conceptions but also, and most importantly, to the students' current (assimilatory) conceptions.

4) *Selecting tasks.* The teacher needs to decide on an initial problem situation and follow-up prompts/questions. Such a decision is based on reasoned hypotheses as to (a) how students may assimilate the tasks into and use the current conceptions and (b) notice intended (new) effects of their activity and relate those effects to the activity. As in #3 above, the activity of task selection requires explicitly linking current and intended conceptions (Karp, 2007).

5) *Engaging students in the tasks.* A teacher needs to introduce the problem situation and make sure students are involved in solving it while using *their* goal in regulating *their* activity. This is a delicate challenge, because it involves continual negotiation of students' interpretation to the task itself, so that their goal is compatible with the teacher-intended goal, without sliding into a futile attempt to govern students' (necessarily internal) goals. Once students are working on the initiating task, the teacher interjects follow-up questions and prompts that enable/challenge student thinking (see Sullivan, Mousley, & Zevenbergen, 2003; Watson & Mason, 1998), hence promote reflection and anticipation. A particularly useful type of follow-up prompt is the introduction of constraints to the ways in which students solve the task. Such constraints assist the teacher in replacing direct attempts to 'show' new concepts to the students with indirect guidance for constructing the new conceptions via transformation of conceptions within the students' existing capacities (Pirie & Kieren, 1992), that is, for using their familiar actions in novel ways.

6) *Monitoring students' progress.* Guided by products of activities 1–4 above, a teacher needs to repeatedly examine students' actual work: which goals seem to regulate their activities, which (mental) activities they seem to employ for accomplishing those goals, which effects of the activities they seem to notice anew, and which reflections they seem to undertake.

7) *Introducing follow-up questions/prompts.* Based on activity #6 above, a teacher needs to interject planned and/or adjusted follow-up prompts/questions (e.g., constraints). Such follow-ups constitute the teacher's chief means of *indirectly* orienting students' attention so they reflect on intended effects of their goal-regulated activity and relate those effects with the activity in anticipation. Information gained through

principal activities #6 and #7, along with data from additional assessment events, feed back into the first activity, that is, into a new cycle.

Attributes of Teaching Rooted in PALP. A dialectic stance toward teaching marks the entire, 7-activity cyclic process. On the one hand, teaching cannot determine or control student learning of intended conceptions nor can it directly engender such learning (Bereiter, 1985; Pirie & Kieren, 1992). On the other hand, student learning does not happen in a vacuum – a point so frequently mistaken by both opponents and advocates of constructivism. Rather, teacher intentions for student learning are attained via activities that indirectly prompt learners to educate their awareness – goals that students may set, activities they may bring forth and employ, effects they may notice, and anticipations they may form anew. Indeed, attaining such intentions is a highly complex endeavour as it occurs in the very dynamic, and contextually challenging situations of young people (children, youth) epistemologies.

This dialectic stance implies acceptance of uncertainty, in fact welcoming it, because of three interrelated reasons (Tzur, 2007b). First, a teacher can never be certain of student conceptions – he or she can only form inferential models of such conceptions on the basis of student observable actions and language. Second, figuring out whether one's teaching activities effected the intended learning, that is, the desired change in student anticipation, cannot be accomplished in advance but only via the teacher's retrospective (and inferential) reflection on students' solution to tasks and follow-up prompts. Third, and most importantly, a teaching plan rooted in PALP can only indirectly and inconclusively steer student learning. The reason I propose not only accepting but also welcoming this uncertainty is that such a mindset opens a window-of-opportunity for promoting and studying mathematics teacher and teacher educator learning through reflection on activity-effect relationships – activities of teaching, effects of student conceptual change.

Advantages of Teaching Rooted in PALP. Epistemologically regulated pedagogies such as the one presented here have three key, interrelated advantages over teaching intuitions of traditional or reform-oriented approaches. First, such pedagogy acknowledges and copes, head-on, with the learning paradox (LP). Thus, teacher-learner interactions are less likely to fall prey to this impeding force. In practice, this is accomplished when mathematics teachers and/or teacher educators use the LP as a 'litmus test' for judging the worth of each task prior to, during, and/or after students' engagement with it (particularly when students do not 'see' what is obvious to the teacher).

Second, PALP-rooted teaching is better suited for moving students from what they know to what they are expected to learn, because it purposely and systematically tailors activities/tasks to *both*. The key difference is with reform-oriented approaches (perception-based perspective). Reform-oriented approaches are clearly advantageous over a traditional approach, because of the explicit focus on student activity. However, reform-oriented tasks/activities are designed and implemented without explicit attention to how they *fit with and proceed from* students' available conceptions, that is, how they capitalise on students' natural and developing powers. Consequently, only students who have already formed conceptions into which such tasks/activities can be assimilated are likely to benefit from the tasks. Students whose available assimilatory conceptions do not afford assimilation of the tasks/activities are not likely to benefit. When active engagement in tasks is employed within a perception-based perspective these students may, too often, be left behind. In this sense, reform-oriented pedagogies are susceptible to increasing achievement gaps.

Third, any mode of teaching may, indeed, fail to bring forth students' learning of the intended mathematics, yet teaching rooted in PALP is better suited for figuring out why and systematically modifying tasks. More often than not, such failures happen when teachers intend to foster students' advancement to abstract levels of notoriously hard-to-grasp mathematical ideas. Any teacher who engages students in actively working on tasks is likely to conclude that when a task fails to foster the intended learning it should be adjusted or replaced (see Karp, 2007; Zaslavsky, 2007). My work with and studies of teachers who used intuitive, reform-oriented pedagogies (Heinz, Kinzel, Simon, & Tzur, 2000; Tzur, Simon, Heinz, & Kinzel, 2001) indicated that in such situations their level of frustration is extremely high. They see essentially three alternative courses of action – particularly when students' activity with manipulatives showed some success whereas removing these aides did not. They may (a) repeat the same task, but their rich past experience of the futility of such an attempt is usually confirmed. Consequently, they either (b) go back to using tangible instructional materials/tasks for directly 'showing' the intended concept to students, or (c) revert to traditional, show-and-tell methods.

In contrast, teaching rooted in PALP provides a naturally frustrated teacher with an additional, powerful alternative. When a task fails (students did not come to 'see'), the teacher subjects it to the LP 'litmus test' as a first, crucial step in carrying out a new, 7-step teaching cycle. That is, the teacher re-analyses the students' available conceptions, using the very task that failed as lenses for examining what in her/his previous inferences could have been at fault. In this sense, the alternative course of action is a vital, meta-cognitive tool for educating teachers' awarenesses (Mason, 1998, this volume) through attentive reflection on their own and/or others' teaching (see also Tzur, 2007b). Because learning to use such a process is initially labour and time intensive it must be gradual. Eventually, it can become second nature for teachers similar to their current practices of reflection/planning, with one advantage – it will more often than not fulfil teachers' paramount intention of promoting their students' mathematical powers, hence may be decreasingly needed.

MY JOURNEY TOWARD PALP AND CONCEPTION-BASED PERSPECTIVE

In this section I describe some key experiences that steered my journey toward the aforementioned epistemological awarenesses and pedagogy. I first analysed this journey by postulating four foci through which one matures: mathematics learner, mathematics teacher, mathematics teacher educator, and mentor of mathematics teacher educators (Tzur, 2001). Similar layers were proposed by Zaslavsky and Leikin (2004) in their articulation of MTEs' development. The overlap between their layers and my foci was enlightening, because the two sets of constructs grew out of distinct-but-complementary perspectives. Their analysis focused on social-cultural aspects of varied communities of practice in which MTEs' interactive participation is situated. My analysis was rooted in a constructivist framework, followed Polkinghorne's (1988) method of narrating and interpreting fragments of personal experience, and focused on processes that underlie shifts in MTEs' cognition. To situate these fragments within the larger context of my journey, I first recap the ones reported on in the previous article (Tzur, 2001).

1. *Learning of mathematics through teaching it.* I described experiences of tutoring my peers how to solve linear equations and of promoting my high school students' understanding of the Cartesian coordinate system. In the former, in spite of my studies of theories of learning that emphasise assimilation, I did not consider what my peers could not 'see.' In the latter, I gained mathematical understandings of the coordinate system that markedly differed from my students' understandings due to the different goals that regulated their and my reflections.

2. *Learning pedagogy through teaching teachers.* I described experiences of working with teachers who were in the process of transition to reform-oriented practices. In my work, I identified 'mistakes' in the teachers' use of the new curriculum but overlooked, in spite of my studies of teacher learning, the assimilatory conceptions that afforded and constrained their way of interpreting and implementing it. For example, I attempted and failed to straightforwardly convince teachers to let students solve problems on their own before the teacher demonstrates a solution. Through my reflections on those attempts, I developed as a mathematics teacher, but made minute progress in educating my awareness of educating teachers.

3. *Learning teacher education through teaching prospective MTEs.* I described experiences of working with graduate students to promote their development as mathematics teacher educators. In this work, as a result of my growing engagement in constructivist-informed research, I gradually became aware of and explicitly analysed their assimilatory conceptions. Those analyses brought about substantial changes to courses they took with me when I realised that reading and discussing literature alone could not foster their deeper understandings. Most importantly, my goal of teaching them about teacher learning promoted further education of my awareness of and attention to teachers' awarenesses, a transition I elaborate on below.

4. *Learning to mentor MTEs through scholarly collaboration and continual teaching.* I described experiences of working with colleagues, my individual struggles to write about my own research, and my continual work with prospective MTEs. In this work, which included writing papers about teachers' perspectives and the paper about my own growth as a MTE, at long last I was explicitly and systematically utilising a conception-based

perspective for teaching at any of the other foci (mathematics, teaching, teacher education).

At the time, I identified the four foci based on the object of reflection in one's learning. In learning mathematics the objects of reflection are quantifying/spatialising activities. In learning mathematics teaching the objects are both activities of doing mathematics and of teaching it. In learning mathematics teacher education the objects are activities of doing mathematics, of teaching it, and of teaching mathematics teachers. Finally, in learning to mentor MTEs the objects are all the above and activities of teaching, studying, and theorising about prospective MTEs' development. Having distinguished the PALP as an underlying construct of epistemologically regulated pedagogy to be promoted in and embraced by MTs and MTEs, I now use this distinction to further examine the four foci.

Clearly, the first two foci were marked by the lack of PALP. In the first I mainly used traditional methods; in the second I gradually shifted to reform-oriented pedagogy in my own teaching of mathematics to students but not for educating mathematics teachers. In the fourth focus, PALP became the basis for my way of operating when fostering others' learning of mathematics, of teaching it, and of teacher education. That is, I was finally using analysis of MTs' and MTEs' conceptions as the invariant, first activity of 7-step teaching cycles.

The third focus calls for closer attention, as it involved my transition toward PALP. Initially the PALP guided only my own teaching of mathematics – analysing my students' conceptions (i.e., practicing and prospective teachers). Putting it to work also in my mathematics teaching education endeavours took longer. Because it constitutes the heart of this chapter, I go into more detail – fragments of experience – regarding this process.

First Encounters with the Learning Paradox as MTE

I first came across a teacher's disbelief when co-creating, with an elementary teacher named Nevil, a plan for teaching fractions to his fifth graders. He was genuinely interested and highly engaged in the process, because of his frustrating past experience of teaching this topic. He was also very intrigued by my suggestions and, several times, honestly stated that he (a) could not see why students would learn the intended conceptions and (b) entrusted me with using this plan for co-teaching his class.

As we progressed with implementing the plan, Nevil made several comments about how satisfied he was with his students' progress and, at the same time, perturbed by his own lack of understanding what made this progress possible. At the time, my team of mathematics educators had not yet distinguished between the perception-based and the conception-based perspectives. Thus, in a mix of traditional and reform-oriented practices, I attempted, straightforwardly, to share with Nevil key notions of the framework that guided my planning. Nevil could follow my talk, but admittedly could not fully understand, let alone use the framework himself.

In retrospect, I can now say that Nevil and I repeatedly confronted the LP while planning, implementing, and/or reflecting on our work. Moreover, I can now point out that my lack of explicit attention to the LP was detrimental to my attempts to orient Nevil's education of his awareness. This lack of attention was partly due to my work with Nevil as part of a research project on mathematics teacher development. I was actively engaged in (a) his fifth grade classroom, (b) the cohort of 20 prospective and practicing teachers who took 5 courses on reform-oriented teaching with us, and (c) the community of our research project team. At issue are the activities upon which our reflection focused in each context. With the fifth graders, we focused on promoting their conceptual understanding of fractions. With the teachers, we focused on activities (theirs, ours) that seemed to promote shifts in their thinking and practices. With the research team our focus was on identifying and explaining those shifts. Through the last two, the distinction between perception- and conception-based perspectives (Simon et al., 2000) slowly evolved as a result of the team's focus on what in the teachers' practices *did* make sense to the teachers. This focus contributed to my lack of examination of what *did not* make sense to teachers in general and to Nevil in particular. Thus, although I was acutely aware of the gap between my intentions for his learning and the perturbingly poor effects of my activities to accomplish these goals, I did not yet identify his disbelief as a crucial object of study.

Further Encounters with the Learning Paradox as MTE

In 2000 I began a 2-year research project on how the reflection on activity-effect relationship framework might inform teaching of mathematics in a whole classroom setting (Tzur, 2007a). To strengthen the possible claims of that study, I asked the principal of a public school in Israel where the teaching experiment was to be conducted to select what the staff considered to be one of the toughest to teach grade-3 classrooms. Of course, I also asked him to make sure that the homeroom teacher agreed to let me work with her. This was not a trivial demand, because the design of the study required that I teach fractions to the class on two of the 6 days children attended school every week. Fortunately, Edna's class matched the request and she welcomed the opportunity to collaborate with me.

Edna's 28 third graders were still struggling with addition and subtraction and were scheduled to begin learning multiplication within 2-3 months. Before I began teaching Edna's class, I presented my plan to her. At the time, she did not believe that such a plan could be implemented in her notoriously tough-to-teach classroom, let alone foster students' learning of fractions. Yet, Edna kept the disbelief to herself and did not give me any indication of it. As the plan unfolded, students began demonstrating understanding and mastery of fraction conceptions she said were never accomplished by over half of her previous, grade-5 students, in spite of her best, months-long and highly frustrating attempts. In one of our weekly meetings, two months into the study, Edna finally shared her initial disbelief. From the start, I learned, she courageously decided to trust me although she simply could not 'see' how my planned teaching activities would bring about the intended effects of student learning. Subsequently, she asked if I could provide her with a similar plan to foster her students' understanding and mastery of multiplication – the topic she struggled to teach on days I did not teach fractions. I first asked Edna to plan lessons she anticipated would work. Then we both revised her plan as I struggled to make my reasons for the adjustments as transparent as possible for her. This was a formidable struggle for me as a MTE, because under the constraints and stresses of Edna's daily obligations, avoiding my own 'running into the learning paradox' (i.e., attempting to straightforwardly communicate with her my own thought processes regarding student thinking and its relation to task design) proved extremely difficult.

When we completed planning, Edna again expressed both her (a) doubts that the plan would work (e.g., "my students are likely to consider the activities childish") and (b) willingness to try anything different from her hitherto frustrating, unsuccessful attempts. I was unable to observe Edna's teaching of the planned lesson the following day. Still, later in the afternoon I received her excited phone call. Not only was she surprised how much the kids loved and were highly engaged in the activity, but within 20 minutes they also seemed to have formed solid understanding and grasp of the entire fact-family for 4's. For example, she described how children could reason that "4x7=28 because 4x6=24 and there's one more group of four."

As in Nevil's case, my collaboration with Edna repeatedly brought us face-to-face with the learning paradox, as it pertained both to students' and to Edna's learning. However, the work in both cases focused my attention on creating and/or adjusting activities for promoting student learning of fractions. In the planning meetings we openly shared with one another the reasoning behind our suggestions. This crucial method was used proactively to foster the education of my researcher awarenesses of ways in which practice could be informed by the conceptual framework. Accordingly, I *did not* focus on issues of educating her teacher's awarenesses or, particularly, on what in my plans did not fit with her intuitive anticipation. It would take four more years, and another research project on teacher professional development, for the "Aha" moment of noticing and reflecting on teachers' recurring disbeliefs to become the focus of my attention and the source for restructuring my MTE awarenesses.

Restructuring My MTE's Awareness of the Learning Paradox

I finally distinguished and learned to regularly take into account traditional/reform-oriented teachers' lack of PALP through encounters with disbeliefs expressed by Chris and her peers. A grade-1 teacher and avid proponent of reform-oriented teaching, Chris volunteered to be videotaped as she taught a lesson on counting-on. We chose to focus on this strategy because the majority of the teachers in our project identified it as a serious impediment for their students. Chris organised the class into "centres" – small groups engaged in working on different tasks while using manipulatives and discussing their solutions. She worked closely with one group (about 10 students), essentially following a detailed lesson plan provided in a *reform-oriented curriculum* adopted by and used across the school district. In essence, this curriculum's task for teaching counting-on is typical of reform minded lessons. Children have an active role in solving realistic problem situations while using manipulatives, communicating and reasoning in small groups, whereas the teacher mainly facilitates the interactive discourse. Using assessment embedded in the curriculum at the end of the lesson, Chris judged it to be yet another instance of dismal failure to teach this hard-to-grasp concept, in spite of her enthusiasm, thorough preparation, and a great deal of personal attention to students.

The processes and outcomes of Chris' lesson, as she openly shared the videotape and her frustrating experience with the teachers, did not surprise me. In fact, having analysed the curriculum a few weeks earlier, without setting foot in Chris' classroom, I anticipated such a turn of events. In a nutshell, the source of the problem was the tenacious-but-futile attempt to use a task for directly teaching counting-on to students who did not yet form a conception for assimilating the task and make sense of the new (to them) strategy for adding two numbers. That is, Chris' vigorous attempt fell prey, head-on, to the LP. Both the features of the task for teaching counting-on and the way Chris used it entailed student engagement in activities that required they already knew... counting-on! To use Tahta's (1981) task characterisation, the outer task (what students were invited to do) was not connecting to a constructive inner task (what they might have encountered when doing it). While they could 'successfully' achieve the outer task (execute the taught procedure), they failed to experience and resolve the inner task.

Over the weekend, Chris and I met to co-create a lesson on counting-on that would be informed by the reflection on activity-effect relationship framework. As in Edna's case, prior to our meeting I asked Chris to prepare what she thought would be a productive task. I anticipated, and found clear evidence of the LP in her plan, and suggested some substantial changes. In this context of being acutely aware of the LP as it pertains to teaching the students, I again encountered a teacher's disbelief as to the likelihood for success of my lesson. Like Nevil and Edna, due to her extremely high frustration and to the growing trust in my experience, Chris agreed to implement the altered lesson I proposed and committed herself to closely studying the plan. On the following Monday she taught the lesson, looking increasingly pleased with students' progress and extremely relieved and satisfied with the wished-for outcomes (indicated for all students in the previous small group by the same curriculum-embedded assessment). Through my subsequent reflection on a post-lesson debriefing with Chris, which led to my recalling of the experiences with Nevil and Edna, I finally distinguished the epistemological source of the difference between their and my intuitions. I realised that to each of these reform-minded, student-dedicated teachers my lesson planning (and teaching) seemed counter-intuitive mainly due to the lack of PALP.

To further examine my MTE evolving "Aha" awareness of teachers' (lack of) awareness, I asked Chris to join me in co-planning and co-leading a session with our entire teacher group. My purpose was to turn Chris' twofold experience of initial frustrating failure followed by a triumphant success into an object for their further reflection, because it is through such reflections that teachers transform their thinking about practice (Cooney, 1994; Jaworski, 1998, 2007; Krainer, 1999). In the process, I wanted to examine the teachers' intuitive responses to a PALP-rooted lesson before and after they observed the videotape. Thus, the teacher session began with hearing Chris' story of how she created her plan and the changes we both made to it, including the reasons *she* saw for these revisions. Next, I asked the teachers to write their individual responses and then discuss, first in small groups and finally in the entire group, the following question: Do you anticipate that such a lesson would bring about the desired student learning of counting-on and why? The responses of all 19 teachers in the group were analogous to Chris' initial disbelief – the revised lesson plan seemed counter-intuitive to them.

Once all teachers had several opportunities to form an anticipated image of the lesson independently, I engaged the entire group in *active* observation of Chris' videotaped, successful lesson. Specifically, to orient their reflection I asked them to take detailed notes of Chris' teaching activities, the ensuing student behaviours, and one's own emerging inferences into changes in the students' mathematical understanding. I oriented the teachers' reflection in advance that the observation notes should revolve around the challenge of identifying turning points in the children's counting strategies, that is, data-supported claims of a child's shift from counting-all to counting-on. Finally, Chris and I co-led a whole group discussion, first about changes in individual student understandings and then about changes in the teachers' anticipation for success of a similar lesson in the future (including reasoning). The productive results of that session for teachers – rudimentary shifts in their anticipations for task impact on student learning – go beyond this chapter's scope. At issue are the productive results in terms of MTE development, which are discussed further in the concluding section below.

CONLCUDING REMARKS

Consistent with reform visions of mathematics education, Ma (1999) introduced the construct of profound understanding of fundamental mathematics (PUFM). This construct portrays the robust nature of desired mathematical capacities teachers need to develop. In recent years, a growing body of literature has focused on teachers' Mathematical Knowledge for Teaching (MKT) as a necessary, albeit insufficient condition for successfully fostering such capacities (Ball, 2000; Doerr & Thompson, 2004; Tsamir, 2005). The position is plain and sound: one must understand something deeply in order to teach it effectively. In this chapter I argued that an equivalent position holds for epistemological stances. That is, alongside MKT, mathematics teaching is empowered when teachers develop PALP – learning to embrace explicitly, intentionally, and consistently the fundamental role of assimilation in students' learning. Using my own journey towards this awareness, I demonstrated how such development might take place as transformation in a teacher's current (intuitive) conceptions.

Applied to MTEs, then, the equivalent position entails that they need to develop PALP in order to foster its development in teachers. It goes without saying that, first, MTEs need to develop and become familiar with using PALP when they teach mathematics. This is not a trivial pursuit, because, quite often, MTEs who have invested in reform consider themselves to have already developed such awareness. Yet, further scrutiny of their practices frequently reveals that they do not systematically and explicitly consider how their students' (i.e., teachers') assimilatory conceptions afford and constrain the learning of intended mathematical ideas.

Transforming MTEs' epistemological stance is important because of the role they play when teaching mathematics to teachers. MTEs frequently do so as a means to provide teachers with examples of desired experiences of mathematics learning and teaching, which the teachers are expected to then use with their students. To this end, the MTE needs to analyse teachers' available mathematical conceptions as part of 7-step cycles geared not only toward teachers' learning of the intended mathematics, but also toward the more difficult processes of re-learning it. For example, almost all the elementary teachers with whom I worked over the years identified any number system that uses grouping by ten as a place-value, base-ten system. This included colour-coded systems that assigned no value to the location of units, or systems where units were grouped inconsistently (e.g., 1's, 5's, 10's, 50's, 100's, and 500's). As a MTE, I needed to learn how to foster these teachers' profound understanding of place-value systems, be it base-ten or other bases, which required consideration of re-learning that proceeds and is then differentiated from their deeply entrenched conception. That is, teachers needed to be engaged in tasks that both (a) allow them initially to use in novel situations their conceptions (familiar actions) of grouping-by-ten and (b) promote a new distinction between and coordination of the location of symbols in a socially agreed upon string (e.g., third digit from the right) and a magnitude determined by how many times units were recursively grouped (that digit 'counts' units produced via grouping 10 groups of 10 groups of 10 'ones').

On top of developing PALP for mathematics teaching, aims for MTE development include the use of PALP for fostering teacher learning of pedagogy. As my own journey indicates, this mathematics teacher education endeavour is much more complex than developing PALP for teaching mathematics, because it requires highly coordinated, two-tier understandings of learning. That is, in teaching pedagogy the MTE needs to understand how teachers: (a) develop understandings of student mathematics learning and (b) learn to enact such understandings within assimilation-regulated pedagogy. In particular, the desired two-tier understandings that serve as aims for MTEs' development include the following five capacities (capacity in the sense of commitment/ability to think about and carry out teacher education endeavours).

1. Analysing teacher development of facility with inferring student assimilatory conceptions and promoting this capacity through analysis of student behaviours. This includes selection of problem situations for conducting interviews with children, analyses of students' work such as error patterns and/or creative solutions, and extensively examining available literature on mathematical thinking of students. The latter can

include a focus on what Simon (2006a) called key developmental understandings (KDUs), that is, mathematical big ideas (Schifter, 1998; Schifter, Russell, & Bastable, 1999) that provide conceptual foundations for future learning, such as the reflective work on teaching counting-on in which I engaged Chris' peers.
2. Analysing teachers' evolving capacity to conceive of mathematics developmentally, that is, to psychologise (Dewey, 1902) the meaning of mathematical operations and strategies, and using PALP for promoting this capacity. To this end, MTEs can teach teachers how to use the constructivist conceptual framework similar to how I engaged Chris and her peers in analysing student conceptions of counting-all and counting-on.
3. Analysing available curricula and tailoring curricular components to both the intended and student available conceptions; analysing teachers' development of the same capacity and using PALP for promoting it.
4. Analysing teachers' development of pedagogical perspectives that coordinate cognitive and socio-cultural components (Cobb & Yackel, 1995; McClain & Cobb, 2001) and using PALP for promoting such development. The cognitive component can revolve around an epistemological stance that recognises the LP and addresses it via student goal-regulated activity. The socio-cultural component can nurture community establishment and negotiation of taken-as-shared norms and practices.
5. Using PALP as a mode of operation for interpreting, designing, and conducting research on how people learn. MTEs' engagement in such research endeavours is central to their teacher education mission, because it provides them with deep appreciation of the teaching-learning process problematics, particularly as it pertains to the learning of mathematics teachers.

The fragments of experiences that contributed to my development of the aforementioned PALP capacities point to three, interrelated but distinct types of reflection that mentors of maturing MTEs can promote. The first type consists of the MTE's reflection on her or his own activities when doing and/or when teaching mathematics. The second consists of reflection on plausible activities that underlie others' actions and language, that is, producing second order models (Steffe, 1995) of others' thinking. This type includes, specifically, reflection on teachers' disbelief regarding PALP-rooted lessons. The third type consists of the MTE's comparisons between the first and the second types, that is, an intentional attempt to link and bracket one's first and second order models. The key is that, when mentoring maturing MTEs, one continually fosters application of all three types of reflection to the learning of mathematics, of mathematics teaching, and of mathematics teacher education.

A characteristic task that mentors of MTEs can use to nurture the three types of reflection in each focus involves analysis of different solutions that different people (children, teachers) find to the 'same task' (Simon, 2006a). Such an analysis focuses on plausible sources for those differences, including the MTE's introspection of her or his own solution. Another example of how all three types of reflection can be applied is the very activity with Chris and her peers that brought about my distinction of PALP. MTEs can be engaged in observing videotapes or actually working with teachers who are presented with a task like the one Chris and I co-designed on the basis of PALP. Such videotapes will include teachers' expressions (disbelief) about the likelihood for promoting student understandings via the task prior to observing the teaching, their expressions when observing its implementation, and their reflections after witnessing student progress. Similarly, MTEs can observe teachers (live or on video) talking about their plan for, and then implementing, an insightful task given to them by a mathematics educator. It is most likely that the teachers' use of the task would differ from what MTEs' anticipate would be done (Zaslavsky and Leikin, 2004), and the difference can become a key component in the MTEs' reflection.

Eventually, maturing MTEs can be engaged in a meta-level discussion of their inferences into teachers' conceptions and what role these conceptions should play in educating the teachers. In particular, the MTEs' reflect on the differences between theirs and the teachers' understandings of the task and its potential impact on student learning. They then consider why particular activities with teachers may or may not promote teachers' transition toward PALP. The key question to be discussed is why teachers are likely to begin noticing assimilation (and learning paradox) when, obviously, during their career the teachers have already seen, and overlooked, many instances of different solutions to the 'same' task. As my experiences with Nevil, Edna, and Chris indicated, addressing this question is well served by observing teachers' reactions to teaching that succeeds where teachers articulately anticipated failure. I anticipate that, similar to my journey, addressing this question can open the way for MTEs' understanding of how teachers' reflection on the difference between anticipated and actual outcomes of teaching mathematics may lead to teachers' reflection on and noticing of the specific role that students' available conceptions play in learning. Most importantly, addressing this question is likely to foster the MTEs' development of PALP and corresponding, epistemologically regulated pedagogies.

REFERENCES

Australian Council of Deans of Education. (2001). *New learning: A charter for Australian education.* Canberra, Australia (http://www.jennylittle.com/Resources/CharterforAustralianEd.pdf).

Ball, D. L. (2000). Bridging practices: Interweaving content and pedagogy in teaching and learning to teach. *Journal of Teacher Education, 51*(3), 241–247.

Bauersfeld, H. (1988). Interaction, construction, and knowledge: Alternative perspectives for mathematics education. In *Effective mathematics teaching* (pp. 27–46). Hillsdale, NJ: Lawrence Erlbaum.

Bereiter, C. (1985). Toward a solution of the learning paradox. *Review of Educational Research, 55*(2), 201–226.

Cobb, P., & Bauersfeld, H. (1995). *The emergence of mathematical meaning.* Hillsdale, NJ: Lawrence Erlbaum.

Cobb, P., Wood, T., Yackel, E., & McNeal, B. (1992). Characteristics of classroom mathematics traditions: An interactional analysis. *American Educational Research Journal, 29*(3), 573–604.
Cobb, P., & Yackel, E. (1995). Constructivist, emergent, and sociocultural perspectives in the context of developmental research. In D. T. Owens, M. K. Reed & G. M. Millsaps (Eds.), *Proceedings of the 17th Annual Meeting of the North American Chapter of the International Group for the Psychology of Mathematics Education*. Columbus, OH.
Cooney, T. J. (1994). Teacher education as an exercise in adaptation. In D. B. Aichele & A. F. Coxford (Eds.), *Professional development for teachers of mathematics: 1994 yearbook* (pp. 9–22). Reston, VA: National Council of Teachers of Mathematics.
Dewey, J. (1902). *The child and the curriculum*. Chicago: The University of Chicago.
Dewey, J., & Bentley, A. F. (1949). *Knowing and the known*. Boston: Beacon.
Doerr, H., & Thompson, T. (2004). Understanding teacher educators and their pre-service teachers through multi-media case studies of practice. *Journal of Mathematics Teacher Education, 7*(3), 175–201.
Heinz, K., Kinzel, M., Simon, M. A., & Tzur, R. (2000). Moving students through steps of mathematical knowing: An account of the practice of an elementary mathematics teacher in transition. *Journal of Mathematical Behavior, 19*, 83–107.
Jaworski, B. (1998). Mathematics teacher research: Process, practice, and the development of teaching. *Journal of Mathematics Teacher Education, 1*(1), 3–31.
Jaworski, B. (2007). Theory and practice in mathematics teaching development: Critical inquiry as a mode of learning in teaching. *Journal of Mathematics Teacher Education, 9*(2), 187–211.
Karp, A. (2007). "Once more about the quadratic trinomial...": On the formation of methodological skills. *Journal of Mathematics Teacher Education, 10*(4), 405–414.
Krainer, K. (1999). Teacher growth and school development. *Journal of Mathematics Teacher Education, 2*(3), 223–225.
Lave, J., & Wenger, E. (1991). *Situated learning: Legitimate peripheral participation* (1 ed., Vol. 3). Cambridge, UK: Cambridge University Press.
Lerman, S. (2006). Socio-cultural research in PME. In A. Gutiérrez & P. Boero (Eds.), *Handbook of research on the psychology of mathematics education: Past, present, and future* (pp. 347–366). Rotterdam, the Netherlands: Sense.
Ma, L. (1999). *Knowing and teaching elementary mathematics: Teachers' understanding of fundamental mathematics in China and the United States*. Mahwah, NJ: Lawrence Erlbaum.
Mason, J. (1998). Enabling teachers to be real teachers: Necessary levels of awareness and structure of attention. *Journal of Mathematics Teacher Education, 1*(3), 243–267.
McClain, K., & Cobb, P. (2001). An analysis of development of sociomathematical norms in one first-grade classroom. *Journal for Research in Mathematics Education, 32*(3), 236–266.
National Council of Teachers of Mathematics. (2000). *Principles and standards for school mathematics*. Reston, VA: Author.
Noss, R., & Hoyles, C. (1996). *Windows on mathematical meanings*. Dordrecht, the Netherlands: Kluwer.
Pascual-Leone, J. (1976). A view of cognition from a formalist's perspective. In K. F. Riegel & J. A. Meacham (Eds.), *The developing individual in a changing world: Vol. 1 Historical and cultural issues* (pp. 89–110). The Hague, the Netherlands: Mouton.
Piaget, J. (1971). *Biology and knowledge* (B. Walsh, Trans.). Chicago: The University of Chicago.
Piaget, J. (1985). *The equilibration of cognitive structures: The central problem of intellectual development* (T. Brown & K. J. Thampy, Trans.). Chicago: The University of Chicago.
Pirie, S. E. B., & Kieren, T. E. (1992). Creating constructivist environments and constructing creative mathematics. *Educational Studies in Mathematics, 23*(5), 505–528.
Polkinghorne, D. E. (1988). *Narrative knowing and the human sciences*. Albany, NY: SUNY.
Schifter, D. (1998). Learning mathematics for teaching: From a teachers' seminar to the classroom. *Journal of Mathematics Teacher Education, 1*(1), 55–87.

Schifter, D., Russell, S. J., & Bastable, V. (1999). Teaching to the big ideas. In M. Z. Solomon (Ed.), *The diagnostic teacher: Constructing new approaches to professional development* (pp. 22–47). New York: Teachers College Press.

Simon, M. A. (1995). Reconstructing mathematics pedagogy from a constructivist perspective. *Journal for Research in Mathematics Education, 26*(2), 114–145.

Simon, M. A. (2006a). Key developmental understandings in mathematics: A direction for investigating and establishing learning goals. *Mathematical Thinking and Learning, 8*(4), 359–371.

Simon, M. A. (2006b). Pedagogical concepts as goals for teacher education: Towards an agenda for research in teacher development. In S. Alatorre, J. L. Cortina, S. Mariana & A. Méndez (Eds.), *Proceedings of the Twenty Eighth Annual Meeting of the North American Chapter of the International Group for the Psychology of Mathematics Education* (Vol. 2, pp. 730–735). Merida, Yucatán, Mexico: Universidad Pedagógica Nacional.

Simon, M. A., & Tzur, R. (2004). Explicating the role of mathematical tasks in conceptual learning: An elaboration of the hypothetical learning trajectory. *Mathematical Thinking and Learning, 6*(2), 91–104.

Simon, M. A., Tzur, R., Heinz, K., Kinzel, M., & Smith, M. S. (2000). Characterizing a perspective underlying the practice of mathematics teachers in transition. *Journal for Research in Mathematics Education, 31*(5), 579–601.

Steffe, L. P. (1990a). Mathematical curriculum design: A constructivist's perspective. In L. P. Steffe & T. Wood (Eds.), *Transforming children's mathematics education* (1 ed., pp. 389–398). Hillsdale, NJ: Lawrence Erlbaum.

Steffe, L. P. (1990b). On the knowledge of mathematics teachers. In R. B. Davis, C. A. Maher & N. Noddings (Eds.), *Constructivist views on the teaching and learning of mathematics* (pp. 167–184). Reston, VA: National Council of Teachers of Mathematics.

Steffe, L. P. (1992). *On the nature of learning theory*. Paper presented at the International Congress of Mathematics Education (ICME-7), Quebec, Canada.

Steffe, L. P. (1995). Alternative epistemologies: An educator's perspective. In L. P. Steffe & J. Gale (Eds.), *Constructivism in education* (pp. 489–523). Hillsdale, NJ: Lawrence Erlbaum.

Steinbring, H. (1998). Elements of epistemological knowledge for mathematics teachers. *Journal of Mathematics Teacher Education, 1*(2), 157–189.

Sullivan, P., Mousley, J., & Zevenbergen, R. (2003). Being explicit about aspects of mathematics pedagogy. In N. A. Pateman, B. J. Dougherty & J. Zilliox (Eds.), *Proceedings of the 2003 Joint Meeting of PME and PME-NA* (Vol. 4, pp. 267–274). Honolulu, Hawai'i: University of Hawai'i.

Tahta, D. (1981). Some thoughts arising from the new Nicolet Films. *Mathematics Teaching, 94*, 25–29.

The Royal Ministry of Education, R., and Church Affairs. (1999). *The curriculum for the 10-year compulsory school in Norway (L97)*. Oslo, Norway: Author.

Tsamir, P. (2005). Enhancing prospective teachers' knowledge of learners' intuitive conceptions: The case of same A–same B. *Journal of Mathematics Teacher Education, 8*(6), 469–497.

Tzur, R. (2001). Becoming a mathematics teacher-educator: Conceptualizing the terrain through self-reflective analysis. *Journal of Mathematics Teacher Education, 4*(4), 259–283.

Tzur, R. (2007a). Fine grain assessment of students' mathematical understanding: Participatory and anticipatory stages in learning a new mathematical conception. *Educational Studies in Mathematics, 66*(3), 273–291.

Tzur, R. (2007b). What and how might teachers learn via teaching: Contributions to closing an unspoken gap. In J.-H. Woo, H.-C. Lew, K.-S. Park & D.-Y. Seo (Eds.), *Proceedings of the 31st Annual Meeting of the International Group for the Psychology of Mathematics Education* (Vol. 1, pp. 142–150). Seoul, Korea: The Korean Society of Educational Studies in Mathematics.

Tzur, R., Simon, M., Heinz, K., & Kinzel, M. (2001). An Account of a teacher's perspective on learning and teaching mathematics: Implications for teacher development. *Journal of Mathematics Teacher Education, 4*(3), 227–254.

von Glasersfeld, E. (1995a). A constructivist approach to teaching. In L. P. Steffe & J. Gale (Eds.), *Constructivism in Education* (1 ed., Vol. 1, pp. 3–15). Hillsdale, NJ: Lawrence Erlbaum.

von Glasersfeld, E. (1995b). *Radical constructivism: A way of knowing and learning.* Washington, D.C.: Falmer.
Watson, A., & Mason, J. (1998). *Questions and prompts for mathematical thinking.* Derby, UK: Anne Watson and John Mason.
Wenger, E. (1998). *Communities of practice: Learning, meaning, and identity.* NY: Cambridge University.
Wertsch, J. V., & Toma, C. (1995). Discourse and learning in the classroom: A sociocultural approach. In L. P. Steffe & J. Gale (Eds.), *Constructivism in education* (1 ed., Vol. 1, pp. 159–174). Hillsdale, NJ: Lawrence Erlbaum.
Zaslavsky, O. (2007). Mathematics-related tasks, teacher education, and teacher educators. *Journal of Mathematics Teacher Education, 10*(4), 433–440.
Zaslavsky, O., & Leikin, R. (2004). Professional development of mathematics teacher educators: Growth through practice. *Journal of Mathematics Teacher Education, 7*(1), 5–32.

Ron Tzur
Curriculum and Instruction Department
Purdue University

RAZIA FAKIR MOHAMMAD

8. BECOMING A MATHEMATICS TEACHER EDUCATOR

Processes and Issues

Two roads diverged in a wood, and I –
I took the one less traveled by,
And that has made all the difference
Robert Frost.

This chapter discusses particular aspects of my real experiences of becoming a mathematics teacher educator in Pakistan; these experiences have been selected carefully to mark the key stages of my development process. Becoming a teacher educator involved having a strong ethical and moral stance; opportunities and exposure to formal learning; on-going reflections and experiences of 'real' work with teachers in their classrooms, and a self-inquiry approach. My self-involvement in on-going critical reflection also allowed me not only to comprehend my own actions and underpinning philosophy but to be able to revisit and modify them. The analysis of my experiences suggest that mathematics teachers and teacher educators need to build a flexible, realistic, moral, ethical and critical approach towards their practices as it opens new avenues of learning and professional development for them.

INTRODUCTION

Using a self-study approach, I have documented in this chapter my journey towards becoming a mathematics teacher educator, highlighting significant turning points and factors contributing to the process of my growth and development. The chapter begins with a narrative account of the growth of my philosophical orientation towards teaching, learning and teacher education. I have shared some examples of my work with mathematics teachers which reflect the stages in my development as a teacher educator, involving discussion of the processes involved, highlighting issues and challenges, and suggesting a way forward in terms of the future and scope of Mathematics Teacher Education (MTE) in Pakistan. The examples discussed in this chapter come from my research studies in the area of mathematics education, my experience of working with teachers at school level, and my involvement in the teacher education programmes offered at the university level.

The chapter, thus, highlights key stages in my journey towards becoming a mathematics teacher educator and also provides significant insights into the processes and issues involved in developing as a teacher educator. Although the specific examples come from a Pakistani context, the processes and issues, I believe, have relevance for mathematics teacher education in other contexts as well.

MY PHILOSOPHICAL ORIENTATION

In this section, I discuss the philosophical and conceptual framework that provided the initial foundation for my work as a teacher educator. The discussion reveals that my ethical and moral perspective provided the impetus for my professional journey while the new avenues of learning and opportunities for growth provided the philosophy for my work with mathematics teachers and learners, which was further reshaped as my work with teachers in the reality of their classrooms progressed. My philosophy and perspectives were further enriched through an on-going process of: self-questioning and self-reflection on former images; gaining new experiences; and engaging myself in examining teachers' involvement in a change process.

THE FORMATIVE YEARS

I was brought up in a family where values such as 'love' and 'care' for humanity were inculcated in me during my formative years, helping me to develop the moral and ethical values that impelled me to realize that the true purpose of my existence was to serve humanity. For example, when, through my religious education, I realized the significance and power of prayer, I started feeling responsible to pray for everyone's safety and protection. This illustrates how my general feeling of care transformed into a moral responsibility to 'take action' for improvement. Similarly, during my early school days, I read a story that affected me deeply. The story, which was in my Urdu[1] textbook for class III, was about a child whose father advised him that he should not throw a banana skin on to the street, as it could cause a serious accident to any passer-by. The moral was that one should avoid practices that could harm others. I realized later that reading such texts could be a powerful tool for raising one's awareness about the basic values of life. This awareness and my notion of care for humanity encouraged me to commit myself to the profession of teaching with an aim to raise awareness through education; I viewed education as a powerful tool to help build a safe society and enhance the quality of life.

My experience of public schooling suggests that Pakistani schools are a reflection of the hierarchical and restricted structures of the society, where respect is viewed as uni-directional; for example, from weak to strong, from poor to rich , from student to teacher, from youngsters to elders, from women to men, and from

---

[1] Urdu is the national language of Pakistan.

teacher to headteacher. Schools employ traditional modes of teaching; the students spend 5-6 hours each day listening and following instructions and doing assigned work quietly. Teachers are compelled to follow the instructions of their school heads and other professional seniors, usually suppressing their individual potentials and eschewing creative thinking. Parents send their children to public schools because of its low and affordable fee structure in comparison to private schools; also, when it comes to the education of their daughters, because they do not see much point in spending money on their daughters' education.

I was sent to a girls public school, being run in a small two-storey building where we (the girls from the nearby community) were expected to learn to acquire reading and writing skills for becoming effective future wives/family members. I was unaware of what I was gaining or could gain from each class; however, I was able to read, complete the assigned tasks and also pass examinations and move to further classes. As I did not have any other framework except my own ethical and moral belief to perceive the quality of learning, I was satisfied with my own learning outcomes. Nevertheless, I experienced that the school environment encourages students to grow as obedient and passive members of community in order to maintain/sustain meaningless actions, power imbalance and injustice in society.

Guided by my ethical and moral purpose, when I started my formal teaching as a Secondary School Mathematics Teacher in a private school in Karachi, I was kind to the children; sensitive towards their feelings and behaviour; and respectful of their self-esteem. In fact, 'individual attention' and 'care' were two distinguishing characteristics of my teaching. In other words, I viewed teaching in the same way as Easley (2005, p. 167) defined it as a "constitution of care and trust", where the purpose of teaching is to make a positive difference in the lives of the students.

However, I realized later (as detailed below) that there was a gap in my teaching approaches, which were mainly driven by my earlier experiences of having learnt through rote memorization and routinized practice and drills which did not help learners develop a critical stance or logical reasoning (Skemp, 1976). My aim was to help develop my students' moral and critical stance for which I would provide them moral lessons about life in general; however, these lessons were isolated and not consistent with my teaching approach. My teaching approach was encouraging their dependence on me; I was presenting knowledge and morality as two separate concepts to my students instead of providing them with a favourable environment to learn through experiences and reflection. Thus, I was contradicting my own moral perspective by directing the students and their learning instead of empowering them.

TEACHING/LEARNING MATHEMATICS – DEVELOPING PHILOSOPHY

As Polanyi (1958) states, 'development is a transformational process'; my professional development began when I was offered an opportunity to challenge, understand and deepen my practices as a teacher through a two-year master's programme in teacher education at a private university in Karachi, Pakistan. These two years marked a crucial stage in my professional development process. My participation along with other programme participants in learning through problem-solving tasks, questioning, arguing, reasoning, seeking alternative solutions, peer discussions, and sharing differences of perspectives resulted not only in the growth of my mathematical knowledge and positive attitudes towards learning mathematics but also acted as powerful tools to help me become an independent learner and critical thinker. We explored and examined alternate perspectives of mathematics teaching and learning based on the philosophy of 'social constructivism' (cf., Cobb, Wood & Yackel, 1991; Ernest, 1991; Jaworski, 1994; Vygotsky, 1981) and their consequences for students' intellectual and social development.

Based on this new learning, I started questioning my former teaching practice of developing rote memorisation. I now understood the limited cognitive outcomes of this practice, as expressed by Skemp (1979, p. 33): "Certain actions are reinforced as a result of their outcomes, so learning follows action. And what is learnt is action: the cognitive element is small". Individuals develop their cognition from their interaction with the environment, when they negotiate meanings, resolve conflicts, make sense of peers' and teachers' perspectives, and construct their own interpretations. Therefore, the context (including other participants) of an individual's learning, as discussed by Vygotsky (1981), plays a fundamental role in his or her development. I understood that meaningful learning and development of students thinking is a process whereby a teacher facilitates the meaningful acquisition of an idea by a learner through active participation and engagement.

The new philosophical underpinning of mathematics learning fitted well with my ethical and moral perspective: to promote students' autonomy, imagination, innovation, spontaneity, enquiry, and flexibility. My teaching approach in mathematics classrooms based on my new learning also affirmed that when students construct new knowledge, they develop high self-esteem and a sense of empowerment; this, in turn, empowers them to believe that they have the capacity to learn, resulting in increased confidence, motivation and openness. From this platform, they are able to tackle difficulties in flexible and reflective ways. Thus, the social constructivist philosophy of learning became an integral part of my philosophy of teaching, providing the foundation of my work as a mathematics teacher and teacher educator, as I could now view teaching not only as a moral but also as an intellectual enterprise.

WORKING WITH MATHEMATICS TEACHERS – THEORETICAL POSITION

My role changed from being a teacher to a mathematics teacher educator. My view of teaching is consistent with Pearl Buck's view, expressed as "sacred as priesthood. If one has not the concern for humanity, the love of living creatures, the vision of the priest and the artist, he must not teach" (cited in Small, 2005, p. 2). Likewise, I took on the role of a teacher educator as a moral and ethical responsibility, since I believed it to have a key role in influencing teachers' perceptions and practices to help them make a difference in the lives of learners.

The university, where I work, has been at the forefront of efforts to improve the quality of education, conceptualizing and implementing teacher education according to the theoretical perspectives discussed above (philosophies of social constructivism, reflective inquiry, etc.) for more than a decade. The university offers a variety of programmes for educational improvement of practicing teachers, with an aim to bring reform at the classroom and school level. It is expected that after the practicing teachers have completed their training courses, they will be leaders in the 'change processes' in their schools (Farah & Jaworski, 2007).

However, it is significant to note here that in the Pakistani context, it is a common practice that teachers hold teacher-educators and/or supervisors in a higher position of respect, viewing themselves as inferior or less knowledgeable, thus, feeling obliged to listen to whatever the supervisors have to say. Most teacher-education institutions employ 'top-down' methods and approaches. Even when they are advocating more creative and innovative ideas and methods, the teacher educators' approaches are likely to be formal and transmission-based. The presence and authority of inspectors and/or teacher educators and management is experienced by the teachers with anxiety and/or hostility and creates a negative impact on teachers' self-confidence. This is due to the overall culture, which does not allow teachers to develop a sense of ownership, thus, they become dependent, resulting in a submissive attitude of obedience and compliance to higher authority.

Based on my philosophy of how meaningful learning takes place, I wanted to achieve equality in our working relationship together in order for teachers to feel more committed, responsible and capable of producing meaningful and practical knowledge about teaching and improving themselves (e.g., Fullan, 1999). Therefore, a big concern for me was how I could establish a working together environment in their own school setting where both participants (teachers and myself) could accept and recognise some kind of accountability and equality in the learning process and teachers could feel free of constraints of my perceived domination. The global research literature suggests that a culture of collaboration, focusing explicitly on nurturing teachers' thinking and practices, can be promoted when both teachers and teacher-educators commit themselves to learning. This commitment towards co-learning promotes mutual dialogue, which is the product of self-reflection and/or a trust relationship (Wagner, 1997 & Jaworski, 1994, 2002). The outcome is reduced external control and authority over teachers' learning, whereas, at the same time, intensifying the responsibility for self-improvement (Frost & Durrant, 2002). In this regard, collaborative practice becomes a powerful force, which is similar to the way Carr (1995) views practice – a 'dynamic force both for social continuity and for social change'. Teacher education, viewed thus, fitted well with my general feeling of care transformed into a moral responsibility to take 'collective/collaborative' action for social improvement.

Thus, I adopted a new theoretical position – establishment of a co-learning partnership that focuses explicitly on empowerment of teachers and nurtures the thinking, practice and commitment towards learning of both teachers and teacher educators; it also promotes mutual dialogue resulting from self-reflection and/or a trusting relationship. However, critical reflection on my role, as a teacher educator and as a co-learner, in the context of the teachers' workplace, provides evidence that my teacher educator's perspective was presented throughout as an authoritative one, which was in contradiction with the co-learner concept. I felt strongly that the teachers were unique and worthy of respect; I tried to avoid a deficit model in analyzing the teachers' accounts (Brown & McIntyre, 1993). I attempted to refrain from asking any question or judging the teachers' actions in ways which could disturb the teachers' emotional, social or professional status in the school. Nevertheless, there were ethical issues that arose during my interaction with the teachers. I realise that professionals cannot take a completely neutral stance in relation to implementation of a theory (in this case the co-learning partnership that I adopted from the literature); there is bound to be a personal element of an interchange of perceptions, judgments, feelings, biases, and so forth – all depending on the nature of reality.

EXAMPLES OF WORKING WITH TEACHERS IN THEIR SCHOOLS

In this section, I present some examples of my work as a mathematics teacher educator in the schools. These examples highlight issues of mathematics teacher education in the context of Pakistan, and also illustrate the emerging tension in my work with the teachers in which my philosophical position was questioned and challenged by the contextual realities. The examples come from my work with mathematics teachers who had resumed their teaching after their participation in an eight-week mathematics teacher education programme at the university in which I was one of the teacher educators. The teachers were all committed and motivated to bring reform in their mathematics teaching and learning processes through implementing the newly learnt strategies; their aim being to promote learner-centred approaches and respect students' individuality. An examination of my work with these teachers, in terms of issues affecting teachers' implementation of learning from the teacher education programmes, suggested that teachers had serious difficulties in initiating their new learning in the real classroom situations. The teachers raised questions as to whether it was possible to bring about change in the prevailing poor conditions of the schools. Some had also expressed the need for a professional colleague to support their professional development in the reality of classroom (Mohammad, 2007). I offer below some examples that illustrate and discuss the nature of support provided as well as the processes of a teacher educator's work with teachers, its outcomes, related issues and challenges. The issues highlight the tension between my approach to teacher learning and my ethical and moral perspectives (my feelings of care and concern, both for the teachers and for children's learning).

Example 1

This example is taken from my work with a government school[2] teacher, Sahib[3]. One day, when I arrived at his school, I saw that the students had queued up in front of the head teacher's room and were going inside one by one. I was informed that money had been stolen from a student's bag – something that had been happening on a regular basis – and in order to catch the thief, the teachers had decided to use the 'holy water' ruse. They mixed an egg in a jug of water and told the students that it was special 'holy' water, which would make the thief sick after he drank it. I was concerned that this 'trick' would have an adverse effect on the children's moral growth and development, i.e., in terms of the implications of the teachers' lying to their students or the psychological implications of the fear of 'holy water'.

Initially, Sahib did not share the same concerns nor did he raise any such questions – a situation that seems to reflect the prevailing culture in Pakistan where one is not expected or educated to question authority or one's own actions and their intended or unintended consequences. Therefore, it was difficult for Sahib to analyse his own actions and practices, as they were embedded in the social fabric of the school. However, my philosophy of learning and understanding of this culture led me to believe that actions such as these contributed to the low motivation and morale of students. I encouraged Sahib to think about the effects of the 'holy water' ruse on the students' mental, physical and emotional health; however, he did not think that there could be any harmful effects, as is evident from the piece of conversation between Sahib and myself in Figure 1 below.

My point of view was that a teacher needs to take a more holistic approach towards teaching, integrating content and methodology with his/her overall moral and ethical stance. This implies that if, for example in the above case, the teacher aimed to improve his students' confidence in doing mathematics, he also needed to be sensitive towards their holistic growth, their motivation, morale, academic self-concept and self-esteem since, for me, teaching entails moral responsibility.

---

[2] The school had very basic resources: desks and benches for students (with 3 students sharing the bench at one small desk), table and chair for the teacher, a blackboard and a few pieces of chalk. There was no playground or sports equipment available in the school.
[3] Teachers' names are pseudonyms.

> Translated from Urdu
> T = The teacher (Sahib)
> TE = The teacher educator (Razia)
> 1 TE There must be some effects of this water on the students' feelings?
> 2 T I don't think so. The water was not very harmful.
> 3 TE How do you think the students would be feeling?
> 4 T I don't think they'd have taken it seriously.
> 5 TE What would happen if somebody is innocent and feels sick after drinking it?
> 6 T No it won't happen, nobody is going to be sick. It was just a trick.
> 7 TE What do you think about it? I mean, what will be the attitude of the student who has done something wrong? Do you think that student could get more confidence to steal again if he is not caught through this trick?

Figure 1. Conversation with Sahib

Besides, there were very important questions for me: What would be the consequences of teacher's ignorance if similar issues were to arise in future? This was a local issue, but what about the implications of many such local issues adding to the wider issue of teachers' learning and growth and their implications for many other students in the future.

My belief was that no one should get harm from anyone's action (neither students from teachers nor teachers from teacher educators). I was also worried about the teacher's perception of me as an evaluator or inspector due to my interference in the school issue. I did not want to subscribe to the current authoritarian practice, to be just another layer in the hierarchy of control within the Pakistani context. However, I realised that simply challenging Sahib's views was not sufficient; he needed more help to be able to analyze the consequences of his actions. Therefore, I suggested to him some ways to address similar issues in future, e.g. I suggested that story-telling with powerful moral messages could also help students understand the implications of their actions and develop a moral perspective. Later, during our conversation, Sahib did acknowledge that he had not considered the negative implications of his approach of dealing with a moral issue and suggested that, to improve the students' moral perspectives, he could discuss with them the problems that such kind of incidents of theft could cause for the victim and his family. Although it seems from the given example that the teacher learned some ideas about addressing the moral concerns, the question remained whether he genuinely considered those alternative perspectives or was agreeing as a way of ending the discussion by accepting the teacher educator's authority.

Example 2

Sometimes, I took over teaching during the lessons when the problems of the teachers' limited understanding of mathematics were revealed to me. For example, Sahib taught a revision lesson one week before the final examinations in the school. According to Sahib, it was a very important topic from the examination point of view, because a number of questions from that topic were included in the

BECOMING A MATHEMATICS TEACHER EDUCATOR

examination paper. He, therefore, wanted the students to have conceptual understanding so as to be able to solve the related questions and pass the examination. In order to discuss my dilemma as a teacher educator and the teacher's constraints in this event, I provide below some details from the lesson in Figure 2 below.

- The topic of the lesson was 'division in algebra'. Sahib began the lesson by testing the students' knowledge of basic algebra; for example, definitions of *variable, constant* etc. and then he asked the students to divide x^3 by x. The students were not able to solve the presented problem correctly; the teacher solved it himself and gave another question of division:

$$\frac{xy}{x}.$$

- None of the students raised hands. When the teacher observed passive participation from the students, he drew the following table and explained the rule of 'powers of two'.

2^4	2 x 2 x 2 x 2	16
2^3	2 x 2x 2	8
2^2	2 x 2	4
2^1	2 x 1	2
2^0	—	1
2^{-1}	—	$\frac{1}{2}$
2^{-2}	—	$\frac{1}{4}$

- After that explanation, the teacher went back to the question $\frac{xy}{x}$? He solved the question in this way: $x^{1-1}y = y$. However, he did not provide any linkages between his explanation of the rule involving powers of two and his solution to the question. Then he gave another question and invited the students to solve this on the blackboard. The whole class sat listening to the teacher; none of them raised their hands.

Figure 2. Details from Sahib's lesson on division in algebra

The teacher's intention in the above lesson seemed to be to help the students to generalise the rule of exponents from that example of 'powers of two', and to apply the rule in the presented task. However, he did not provide adequate explanations to support the students' ability to understand such questions. I observed silence in the class. The teacher himself did not seem to understand the barrier of his own way of presenting the material impeding achievement of his aims of helping the students to understand the questions. Partly, I was aware, the problem was with the teachers' own mathematical background, and his reliance on the standard text book. The question this raised for me was that if the teacher does not have adequate mathematical background, how could he help students to learn the concept? Should I leave children alone with the conceptual gaps despite having relevant expertise and mathematical knowledge to address the issue?

From a co-learning perspective, I could have discussed that issue with the teacher in the feedback session, as I believe that dialogue promotes shared understanding. However, the moral dilemma for me here was that how would this post-discussion with the teachers benefit the students? The final examination was imminent; the teacher might not have time to clarify the students' concepts and there were possible consequences of students failing the examination. Though from my theoretical assumptions of meaningful learning I assumed that by interfering, I might humiliate the teacher, I felt morally obliged to take care of students' learning and help them gain conceptual clarity of the topic. Therefore, I decided to take charge of the lesson to support the students' as well as the teacher's learning. I phrased my intervention in this way; I asked him if I could continue his teaching as I developed interest in this topic and Sahib accepted my request. This nevertheless leaves a question of what happens to these students when a teacher educator is not there to mediate the teacher's presentation of material.

Example 3

There were events and situations in the schools when I judged the situations in terms of my own ethical and moral perspectives and led the discussion and decided the agenda for working together on those occasions. For example, during my meeting with Naeem, another government school teacher, at beginning of the new school year, I noticed that classes were not being conducted and the students were roaming around. Upon inquiry, I learned that this was due to the fact that the head teacher had not prepared the new timetable as yet; the teachers did not have a schedule of classes and were, therefore, enjoying their 'free' time. When asked about his responsibilities during this 'interim period', the participant-teacher (Naeem) replied,

> Before you arrived, I was just chatting with the other teachers. There is no work. Teachers have to come to school regularly because they [the school management] do record our attendance but teaching has not begun yet. It is a relief time for us. ... We read newspapers, discuss politics [country's current political situation] and [do] nothing else.

From my ethical and moral stance, this could have been used as a 'productive' time for students' and/or the teachers' learning, however, the school management or the teachers did not seem to be concerned about this. I wanted the teacher to realize the value of this time and learn about ways to utilize it properly himself in his role as a teacher, to take initiatives in planning his time in a better way rather than relying on the school management or some authority figure to lay out a plan for him. I suggested to Naeem that he could use this time to plan teaching lessons in preparation for our collaborative work in the forthcoming weeks. In order to see the reason behind why I was making the suggestion, I explained to him that by doing so (a) the time could be used productively and (b) he might get some ideas about how to use 'slack' periods in future. Naeem agreed and identified some topics for which he felt the need to do advance planning and he planned several lessons and resolved issues related to his own mathematical understanding. Analysis of the data provides evidence that working together with the teacher educator directed the teacher from a passive submissive role into taking a more constructive role and positive actions for improvement. Naeem also realized that improving upon one's practices was possible not only through his understanding of mathematics but also by realizing his moral responsibility towards his profession that implied effective use of time and resources for learning.

From my perspective of co-learner in this collaboration, I felt that I had imposed my decision on the teacher through advising him to use his free time in planning. This raised various questions for me to reflect and ponder on, e.g., what were the unintended outcomes of my suggestion? In taking a moral stance, have I promoted a situation where teachers felt submissive and powerless, and experienced a lack of professional autonomy? These were important questions for me in consideration of the fact that most teachers in Pakistan regard teacher-educators as superiors whose words and actions should not be questioned, which inclines the teachers towards dependence, thus, making them less inclined to take initiatives. The imposition was justified from a moral perspective (i.e., teachers should use their time effectively for improved learning outcomes); however, it could be questioned if viewed from the philosophy of co-learning partnership; thus, creating a dilemma for me as to which stance to take.

Ethical and Moral Dilemmas of My Role: Issues and Challenges

The discussion in the above examples referred to issues and tensions related to my taking on a lead role in collaborative work with teachers. Theoretically, I believed that teachers would learn best by exploring their own issues, identifying their own needs, and taking decisions on their own to address the issues concerning them. However, as evident from the above examples, my own moral stance made me take on a role that was relatively more directive and leading with respect to teachers' own learning as well as addressing critical issues related to the children and their learning. I recognized that my response was driven by the need of the situation, where I could not stop myself from directing the teachers' thinking and behaviour. From my analysis, the following issues hindered the teacher educator in developing a shared sense of equal participation in the collaborative partnership:

- Intellectual/ Conceptual Constraints
- High Distance Culture

– Moral obligation of being a teacher educator

INTELLECTUAL/ CONCEPTUAL CONSTRAINTS

My analysis is that the teachers were unable to visualise, analyse and cope with the issues related to their knowledge of mathematics, mathematics teaching, and teaching and learning, in general. More importantly, the teachers did not see them as a matter of concern and could not visualise consequences arising from those situations. For example, it was evident from various experiences that teachers' knowledge was based mostly on the exercises given in the textbook, for which they lacked adequate conceptual understanding, and this limited their efforts to move from traditional to student-centred learning approaches, and their ability to reflect on their practice and learn from it. Lampert (1988), Ball (1993) and Ma (1999) have raised similar issues of mathematics teachers' transition in terms of the limitations of their mathematics content knowledge.

As evident from Example 2 of Sahib, the teachers were unable to review, clarify and rationalize their own mathematical assumptions alone; they were also unable to extend students' ideas, to help them formalise intuitive understanding and challenge incorrect notions. The teachers imposed on students their own idiosyncratic way of thinking, mainly derived from the textbook explanation without making any logical or mathematical sense of the topic, or linking their own activities with appropriate mathematical explanation. Moreover, it was difficult for them to reduce their own domination in the lessons, to stop telling the students what to do or to provide the students with the space to organize their thinking. The teachers' actions did not allow the students to step back from dependent modes of behaviour, although the teachers intended to move away from their routine teaching. This, in turn, also resulted in routinization of innovative strategies, learned in their university course, at a low level of competence. The question remained: What would be the consequences of teachers' limited knowledge for the students' learning?

'HIGH DISTANCE' CULTURE

In addition to teachers' intellectual limitations, another issue was the prevailing 'high distance' culture in Pakistan, which encourages obedience and submission (Simkins, Sisum & Memon, 2003). As discussed earlier (Examples 1 and 3), in the context of Pakistani government schools, teachers are in the habit of looking and thinking from an external perspective rather than rationalizing and judging self-practices. Since authority lines are maintained and limitations are unchallenged, teachers do not develop a habit of looking explicitly at their practice and maintaining an open vision of their teaching. In such a culture, there is a belief in the 'naturalness' of hierarchy, which is defined and respected on the basis of professional and personal status; subordinates exhibit a strong sense of dependence on their superiors, while the superiors see things from the perspective of their own agendas and benefits. Being part of this culture, the participants found it difficult to reflect on the implications of their decisions. This, in turn, results in teachers' limited moral and ethical decision-making, which reflects another serious aspect of constraints on teachers' thinking.

The examples discussed in this chapter also reveal that schools were not being used as places for encouraging teacher or students' intellectual thinking nor was there any mechanism or culture of professional dialogue or discourse among teachers through which they could reflect on or examine their practices. It is evident that years of working in a culture and environment that does not promote self-reflection or a practice of questioning one's perspective or actions, does not help teachers to develop a strong moral and critical stance towards their practice. Furthermore, there are examples in my data set which suggest that due to an unsupportive school culture and in the absence of any professional support available, teachers perceive routine teaching as a secure, convenient and compensated option, since they assume that it protects their time, position and promotion in the school and also saves them from any additional stress. For example, Naeem mentioned that in his class, the majority of the students failed in the half-yearly examination. Naeem wanted to work on the students' former weaknesses in order to improve their foundations. But the headteacher asked him to ignore the students' needs and to complete the remaining part of the textbook as a requirement for the final examination. Thus, teachers, on their own, were not able to raise and address issues related to their intellectual constraints.

MORAL OBLIGATION OF BEING A TEACHER EDUCATOR

My judgment was that within the intellectual limitations and contextual constraints, the teachers could not think differently from their routine practice. The teachers appeared unable to take notice of their conceptual problems or to cope with them unaided in trying to implement practical change at this initial stage. I felt it my moral responsibility not to leave teachers alone with their constraints and their negative consequences for the students. Similarly, since I was concerned about teachers' limitations in terms of content knowledge and its negative impact on the students' learning I felt responsible for directing the teachers' actions and discussion in many instances. This again made me question my philosophy of working with teachers: should I take the initiative to challenge or leave the teachers to continue practices that I perceived as damaging to students' mathematical learning and overall growth? Do the teachers have resources, expectations (for example, from the senior management and colleagues at school) and support to advance their knowledge and critical thinking at the school? It was difficult for me to cope with my perceptions that my input would create some kind of dependency. I questioned my assumptions: How could my philosophy fit in this context of constraints? Was the philosophy flawed? My judgment was that working with my own theoretical positioning would cost teachers' time, motivation and energy, which could result in lower student learning outcomes. As a teacher educator in this context, being aware of the issues and having the relevant knowledge and expertise, I feel an enormous responsibility to take some action to address the enormity of teachers' issues. Thus, it remained impossible to follow the theoretical perspective of collaborative learning and practice in the face of the realities teachers encounter in schools, resulting in my taking on a leading role wherever and whenever it was necessitated by the circumstances.

The teachers' behaviour seems to reflect the influence of a culture where power (superiority in relation to age, experience, status, etc) results in acceptance rather than negotiation or questioning. It seemed that my acts or decisions created an imbalance in terms of power relations; instead of taking an equal role, my advice put me in an authoritative position. I was not able to analyze issues related to the practical implications of such theory until I inquired into and reflected upon my role as a co-learner.

REVISITING THE PHILOSOPHICAL POSITION: A TEACHER EDUCATOR'S ROLE IN PAKISTAN

My analysis suggests that the teachers' needs in schools and my own judgement of the situations had influenced my actions and my role in collaborative work, which seems contrary to my earlier theoretical understanding of a co-learning partnership. Given the contextual and conceptual constraints of teachers in a developing country, my judgment is that these teachers need explicit input in understanding mathematics and pedagogy in order to challenge their misconceptions and improve practice – reflection alone, without having relevant knowledge and skills would not help them move beyond their routinized thinking and practices. Developing an attitude in which teachers see and experience questioning as learning can be integrated with the provision of adequate interventions and relevant input. In addition, both teachers and teacher educators may have different levels of knowledge, experiences and concerns and their own ways of viewing issues. As a teacher educator in this context, I have felt an enormous responsibility to take some action to address the extent of teachers' issues. This resulted in my taking on a leading role wherever and whenever it was necessitated by the circumstances. I believe that ignoring teachers' needs and failing to provide appropriate support negates the notion of true equality, as Macmurray (1961, p. 120) suggests, "It is to be noted that the moral rightness or wrongness of an action resides in its intention". I also refer here to a teacher's comment at the end of a partnership day as an example of the issues raised by the collaborative partnership.

> You are leaving me at the wrong time. With you, I understood my role in my improvement. I am becoming confident about how that learning situation could be improved in the inconvenient situation of the classroom.

My learning is that a dialogic relationship was achieved, which is a big progress in this situation, even if the authority/acceptance relationship was not entirely overcome. I learned that resolving power relationships in terms of equality in role is a difficult aim to achieve in the Pakistani context. From the examples cited above, it is also clear that if teacher educators are to help liberate teachers from the constraining impositions of their schools, to support the teachers as they work, to help teachers develop the teaching practices that they envisage, they (the teacher educators) need to be sensitive to, and understand, the true nature of the difficulties faced by the teachers. Furthermore, teacher educators need to appreciate not only how the constraints affect the teachers' ability to transfer knowledge and ideas from the university to the schoolroom, but also the effects that these constraints have on learning in classrooms.

Therefore, my redefinition suggests that collaborative learning is not the result of a contribution of equal levels of knowledge and understanding. Rather, it is the achievement of a growing relationship where a teacher educator supports teachers morally and practically while trying not to lower their self-esteem. Macmurray (1961, p. 158) supports my notion of "equality" in terms of mutual respect between both teachers and teacher educators in their roles, discussing "equality" in personal relations:

> This does not mean they have, as matter of fact, equal abilities, equal rights, equal functions [...]. The equality is intentional: it is an aspect of the mutuality of the relation. If it were not an equal relation, the motivation would be negative; or a relation in which one was using the other as a means to [one's] own end.

In such relations, a power differential always exists but the threat of power in impeding equality can be reduced if a teacher educator offers support and encourages teacher's thinking and autonomy.

FUTURE AND SCOPE OF MTE IN PAKISTAN

My reflection on my developmental processes suggests that teacher education should not be seen only as an intellectual activity; it should take into serious consideration teachers' realities and constraints. In this regard, there is a need to go beyond the idea of how things should be towards a clearer recognition of what they are. This means there is a need to go from an abstract intellectual dimension to a practical one, that is consideration of teachers' constraints and provision of school based support and also of developing teachers' moral and ethical perspectives of teaching.

In a Pakistani context, and also as Sally and Monk (2000) note from their work in developing countries more generally, attempts to improve teaching without addressing the issues of wider reform in the school can cause additional pressures on teachers, whose professional lives are already full of challenges. Fullan and Hargreaves (1992) suggest a focus on teachers' work environment and on teaching itself for teacher development. Clearly, these are important considerations when trying to bring about reform in a restrictive, authoritarian structure, such as is evident in schools in Pakistan. In addition, in the presence of numerous conceptual and contextual constraints, these teachers need direct discussions, advice, and input on various issues that are related to their teaching in an on-going basis. To support this, I quote Naeem,

> Only thinking alone is a tiring job. I am not in the position to cope with all the mathematical and practical problems of mathematics teaching in the school. Your presence is my assistance, which is reducing my tension. ... If I move back to my previous style [methods of teaching] then there will have to be some reasons and pressure; it will not only be my fault; we need to work as a group if we want improvement. .... We need an environment to 'push' [drive] us. You are here and the work has started again, but what will happen if you leave? ... We need favourable conditions. At the school nothing is certain.

The quotation indicates that the realities in public schools in Pakistan are so discouraging that teachers almost entirely abandon their efforts to change their practices. Inevitably, however, the clash between their desire for change and their failure to practice it has generated pressures and tensions. Lack of support, together with a rigorous emphasis on set routines, and a lack of independent thinking/approach prevent teachers from growing in knowledge within the school context. Clearly, teachers in a Pakistani context cannot be effective change agents if they are not provided with support in their work environments. This is a serious issue because even the most influential and sensitive teachers cannot, within their limited spheres of activity, do much to bring improvement. This is reflected well in O'Sullivan's (2002) comments following:

> The failures are particularly tragic in developing countries that can ill afford the wasted time, resources, and effort, and which urgently need reforms that successfully lead to significant much-needed improvements in their educational systems. (p. 233)

The global literature (e.g., Breen, 1999; Pimm, 1993) as well as the above discussion notes that it seems unproductive to leave teachers alone just when they have the desire to apply new aims and methods. Theory and experience suggest that the greater the degree of change that is required, the more the time and attention that should be given to securing the understanding and collaboration of the teachers and teacher educators. Moreover, reforms are more likely to be successful when they are integrated into a comprehensive plan for improvements across the whole system, with all involved agreeing on what is to be done and how. Cooney and Shealy (1997) sum up the preceding argument/ discussion well where they suggest that reform is the product of an integration of existing realities with new ways of teaching in which the reform unfolds through practical actions.

It is important to recognize here that issues related to teachers' implementation of new learning in the classroom context have been identified for decades; as Pressman and Wildavsky (1973) advise, 'the separation of policy design from implementation is fatal.' Similarly, Larson (1980) considers it unrealistic to set goals without considering one's resources. In addition to all the debate in the literature, there is no mechanism in Pakistani schools for school-based support and follow up in relation to teachers' professional growth. We teacher educators need to understand, identify with, and implement a reform with conviction; we need thoroughly to 'learn' it, in the deep and comprehensive senses of teacher learning. Allix (2003) explains this,

> New knowledge is created by dynamic cognitive and behavioural processes of search, inquiry, trial and error, experimentation, novel association, and intelligent adaptation in finding and generating solutions or resolutions to the demands of particular problematic contexts. (p. 14)

Therefore, teacher educators need to think about how to integrate teachers' learning at the university – the phase when the 'teacher is developed' (Jackson, 1971) – with their implementation phase in school where they are 'developing their profession' (Day, 1999). Whatever the difficulties, the ongoing professional growth of teachers in their schools is vital for (a) maintaining and developing teachers' own commitment and expertise, and (b) preventing the damage that might be done to students if teachers take imprudent risks. It is a central feature of my argument that follow-up support or teacher-teacher educator partnership should be taken into consideration to anticipate and plan for successful implementation. Thus, preparing the teachers for change in teacher education programmes without addressing their needs and providing on-going support at their school does not allow teachers to acquire the depth of improvement of their new practice. However, provision of one teacher educator to each teacher, as my work exemplifies, could be seen as an unrealistic option; there are few educators and it is too costly. So, the question for the community of teacher educators to think about is: What are the options for the future? I suggest strongly that teacher educators at the university need to take very seriously how the questions of theory can be incorporated into reality; that schools would need to take on a more active role than they presently do as partners in school reform rather than recipients of change: since even the best informed, most sensitive, and most consultative teacher education programmes cannot possibly anticipate and incorporate all of the variables that will affect the outcome.

Moreover, the issues emerging from teachers' practice are also of a moral and ethical nature (i.e., developing an attitude of care for students and what are the implications of their actions for pupils' development, the wasting of precious free time, their imposition of their own thinking, and so on). For example, the teachers' imposition of one procedure and rephrasing the students' answers (after inviting the students to bring their own ideas) first encouraged and then discouraged participation. This leads to confusion and sustained dependency on the teacher which may cause intellectual, emotional and affective hindrance of students' growth and raise questions such as: if a teacher does not understand or deal with students' answers, what is the motivation for students to supply their own answers? What are its implications for students' holistic development? Teachers need to have a strong moral and ethical perspective on education as well as on their teaching profession, as Ayers (2006) suggests:

> ... something beyond the instrumental and the linear. We need to understand that teaching requires thoughtful caring people to carry it forward successfully, and we need, then, to commit to becoming more careful and more thoughtful as we grow into our work. (p. 270)

Becoming a teacher educator involved having a strong moral stance; opportunities for and exposure to formal learning; on-going reflections and experiences of 'real' work with teachers in their classrooms, a self-inquiry approach and critical perspective. My self-involvement in on-going critical reflection allowed me to not only comprehend my own actions and underpinning philosophy but to be able to modify and revisit them. I believe that professionals do need to build a flexible, realistic and critical approach towards their practices as it opens new avenues of learning for them. Self-reflective inquiry upon practices undertaken by teachers and or teacher educators enhances their understanding of these practices and the situations in which these practices are carried out (Carr and Kemmis, 1986); however, having a strong moral and ethical perspective provides the very basis to extend individuals' intellectual and professional capacity. And if we want our teachers to be life-long learners, then, we, as teacher educators, have to acknowledge our position as life-long learners too, refocusing teaching and teacher education as an intellectual, moral and ethical enterprise. My conclusion is that reform in practice, therefore, not only requires a conceptual understanding of content and new pedagogy, but teacher educators also need to help teachers to develop an analytical stance towards their practice in terms of its moral and ethical implications as in the absence of this the teachers may not develop an attitude of care towards students' learning and its connection with their own practice.

REFERENCES

Allix, N. (2003). Epistemology and knowledge management concepts and practices. *Journal of Knowledge Management Practice, 4.* http://wwwtlainc.com/articl49.htm

Ayers, W. (2006). The hope and practice of teaching. *Journal of Teacher Education, 57(3),* 269–277.

Ball, D. (1993) With an eye on the mathematical horizon: Dilemmas of teaching elementary school mathematics. *Elementary School Journal, (4),* 373–397.

Breen, C. (1999). Circling the square: Issues and dilemmas concerning teacher transformation. In B. Jaworski, T. Wood, & A. J. Dawson (Eds.), *Mathematics teacher education: Critical international perspectives* (pp. 113–24). London: Falmer Press.

Brown, S & McIntyre, D.. (1993). *Making sense of teaching.* Buckingham: Open University Press.

Carr, W. (1995). *For education: Towards critical education inquiry.* Buckingham: Open University Press.

Carr, W., & Kemmis, S. (1986). *Becoming critical: Education, knowledge and action research.* London: The Falmer Press.

Cobb, P., Wood, T., & Yackel, E. (1991). A constructive approach to second grade mathematics. In E. von Glasersfeld (Eds.), *Radical constructivism in mathematics education* (pp. 157–176). Dordrecht: Kluwer Academic Publishers.

Cooney, T. J., & Shealy, B. (1997). On understanding of the structure of teachers' beliefs and their relationship to change. In E. Fennema & B. Nelson (Eds.), *Mathematics teachers in transition* (pp. 87–110). Hillsdale, NJ: Lawrence Erlbaum Associates.

Day, C. (1999). *Developing teachers: The challenges of lifelong learning.* Norwich: The Falmer Press.

Easley, J. (2005). A struggle to leave no child behind: The dichotomies of reform, urban school teachers and their moral leadership. *Improving Schools, 8(2)*, 161-177.

Ernest, P. (1991). *The philosophy of mathematics education*. London: The Falmer Press.

Farah, I. & Jaworski, B. (2007) *Partnerships in educational development*. Karachi, Pakistan: Oxford University Press.

Frost, D., & Durrant, J. (2002). Teachers as leaders: Exploring the impact of teacher-led development. *School Leadership and Management (2)*, 143–161.

Fullan, M. (1999). *Change forces: The sequel*. London: The Falmer Press.

Fullan, M., & Hargreaves, A. (1992). *Understanding teacher development*. New York: Cassell.

Jackson, W. P. (1971). Old dogs and new tricks: Observation on the continuing education of teachers. In L. J. Rubin (Eds.), *Improving in-service education: Proposal and procedures for change* (pp.19–35). Boston: Allyn and Bacon.

Jaworski, B. (1994). *Investigating mathematics teaching: A constructive enquiry*. London: The Falmer Press.

Jaworski, B. (2002). The student-teacher-educator-researcher in the mathematics classroom: Co-learning partnership in the mathematics teaching and teaching development. In C. Bergsten, G. Dahland, & B. Grevholm (Eds.), *Research and Action in the Mathematics Classroom. Proceedings of MADIF 2, The 2nd Swedish Mathematics Education Research* Seminar (pp. 37–54). Linkoping, Sweden: Linkopings Universitet.

Lampert, M. (1988). The teacher's role in reinventing the meaning of mathematical knowing in the classroom. In M. Behr,.C. Lacampage, & M. Wheeler (Eds.), *Proceedings of the Tenth Annual Meeting of the Psychology of Mathematics Education-North America* (pp. 433–480). Dekalb, IL: Northern Illinois University.

Larson, J. (1980). *Why government programs fail*. New York: Praeger Publishers.

Ma, L. (1999). *Knowing and teaching elementary mathematics*. London: Lawrence Erlbaum Associates.

Macmurray, J. (1961). *Persons in relation*. London: Faber and Faber.

Mohammad, F.R. (2007). Problems of teachers' re-entry in schools after in-service education. In I. Farah, & B. Jaworski, (Eds.), *Partnerships in educational development* (pp.103–107). Karachi, Pakistan: Oxford University Press.

O'Sullivan, M. (2002). Reform implementation and the realities within which teachers work: A Namibian case study. *Compare, 32*(2), 219–236.

Polanyi, M. (1958). *Personal Knowledge*. London: Routledge.

Pressman, J. & Wildavsky, A. (1973). *Implementation*, Berkeley: University of California Press.

Pimm, D. (1993). From should to could: Reflections on possibilities of mathematics teacher education. *For the Learning of Mathematics, 13*(2), 22–27.

Sally, J., & Monk, M. (2000). Teacher development and change in South Africa: A critique of the appropriateness of transfer of northern/western practice. *Compare, 3*(2), 175–192.

Simkins, T., Sisum, C., & Memon, M. (2003). School leadership in Pakistan: Exploring the head teacher's role. *School Effectiveness and School Improvement, 14* (3), 275–291.

Skemp, R.P (1976). Relational understanding and instrumental understanding. *Mathematics Teaching, 77*, 20–26.

Skemp, R.P. (1979). *Intelligence learning and action: A foundation of theory and practice in education*. London: Wiley.

Small, D. (2005). *Education World*. Retrieved Jan 12, 2008, from http://www.education-world.com/a_curr/profdev/profdev114s.shtml

Vygotsky, L (1981) The genesis of higher mental function. In J. V.Wertsch, (Eds.), The concept of activity in Soviet psychology (pp. 144–188). ArmonK, NT: Sharpe.

Wagner, J. (1997). The unavoidable intervention of educational research: A framework for reconsidering researcher-practitioner cooperation. *Educational Researcher, 26*(7), 13–22.

RAZIA FAKIR MOHAMMAD

Razia Fakir Mohammad
Institute for Educational Development
The Aga Khan University

KONRAD KRAINER

9. REFLECTING THE DEVELOPMENT OF A MATHEMATICS TEACHER EDUCATOR AND HIS DISCIPLINE

The chapter gives an insight into a mathematics teacher educator's personal development and its relation to the development of mathematics teacher education as a discipline. The paper presents first a nested model for seven domains of mathematics education research, with mathematics as its kernel. The model is used as a background and structure to describe and discuss the author's own personal development, starting with reflections on his reasons for aiming at and becoming a mathematics teacher, and later a mathematics educator and researcher. The conclusion is a kind of meta-reflection on his journey in mathematics teacher education and on the discipline itself.

INTRODUCTION

The motivation to accept writing a paper on my own development as a mathematics educator is grounded in the belief that our efforts in investigating (prospective and practising) teachers' learning have to be supplemented by efforts in reflecting also on our own learning process. The roots for this belief are manifold. Firstly, this reflection activity helps us in understanding our own development, both as individuals and as a field of practice and research. Secondly, this new knowledge supports us in understanding and interpreting teachers' growth. Thirdly, if teacher educators act as role models in the sense of reflective practitioners, teachers can learn from that. Fourthly, teacher educators' reflection on their own practice might motivate teachers to reflect their learning themselves more deeply. Finally, the self-application of reflecting on one's own learning has also an ethical facet: we do not only demand activities of those for whose growth we are co-responsible, but we demand it also of ourselves. This underlines not only the importance of the activity and the advantage to live democracy in education, but stresses also the genuine responsibility educators have in general.

Personally, I would like to add here that one of my first activities in mathematics teacher education dealt with motivating teachers to write mathematical curriculum vitae. These biographical texts were used as starting points for the collaboration of a group of mathematics teachers within a two year professional development programme in the 1980s (many years later, I wrote a research paper about this approach, see e.g., Krainer, 1994). It was very helpful to the whole process that I, as a beginning teacher educator, also wrote a curriculum vitae, and thus expressed that not only teachers but also teacher educators are learning. This is still the message I think we need to spread. In that sense, I dedicate the chapter to all prospective and practising teachers and teacher educators from whom I have had the honour to learn.

The title of this Volume of the International Handbook of Mathematics Teacher Education is "*The mathematics teacher educator as a developing professional*". I regard this *special focus on teacher educators* as an intermediate zenith of the further development of our discipline. The situation so far was not very promising. In the following, I sketch some thoughts on the development so far.

After the students and the teachers, only recently, the teacher educators themselves get increasingly into the focus of research in mathematics education. There are at least *four possible options* for designing such research.

The *first option* is (apparently) the one where teacher educators themselves reflect on their learning process. The width of approaches is enormous: it spans from a non-intensive self-reflection during and after a single teacher education intervention, over writing a reflective paper on one's own learning process as a leader of a larger professional development programme to publishing an intensive self-biography working out one's own becoming and being a mathematics student, teacher, teacher educator, and researcher.

A *second option* is a survey by a team of researchers on the (recent) development of the field of teacher education, for example sifting out trends, explaining them and working out challenges and possible steps for the future.

A *third option* is that a mathematics teacher educator writes a study on other mathematics teacher educators' growth. This could be done as a meta-reflection by an experienced mathematics educator who educates mathematics teacher educators and investigates their growth during a professional development programme; or a mathematics teacher educator (or another expert near the field) is invited to write a comparative study on mathematics teacher educators, for example, as part of an evaluation of a bigger teacher education project with several mathematics teacher educators involved, or as part of an international project that investigates different careers and general working conditions of colleagues in different contexts, countries, and cultures.

A *fourth option* is that a state or a university is interested in mathematics teacher educators' competencies and mandates a commission or an organisation to gather data and to present results. In principle, some elements of such processes occur in the context of applications for the position of university professors, selection procedures for scientific awards, evaluations of teacher education institutes or larger teacher education programmes. The boundary between evaluation and research is fluid.

Only ten years ago, reports about mathematics teacher educators' learning was to a large extent confined to short notes at the end of studies about teachers. Recently, indicating that also mathematics teacher educators' learning and their self-reflection is a crucial activity, papers are dedicated to teacher educators' growth. The literature is not vast, but increasing. There are examples related to options 1, 2 and 3 mentioned above. An example for option 1 is the self-study by

Tzur (1999, 2001; see also the reaction by Jaworski, 1999) who provides a four-focus model for the development of a mathematics teacher educator through an exercise of self-reflective analysis: learning mathematics, learning to teach mathematics, learning to teach teachers, and learning to mentor teacher educators (see also Tzur, this Volume).

In a combination of options 1 and 3, Zaslavsky and Leikin (1999, 2004; see also the reaction by Krainer, 1999) describe mathematics teachers' and teacher educators' learning within a professional development programme and discuss promoting and hindering factors that influence their practice.

An example for option 2 is the survey by Llinares and Krainer (2006) on relevant research on mathematics teachers' learning produced by the PME community. They conclude that at present, the theoretical references used to explain and to interpret the development of mathematics teacher educators are basically the same as those for mathematics teachers.

The major research question focused on in this chapter is the following: How is my development related to the corresponding development of teacher education as a field of research and practice? Related questions are: Which further factors influenced my development as a mathematics teacher educator? What were critical incidents that had a strong influence on my career and further development? How is my professional growth interconnected with the professional growth of prospective and practising teachers and teacher educators I have worked with?

FRAMEWORK

In the following, starting from *mathematics* as a research domain, I introduce seven extensions of interest concerning the development of mathematics education. These extensions form *seven research domains* related to mathematics education. These domains are used in the following sections as a background and structure to describe my own personal development as a basis to talk about mathematics teacher educators' learning and development more generally.

Several times in my career, I tried to understand the genesis and further development of mathematics education and to describe it in order better to grasp the position of my own work and of my department (e.g., Krainer, 2007). Feedback at two recent international meetings motivated me to develop these ideas further into a *nested model,* which has mathematics in its core.

Focusing on mathematical issues is the realm of mathematicians. Only, when also education-oriented research questions are raised, the field of mathematics education starts. The following model shows the extension of interest concerning the development of mathematics education, having mathematics (M) in its centre and continuously extended research domains of mathematics education (ME1, ME2, etc.) in the outer domains (see Figure 1). Each outer domain contains the *research questions* of the interior domains, however, the major emphasis is put on new questions and thus shows *an extension* of interest and a *new research focus* with corresponding *new questions*. Examples of research will be discussed in the following sections of the paper.

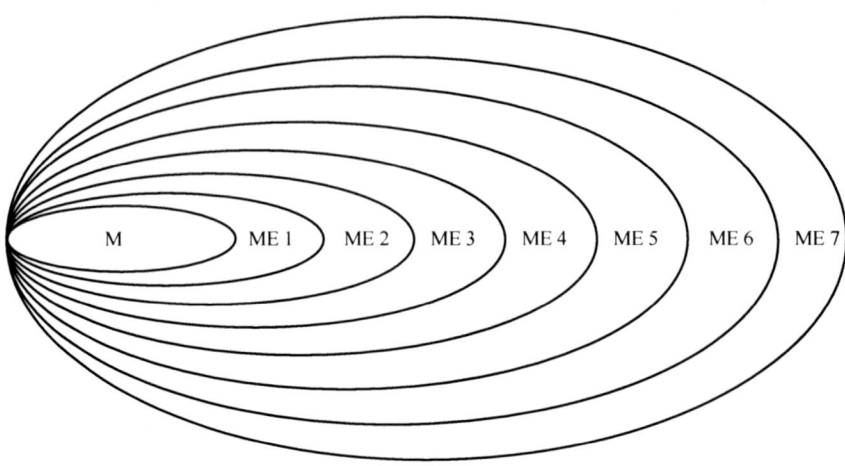

Figure 1. Research domains in mathematics education

Here are the *seven research domains of mathematics education* with *mathematics* as the starting point:

Research domain M: Focus on mathematical content, no educational considerations. This involves mathematicians doing research in mathematics. No issues of mathematics education are considered.

Research domain ME1: Focus on mathematical content, but from an educational point of view. Typical questions include: Which mathematical concepts, procedures etc. should be introduced in school mathematics (curricular issues, thus the question of what, mostly including when and how)? What fundamental ideas should be taught? What kind of introductions into different fields of school mathematics should be generated? Which powerful tasks can be constructed? When and how should different kind of tools and technologies (e.g., calculator, CAS, Internet) be used? In particular, throughout the emergence of mathematics education as a scientific discipline, many researchers entering the field of mathematics education wrote textbooks and developed other curricular material and tools. The development of these sources was often rather based on inner-mathematical and normative pedagogical considerations than on (qualitative and/or quantitative) empirical studies as they are more or less standard nowadays.

Research domain ME2: Focus on applications and history of mathematics. Typical questions include: Which contents can make mathematics more meaningful to learners? How can applications make the picture of mathematics broader and more attractive to learners? How did mathematical ideas (e.g., the concept of fraction) emerge over time (partially assuming that there are parallels between the discipline's and the learner's development)? Who are the people behind the fascinating development of mathematical ideas? Also this kind of research often lacked strong empirical evidence in the early days of mathematics education. The major extension of interest was a conscious effort towards "sense making" of mathematics to the learners, thus asking for answers to students' question of why (why learn mathematics)?

Research domain ME3: Focus on the mathematical learning of students. Typical questions include: How do students learn mathematical concepts? Which errors and misconceptions are typical? What are strategies to overcome such difficulties? How can problem solving and problem posing be promoted? What views of mathematics (teaching) do students have? The new feature of this research was the focus on the learner. In particular, pedagogical and psychological considerations came into play. Empirical research – primarily focused on individual students' cognition – became more prominent. Increasingly, evidence from empirical findings formed the basis for developing instructional material or interventions. In particular, psychological theories were integrated into mathematics education research.

Research domain ME4: Focus on interactions between students and teachers in classrooms and on the epistemological status of concepts. Typical questions include: Which kind of interactions, routines and patterns occur in mathematics classrooms? How are norms negotiated? What is the didactical contract between the teacher and the learners? How can teachers initiate team work by students or classroom discussions in order to promote sharing of ideas and mathematical reasoning? How can students' reflection and meta-cognition be fostered? The research object changed from the individual student to the whole class including the communication with the teacher. In particular, socio-cultural theories were integrated into mathematics education research. Many studies were based on data from audio- and video-taped teaching situations. The focus was not only on the cognitive aspects of learning, but also on the meta-cognitive, the affective and the social ones.

KONRAD KRAINER

Research domain ME5: Focus on the learning of teachers and on the impact of teacher education and school development. Typical questions include: What are teachers' beliefs, knowledge and practices? How do they change through interventions (as e.g., courses for prospective teachers or professional development programmes)? To which extent are learning environments (goals, contents, philosophies, tools etc.) successful? How can mathematics teaching in departments and even whole schools be promoted? What kind of programmes, materials, and tools should be developed and evaluated? The major shift of this research domain is that – at least in intervention studies – the researcher often also has the role of the teacher educator. In contrast to other studies, here the goals of understanding and improving merge. The advantages and disadvantages of researchers' nearness and distance to the research object becomes an issue. Until now, small-scale qualitative studies are predominant.

Research domain ME6: Focus on the learning of teacher educators. Typical questions are the same as for teachers (see above). However, the question of advantages and disadvantages of nearness and distance to the research object becomes even more delicate. Teacher education and its research is about understanding and improving our own practice as teacher educators, and aims at getting particular and general insights. Basically, this research domain can be seen as a self-reflection-based extension of mathematics teacher education. This domain is in its very early stage, only a few studies on teacher educators' learning exist.

Research domain ME7: Focus on mathematical abilities and potentials of education systems, the economy and of the society as a whole. Typical questions include: Which mathematical competencies are regarded as relevant for our society? What do international comparisons on students' achievements, beliefs, interests (e.g., in studying mathematics, science or technology) show? How can countries improve their situation concerning the mathematical competences of its students and citizens? What role can teacher education play? How can studies generate steering knowledge for educational policy? The major shift of this research domain is that the research interest is also strongly coming from outside (economy, policy, society) and that there is a closer focus on the impact of education. What counts as important mathematical competence is no more solely a matter for mathematicians and mathematics educators; it is also a matter for society- and policy-based negotiation. This domain seems to develop very fast. It might be very important for our discipline, to get more deeper-grounded roots concerning ME5 and ME6 before we are driven too fast into issues of ME7.

In the above sketched model, mathematics (including the question of tools) is integrated in all domains, although of course, the focus on it in inner domains is more intensive than in the outer ones. However, this does not carry any implication for the importance of research questions and related results. We need research in all fields, and also, the questions are deeply interrelated. Different mathematics educators surely start(ed) their careers and developed their research questions in different fields. For example, most mathematicians becoming interested in mathematics education take their deep mathematical background as a starting point and ask questions closely related to mathematics. Educational researchers without such a deep background in mathematics may start more likely in outer domains, for example, investigating the impact of mathematics reform programmes on teachers' beliefs and practice at a district or a national level. The research and development interest of people within the field of mathematics education might remain stable for a long period, whereas others might change their focus of attention more often, maybe going only towards the inner domains, or the outer ones, or even changing direction from time to time.

No specific research domain is more important than another one, and no transition from one domain to another is preferable. This equity of different approaches and research interests is an expression of the broadness and complexity of the field of mathematics education. Its specific focus on the teaching and learning of mathematics together with its inter-disciplinary stance makes mathematics education a genuine and autonomous scientific discipline. The start of mathematics education is closely related to mathematics. Therefore, most mathematics educators were originally mathematicians who became interested in educational issues. Thus the succession from mathematics to the outer domains – from a macro-perspective – roughly sketches the historical extension of research interests in mathematics education.

This *macro-perspective* also highlights the *increasing status of mathematics teacher education* within the field of mathematics education. This development is, for example, reflected in the emergence of international handbooks. In 1996, the first *International Handbook of Mathematics Education* (Bishop, Clements, Keitel, Kilpatrick, & Laborde, 1996) was published. The first *International Handbook of Mathematics Teacher Education* is published now. Till about 1990, mathematics teacher education was mainly a *field of practice* (mostly presenting "success stories", more or less theory-grounded and evidence-based), since then it increasingly became also a *field of research*. In 1998, the Journal of Mathematics Teacher Education (JMTE) was launched. JMTE marked the start of a new era since this was the first international journal focusing genuinely on research in mathematics teacher education. It was only four years before, when the founding editor of JMTE (Cooney, 1994, p. 631) stressed: "As a profession we have just begun to recognize the significance of conceptual orientations for guiding research on teacher education".

Parallel to the start of JMTE, also numerous books and articles on problems and progress in describing and interpreting learning processes by mathematics teachers were published (e.g., Loucks-Horsley, Hewson, Love, & Stiles, 1998; Jaworski, Wood, & Dawson, 1999; Krainer, Goffree, & Berger, 1999; Lin & Cooney, 2001; Peter-Koop, Begg, Breen, & Santos-Wagner, 2003; Strässer, Brandell, Grevholm, & Helenius, 2004; Ball & Even, in preparation).

KONRAD KRAINER

CRITICAL INCIDENTS OF MY DEVELOPMENT AS MATHEMATICS EDUCATOR

As mentioned above, the succession of research interests of different mathematics educators might show very different paths. In the following, I reflect – from a *micro-perspective* – my own development within mathematics education taking the above described research domains as a background. As one might see later, the development is not linear and continuous; it has many curves, jumps and links.

Mathematics and ME1: Becoming a Prospective Mathematics Teacher

Starting early in primary school, I was good in mathematics. My primary teachers apparently promoted my mathematical abilities. In addition to that, one of my brothers gave me challenging mathematical tasks. My mathematical experiences in secondary schooling were, on a first view, like in primary school. I was much better in mathematics than in other subjects. However, in upper secondary schooling this broadened up to most other subjects. In mathematics, I loved geometry in particular. I liked to construct triangles and was fascinated that in most cases you need exactly three entities; it challenged me that sometimes three entities led to no solution, and sometimes to two or even infinite solutions. I remember that – although in upper secondary the content was not geometry anymore – I still tried new ways to construct triangles. I succeeded in developing a formula for the angles of the triangle when the three altitudes are given.

Looking back, none of my mathematics teachers in secondary school exceptionally promoted my abilities, none of them deeply impressed me as a teacher either. My motivation to become a mathematics teacher arose primarily from giving numerous private lessons. I started that at the age of about twelve years and learned a lot about students' thinking. In particular, I realized that many students feared mathematics because they did not understand and thus had no experiences of success. Therefore, I would regard private lessons as a crucial element of teacher education for prospective teachers: the one-to-one situation of student to teacher gives a good opportunity to reflect students' growth, from a cognitive and also from an affective point of view. Understanding students' understanding is surely a key factor for prospective teachers. I agree with Wagenschein (1970) who considered *understanding as a human right*, and I want to add that it should also be a human duty.

When I entered university, I already had a good mathematical and pedagogical content knowledge. I decided to become a mathematics teacher for secondary schools, and I chose geography as my second subject since it was the only combination with a natural science at the University of Klagenfurt. Some mathematicians motivated me to study for a mathematics diploma in order to become a mathematician. However, my main interest was in students' learning. The teaching practicum in the middle of my studies was a motivating experience and thus confirmed my desire. In addition, we had well-known mathematics educators at our university. So I decided to keep on track becoming a teacher and wrote my diploma thesis in geometry education.

From my diploma thesis supervisor, I got an interesting reference concerning an *operative approach* to geometry and geometry teaching by Bender and Schreiber (1985). The book strongly formed my insight that mathematics is about constructing and that mathematical learning is about grasping concepts by doing. Looking back, I feel that this was a critical incident, both with regard to my beliefs about mathematics and my beliefs about learning. It shook my predominantly Platonic view of mathematics. Given my experiences with private lessons, I realized once more that teaching gives at best an opportunity to learn, but the crucial point is the active learning process by the learner.

I started my diploma thesis with the concept of the straight line. One major part of my thesis was to develop tasks to be used as learning environments for teaching geometry at Grade 5. The tasks aimed at supporting students to generate the general idea of the "ideal" straight line and purposeful "realizations" of that idea in reality. Realizations of the straight line (and corresponding intuitions) were, for example, a stretched rope (as a representation of the shortest distance between two points), an antenna (using the characteristic that a straight line allows translations in itself), or an edge of a snow shovel (as an ideal matching of a straight line in a plane). Another intuition of a straight line is that it can be turned in such a way that we can see only a single point (orthogonal projection); this is used by carpenters to check whether an edge of a piece of wood is straight.

I got my intuitions from several sources. A lot of my subjective domains of experience go back to experiences out of school, in particular to the work of my father who was a carpenter. My learning was highly situated in specific contexts and activities, carpenters (and their sons) usually do: they plane pieces of wood and check whether edges are straight; they repair snow shovels etc. Another influence came from reading. In particular, the book "Didactical phenomenology of mathematical structures" by Hans Freudenthal (1983) evoked my intuitions. Theories also influenced me. For example, I was fascinated by the idea that in the operative approach (see above) the straight line could be seen as the only geometrical object in "our Euclidian" plane that combines inner and outer homogeneity. Thus, a straight line represents total equality, even more than a circle: The circle is an expression of the idea of "inner homogeneity"; each point on a circular line has the same characteristics; no single point is "outstanding". This is the same for the straight line. But a circle has an inside and an outside, whereas the straight line has two equal sides, expressing also "outer homogeneity". Looking back, I think that this content-related grounding of my didactical knowledge made it easier for me to work with secondary mathematics teachers.

All in all, this means, that my diploma thesis was my entrance to mathematics education. At that time, mathematics education at my university was largely focused on sub-disciplines, e.g., didactics of algebra, calculus and statistics. This situation was not specific to Klagenfurt. In retrospect, mathematics education was to a large extent – in particular in German speaking countries – a design science (see, for example, Wittmann, 1995). Early mathematics educators' thinking – due to their socialization as mathematicians – was mainly *content-related*; the further development was driven by *system-internal questions*. The focus was on mathematical contents and curriculum. Many educators wrote "introductions" into several sub-disciplines or about special challenges, as for example, introducing variables. Mathematics teacher education (for prospective and practicing teachers) was mostly about presenting these elaborations. In particular, due to my experiences in giving private lessons, in the teaching practicum and in writing my diploma thesis, I learned to view active students as the key in mathematics learning and teaching. I realized that teaching does not necessarily promote students' learning. The huge need for private lessons suggested the contrary.

ME2 and ME3: Becoming a Mathematics Teacher, Mathematics Educator and Researcher

Applications of mathematics played an important role for me from the very beginning. In particular, also the diploma thesis dealt with geometry in real life. However, there had been also other impressive experiences with applications, throughout my university study. For example, I remember a seminar where we focused on a research project on applied mathematics at our university. The goal of the project was to develop a programme for a sawmill in order to decide which kind of boards should be cut, taking into account the quantity and quality of the trees to be cut, the storage of the sawmill, the demand and price for specific boards etc. This was about non-linear optimization and helped me later tremendously when teaching linear optimization at a vocational high school. If you teach linear optimization (confined to a few variables) and you don't reflect on the fact that most applications are non-linear and have much more (even hundreds of) variables, then students could get a view of mathematics that each problem has an easy solution, but they do not get a feeling for how complex and demanding it is to build mathematical models. The project showed that the challenge is not only a mathematical one, but also a social-communicative one. The applied mathematician needs to collaborate with the sawmill expert who is responsible for deciding which kinds of boards to cut before such a programme gets introduced (and who of course fears that his work could be useless after the implementation of the programme). I was struck to see that mathematics is not only about content but also about communication, and that mathematics itself facilitates and improves our thinking and communication. I strongly suggest integrating seminars into teacher education where the issues of *content, application* and *communication* are combined with prospective teachers' *joint reflection* on learning. This increases the opportunities to construct rich links between mathematical and didactical knowledge.

In 1981, I worked for several months as a part time study assistant. I became involved in discussion and working groups where *psychological aspects of students' mathematical learning* (e.g., Dörfler, 1987) were discussed. In particular, the work by Jean Piaget, Bob Davis and Kath Hart and the stories by Robert Lawler on his daughter Miriam and by Hans Freudenthal about his grandson Bastiaan deeply impressed me. Increasingly, *interviews with students* got the focus of mathematics education research in Klagenfurt. I realized that tasks are not only a means to represent the mathematical kernel of a topic and to construct "learning trajectories" (as we would say today), but also as a means to sift out students' understanding. This contributes to building more on their strengths ("ideas") rather than on their weaknesses ("errors"). Motivating teachers to do interviews with students and listening to them is a key intervention to change their attitude: you can support students' learning better when you know better their pre-knowledge and strengths.

In 1981, I was asked also to join a group of university staff members and experienced teachers planning a *professional development programme* for practicing English, German, history and mathematics teachers in four parallel courses. The main idea for the programme "Pedagogy and subject-specific methodology" (German abbreviation PFL) came from the pedagogue Peter Posch who brought in ideas of action research (see e.g., Altrichter, Posch, & Somekh, 1993). My role as a teacher educator – being a prospective and later novice teacher among very experienced mathematics teachers – was very demanding. However, my sound mathematical background, my dozens of private lessons and my first research and development experiences helped me a lot. We tried to plan the course in a way that the teachers continually could take responsibility for their active learning within the course, thus trying to live our principles that we regarded as essential for their activities in class (see e.g., Fischer, Krainer, Malle, Posch, & Zenkl, 1985).

During the last phase of the first PFL course, I began teaching mathematics at a High School, firstly full-time, then part-time (the other half working for PFL mathematics). The *double-role as mathematics teacher and teacher educator* was fascinating and challenging at the same time. Having been influenced by action research in the context of PFL, I also reflected on my teaching and presented my experiences among colleagues and at conferences.

When I taught the concept of angle in one class, I felt that the angle is easy to grasp for students. However, this practical experience was in stark contrast to a paper in mathematics education which asserted that this concept is among the most difficult in school mathematics. This irritated and provoked me. Like teachers in PFL that got motivated by us to take their problems as starting points of their studies, I took the concept of angle as the starting point of my doctoral thesis (1983-1989).

Influenced by the local research group, I conducted interviews with students to investigate their understanding of the concept of angle. One important outcome of my doctoral thesis was a "task system" for this concept. It contained tasks showing applications of the angle in the "real world", but also tasks that solely refer to inner-mathematical problems, and tasks where both "worlds" are combined and reflected (see e.g., Krainer, 1993). The *emphasis on students' reflection* was influenced by readings and experiences within the context of action research where the cycle of action and reflection is a crucial basis of teachers' learning (see e.g., Schön, 1983). A "task system" should represent the full complexity and depth of the concept of angle. Therefore, not only a set of single tasks is needed, but also a close interconnection between these tasks; in addition, each task should have its own quality and meaning. This is a first trace of my four dimensions *action and reflection* on the one hand, and *autonomy and networking* on the other (which will be described more detailed in the next sections).

At that time, I put the focus more on the mathematical content of the tasks (as learning environments for human beings) than on the human beings (the students in this case) working with these tasks. I assume that this is due to my deep grounding in mathematics which later was supplemented by also focusing on pedagogical aspects. I think both perspectives are important, and mathematics education has to combine both: mathematics teaching neither isolates mathematics nor isolates students, but it is about students' active confrontation with mathematical activities (see e.g., Fischer & Malle, 1985), meaningful for them but also directed towards curricular goals.

Summing up, I can stress that mathematical applications played an important role in my early research career; partially also links to the history of mathematics (e.g., the concept of angle) occurred. The involvement in research groups as well as in the PFL programme (I was a team member of PFL mathematics throughout the period 1982-1990, later I coordinated the whole programme for many years) brought me to put a focus on pedagogical and psychological issues. In particular, the learners of mathematics became my focus of attention, in my doctoral theses as well as in my professional development activities. In the field of mathematics education in general, a lot of research was done in order to investigate students' thinking in different areas and contexts. Since I got involved in research projects and in the PFL mathematics course which also aimed at integrating new results from mathematics education and related fields, I was more challenged by research and development issues (in particular, with regard to students', teachers' and my own learning) than typical prospective and novice teachers; and also, other mathematics educators might have taken quite other routes.

Nevertheless, some general insights might be sifted out of the first part of my career: it is important to promote (prospective) teachers' engagement in research and development projects. This active confrontation of prospective and beginning teachers with didactical issues increases the likelihood to educate reflective practitioners that continually investigate their teaching, and to win them as research partners or even as future colleagues (as teacher educators and researchers). Conducive factors are a narrow link between theory and practice, and the focus on teachers' learning (which itself raises the question how to promote and understand students' learning). I suggest involving our prospective teachers increasingly in research projects which form the basis of professional development courses for practising teachers. This might contribute to prospective and practising teachers' joint learning, to promote prospective teachers' better induction into their profession, to further develop community building among teachers, and to construct more appropriate bridges between schools and universities.

ME4 and ME5: Promoting and Investigating Individual and Groups of Mathematics Teachers' Growth through Professional Development and School Development

My PFL mathematics experience (1982 to 1990) showed that several teachers had difficulties to keep the role of the interviewer (trying to observe and understand students' thinking, taking an investigative stance) and not to switch to the role of the teacher (trying to influence and change students' thinking, aiming at a desired learning goal). The PFL teachers – and I too – learned that viewing students as experts of their thinking – and not as deficient human beings – is a challenging change of view. The teachers expressed the view that interviewing improved their knowledge about students' mathematical thinking. They also realized that listening to students could be seen as a general attitude helpful for teaching. These interviews were a kind of starting point to other investigative activities by PFL teachers. In particular, the teachers had to write case studies on their own teaching. Evaluations showed that writing was a tremendous challenge for many teachers; however, at the end, when the studies had been written, mostly positive feelings and views remained. Only recently, a doctoral student (Schuster, 2008) began to investigate the impact of mathematics and science teachers' writing on their motivation and competences.

My teaching practice profited from experiences in PFL and from reading research literature. Vice versa, my mathematics teaching was the basis for many questions and problems that made me think didactically. In PFL, I became familiar with research on interactions between mathematics students and teachers in classrooms and on the epistemological status of concepts at the University of Bielefeld. This work can be seen as a contribution to the *sociological and epistemological extension of the field* of mathematics education. In particular, the group around Bauersfeld – putting an emphasis on social interactions in mathematics classrooms (see e.g., contributions in Cobb & Bauersfeld, 1995) – influenced PFL and my thinking. In PFL, we used transcripts of classroom interactions as opportunities to go into deep discussions among and with teachers. It was these experiences that motivated me later to put an emphasis on teaching and learning processes in the Austrian-wide reform initiative IMST, discussed further below.

PFL fostered my attitude of being curious about teachers' and my own learning and the view that researchers and practitioners can meet at eye-level. Teachers' feedback showed that the PFL course changed their attitude and competence in teaching. However, we did not go deeply in investigating the impact of the courses (e.g., concerning teachers' change of beliefs, knowledge or practice). Our main emphasis was on supporting teachers' action research and thus helping them to write down their experiences in a systematic and self-critical way, and not on doing our own research on teachers in addition to their investigations into their own teaching. In general, research on mathematics teacher education was not highly developed at that time. What we did, as I would say today, was predominantly telling success stories, based on evidence by data from teachers' self-study and additional data we gathered and analyzed.

In the 90-ies, we put more emphasis on the social and organizational dimension within the PFL programme. A minor influence came from our perspectives on social interactions in classrooms. However, the new emphasis on school development was primarily a result of the ongoing (and intensified) evaluation of the PFL programme: the evaluation showed that the programme contributed successfully to teachers' individual competences and practices; however, in general, there was little impact on other teachers at participants' schools. Professional development initiatives based on voluntary participation of individual teachers like PFL are usually confronted with problems of realization and dissemination (see e.g., Krainer, 1998).

At my department, we decided to react to these results in three ways. Firstly, we motivated teachers to come in pairs or small groups from one school in order to form a critical mass; we assumed that this might be a better starting point for dissemination of participants' ideas and for bringing about change at their schools. Secondly, the PFL courses put an emphasis on the question of what participants can do during and after the courses in order to get the principal informed about their activities and to get other mathematics teachers involved in activities aiming at improving mathematics teaching at their school. Thus we started to think about strategies in order to give the courses more sustainability (see also Jaworski, and Lerman & Zehetmeier in Volume 3 of this Handbook). Thirdly, we looked for an additional approach that goes beyond offering professional development programmes for individual teachers: a few colleagues at my department attended courses for organizational development and we read relevant literature. The goal was to understand better how schools learn during a whole *school development* process when getting external support.

The experiences with two schools showed us that organizational development has a great potential; however, there was also a tendency that teachers focused on various important issues (school climate, parents-teacher collaboration, corporate identity etc.), but rarely touched the kernel process of schooling, namely students' learning of mathematics (and other subjects). There was a tendency by teachers and schools to non-subject-related issues. This motivated us to regard *working with mathematics departments* as an adequate way: here the subject is the core focus, nevertheless, issues of school development like the interconnection to other subjects or the support by the administration can be raised and worked on.

Our experiences showed that the presence of a principal interested in the development of the mathematics department and at least one highly competent and innovative teacher in this department (e.g., a PFL graduate) were important factors that influenced positively the external support we supplied. We realized that – from a systemic point of view – teachers and schools need to be supported by a combination of at least three measures: through educating teacher leaders with a focus on pedagogical content knowledge, through educating teacher leaders with a focus on content-related school development, and through supporting whole departments and schools in their efforts in improving teaching in mathematics etc.

During this period of putting a focus on the development of teams, departments and schools (mainly in the 1990s), it became clear to me that we cannot reduce the quality of education simply to the quality of teaching, but to see the teachers' contribution to the quality of education in a broader context. I suggested to consider four dimensions of teachers' professional practice: *action and reflection, autonomy and networking* (see e.g., Krainer, 1998). These dimensions can be used for reflecting on mathematics teachers' work at different levels: with regard to their *own further development*, their *students' further development*, and their *school's further development* (e.g., dealing with questions such as: Is there efficient communication among the mathematics teachers in the school? Is there a fruitful collaboration between mathematics teachers and teachers of other subjects? Is mathematics seen as an important learning field at their school? Does the working climate promote innovations in classrooms?) In addition, it can be used to reflect on the *further development of the teaching profession, the education system and its interaction with the society as a whole,* for example by asking: Which role can mathematics, science and technology play in our society and which consequences does this role have for further developing mathematics teaching? What kind of influence do teachers have on regulations (curriculum, assessment, etc.), on standards, or on the status of their profession? Are teachers' reflections on their profession seen as a relevant contribution to the education system? Is professional communication and collaboration among teachers promoted? Do we promote it? Where do the rewarding effects of closer collaboration between theoreticians and practitioners (universities and schools) lie?

These kinds of questions irritated some colleagues within the mathematics education community. They found the approach too general and too far away from mathematics. On the other hand, I was convinced that the challenge of improving mathematics teaching needs a broader view than relying on the initial and further education of individual mathematics teachers. Literature (e.g., Grouws & Schultz, 1996) supported this view.

To some extent, the four dimensions were in contrast to early models of explaining teachers' growth that were mainly taken or modified from psychology as Perry's development scheme used, for example, in Cooney and Shealy (1994). This led to various discussions with Tom Cooney (see e.g., Cooney & Krainer, 2000). These discussions about the notions of reflection and networking helped in further developing my thinking on teachers' learning. In particular, it confirmed my assumption that the individual and the social dimensions are closely interconnected.

In addition, I got involved into the launching of the Journal of Mathematics Teacher Education (JMTE) which started in 1998. This journal gave the field of research on mathematics teacher education a strong impulse, and I profited very much from reading and reviewing. It was not by chance that my editorials for JMTE focused on issues like "teacher growth and school development", "teams, communities and networks", "students' understanding", "good mathematics teaching" or "action research". All these contributions are based on the assumption that besides an emphasis on mathematics and activities by the individual learner, we also should look at the social dimension and should emphasize the role of reflection. I do not know exactly when the notion of "collective reflection" emerged for me, however, I am pretty sure that in was in the PFL context.

All in all, my professional development and school development experiences showed that working with teachers, both promoting them to reflect on their practice and reflecting on our practice as teacher educators are key features for teacher educators' learning. This attitude of being a learner convinced many teachers also to view themselves as learners from learners. Vice versa, the creative and dedicated efforts by many teachers to reflect and improve their practice motivated us to continue our efforts. If I had to reduce my experiences to one suggestion, I would say: The best initiative to promote teacher educators' growth is to involve them in challenging teacher education activities where teachers' and their own learning is a continuous process. You only get a deeper understanding of teachers if you support them in their efforts to change. Then, teacher education is always also a teacher educator education.

ME6 and ME7: Promoting and Investigating the Improvement of Mathematics and Science Teaching in Austria

Following the unsatisfactory results of Austrian high school students in the TIMSS achievement test (see e.g., Mullis et al., 1998; Krainer, 2003), the responsible Austrian ministry ordered an investigation into causes and possible actions for improvement. Based on our department's broad experiences with the PFL programme and other professional development and school development initiatives, the ministry entrusted us with the one-year research project IMST – Innovations in Mathematics and Science Teaching (1998-1999). The analyses showed a picture of a "fragmentary educational system" of lone fighters with a high level of (individual) autonomy and action, however, less reflection and networking. For example, it was astonishing to find how many creative initiatives individuals, groups or institutions carried out, and to realize that these initiatives remained more or less unlinked. This impression is repeated when looking at the whole educational system.

As a consequence of the analyses, the project $IMST^2$ (Innovations in Mathematics, Science and Technology Teaching, 2000–2004) at the upper secondary level was launched. Annually, about 50 projects at Austrian schools were supported. Teachers could apply for participation as individuals or as school teams. At system level, initiatives were started to make these innovations public, discussable and usable to other teachers; recommendations have been incorporated into the new national curriculum for upper secondary general education. The project finally led to IMST3 (2004–2009) which since 2004 includes all secondary schools and since 2007 covers all school levels. In IMST3, more than 150 projects at schools are supported nation-wide; in addition, thousands of teachers take part in meetings of so-called regional networks. Furthermore, IMST3 initiated the setting up of mathematics and science education research centres at regional and national level, the establishment of regional science teacher networks and gradual implementation of science, mathematics and German language subject-related educational managers. More about this national programme IMST can be found in the chapter by Pegg and Krainer in Volume 3 of this Handbook.

In this chapter, I want to focus only on my professional growth in the last eight years, starting with the beginning of the IMST research project 1998–99. To deal with the analysis of an international comparative study was totally new to me. It included discussions about the importance of such studies for school practice, educational research and educational policy. It meant building a wide network of people and institutions all over the country in order to reflect the goals, the implementation, the communication processes, the evaluation and the further development and modification of the programme. My experience within the PFL programme and with organizational development as well as my readings of studies on teacher education initiatives and reform initiatives were an essential basis for my IMST activities which are best described as a mixture of scientific work and the management of science.

In particular, the change from supporting teachers in a professional development programme or in a department to think about bringing about change in a whole country is an enormous one. Also evaluation and research have to be reconsidered (see also Jaworski in Volume 3 of this Handbook). In order to take adequate steps to overcome the "fragmentary educational system", a *systemic approach* to change was needed. It was based on the four dimensions sketched above and integrated systemic approaches to educational change and system theory (Fullan, 1993; Willke, 1999). The approach of a *"learning system"* (see e.g., Krainer, 2005) was taken. This meant adopting enhanced reflection and networking as the basic *intervention strategy* at manifold levels (with different individuals or organisations that practice reflection and networking), for example: in classrooms (students), at departments (mathematics and/or science teachers) and schools (involving also other teachers), in IMST seminars (teachers from different schools and regions), IMST boards and networks (teacher leaders, teacher educators, superintendents, people from various relevant environments as e.g., economy, media, parents association, etc.), mathematics and science education research centres.

From a content-related point of view, IMST tries to overcome the traditional boarders between didactics and school development (see e.g., Rauch & Kreis, 2007). The project aims at designing a *content-related school development* which has in mind, both the quality of teaching and that of the whole school as an organization: having good mathematics teaching in a classroom is one important thing to aim at; having a special school focus on mathematics is another important but different goal. Whereas in the former case the collaboration of mathematics teachers is sufficient, in the second all other teachers should also be involved and be open to collaborations and interdisciplinary work with mathematics teachers.

I learned from IMST that science teachers have quite other views of (the learning of) mathematics than mathematics teachers have, and the same is true for the (learning of) natural sciences. All schools in IMST which put an emphasis on implementing or further developing a mathematics and science branch, established interdisciplinary practical studies and laboratory teaching. However, often mathematics was either not fully integrated or even excluded. This corresponds to an IMST study which showed that science teachers ascribe school development a much higher role than mathematics teachers. What are the reasons for these phenomena? Does the high status of mathematics make teachers (partially) reserved against serving as a complementary science? Do mathematics teachers or mathematicians, more than others, want to stay among themselves? Are there traces of the great Greek history where only mathematical elite had entry to the temple? If we want to claim "mathematics for all" we need to open our doors. These experiences with IMST set me thinking about the image of mathematics and whether there is enough sense for the necessity of changing this image.

A lot of old and new questions arose when working in IMST, for example: What is *good* mathematics teaching? What *relevant* factors promote or hinder good teaching? What is *successful* mathematics teacher education? Telling what is understood by good, relevant or successful goes beyond describing and interpreting things. It is prescribing how things should be seen or expected, it means establishing a *norm*. Do we need such norms, and if so, for what purpose? Who should define these norms, for whom, and with what consequences? How diverse are the opinions? What role can research play? What *relationship between research and norms* should we aim at (thus asking a normative question)?

In particular the last question is decisive for the further development of our discipline. My current point of view is that we as researchers and educators should actively contribute to the *negotiation of norms*. This means not to refuse to think about norms, but also not to try to dominate the generation and dissemination of norms. We should always reflect critically whether our intervention goals (based on assumptions, theories, etc.) are viable. This has also consequences for our approach to research in teacher education. Since teacher education research is mostly done by the same people that are co-responsible for the intervention (Adler, Ball, Krainer, Lin, & Novotná, 2005), this kind of "intervention research" has to balance an *interest in development* and an *interest in understanding* (e.g., investigating which factors had an impact on that development). Intervention research does not only apply knowledge that has been generated within the university, but much more, it generates "local knowledge" that could not be generated outside the practice. Thus this kind of research is mostly process-oriented and context-bounded, generated through continuous interaction and communication with practice. Intervention research tries to overcome the institutionalised division of labour between science and practice.

My experiences within IMST changed my views on mathematics teaching and mathematics education. Firstly, I regard not only mathematics teachers (and principals) as important change agents of mathematics teaching, but also other teachers, in particular science teachers. Secondly, I no longer regard the question of good mathematics teaching as a question limited to teachers, mathematics educators and researchers alone. I am convinced that we need more efforts as a discipline to discuss the future of mathematics with economics experts, educational policy makers, mathematicians, parents, etc. The more qualified opinions we get, the better. If we do not play an active role, other relevant environments – like economy and policy – will take the lead.

CONCLUSION

My teacher education practice and research started with supporting individual mathematics teachers and investigating their growth and the factors that supported or constrained it. Although most courses were designed in a way that teachers – and to some extent including the staff members – form a learning community, the collaboration only in a few cases brought about change at teachers' home schools.

My involvement in school development and national reform efforts underlined my conviction that *sustainable improvement* of mathematics teaching needs single active teachers as change agents; however, I also learned that one single mathematics teacher can at best change teaching in his or her own classes (Krainer, 2006). The assumption that an improvement of mathematics teaching in all classes can be achieved only by trusting the professional development of all individual teachers, centrally organised curricula changes, or formulating standards, is naïve. In my view, educators and representatives of educational policy all over the world are still trusting too much in this kind of top-down efforts. Without communication and collaboration among mathematics teachers we will have no improvement on a larger scale. Only believing in the growth of individual teachers – and generally always the same teachers come to professional development courses – is also too narrow. A group of mathematics teachers or a whole mathematics department, given all people share to a large extent the same vision of mathematics teaching, has the potential to change more. However, if they cannot communicate the importance of mathematics to other teachers (in particular to science teachers who are, in general, more likely to be interested in mathematical thinking), the principal, the parents and other stakeholders, the impact will be limited. A principal who hates mathematics or even a hostile "school climate" against mathematics can hinder or ruin the best competences, attitudes and efforts by mathematics teachers. After a school, the next organizational level is that of the *district*. Schools should share their experiences with improving mathematics teaching at the district level, and, very important, make their efforts, successes but also problems public and thus discussable. The same is true for the next levels – let us name them *regions* and *nations* (see also the contributions by Cobb & Smith, Nickerson, and Pegg & Krainer in Volume 3 of this Handbook). If the status of mathematics at schools and in nation or society is low, it is a hard job for mathematics teachers to preach the importance, applicability etc. of this wonderful science. Thus, although mathematical competences of mathematics teachers are a very important precondition for good mathematics teaching, also other competences are highly relevant.

One competence deals with the ability to *communicate the power and beauty of mathematics to other people*, in particular to students and to teachers of other subjects. It is important to involve the public in reflections about mathematics. Something that is not discussed in the public is not considered as important, particularly nowadays living in an information society. This is true for the media, but also true for schools. If mathematics is not visible at a school (in exhibitions, conferences, competitions etc.), it is not in the centre of school life. If teachers of other subjects get a better understanding of mathematics, and they possibly lose part of their fears, and maybe realize some common issues between their subject and mathematics, it will be much easier to raise the reputation of mathematics at a school.

I totally agree with Alan Schoenfeld when he argued at the Conference "The Future of Mathematics Education in Europe" in Lisbon (Portugal, December 2007) that *mathematics has an image problem*. We – mathematics educators and mathematicians – are too sure that it is totally evident to all people that mathematics is important. I think we should try to communicate the advantages and benefits of mathematics better to the public. I am convinced that mathematics is important, thus the image problem has a *communication problem* in its kernel, but also a *quality problem*. Too many students (at schools, universities etc.) do not get the picture of an enjoyable, fascinating and vivid field of activities that helps them to become freer. In contrast, too many students fear mathematics and find it boring and like a ready-made and even dead body; sometimes students even see mathematics as a discipline where they experience rather limitations than opportunities. Changing this situation affords the improvement of mathematics teaching and making innovations public. Trusting only on top-down and on individual teachers' change is not a sustainable solution. It needs the *efforts of all* and it needs a lot of *reflection and networking*.

To some extent, it seems that my journey in mathematics teacher education brought me more and more away from its kernel, the mathematics and its teaching. However, I have never had the feeling of being so close to contributing to the change of mathematics teaching and its image on a larger scale.

ACKNOWLEDGEMENTS

I gratefully thank Sandy Dawson, Barbara Jaworski, and Dagmar Zois for their very helpful comments on an earlier draft of this chapter.

REFERENCES

Adler, J., Ball, D., Krainer, K., Lin, F.-L., & Novotná, J. (2005). Mirror images of an emerging field: Researching mathematics teacher education. *Educational Studies in Mathematics, 60*, 359–381.

Altrichter, H., Posch P., & Somekh, B. (1993). *Teachers investigate their work. An introduction to the methods of action research.* London: Routledge.

Ball, D., & Even, R. (Eds.). (in preparation). *The International Commission on Mathematical Instruction (ICMI) – The Fifteenth ICMI Study: The professional education and development of teachers of mathematics.*

Bender, P., & Schreiber, A. (1985). *Operative Genese der Geometrie* [Operative genesis of geometry]. Wien: Hölder-Pichler-Tempsky.

Bishop, A., Clements, K., Keitel, C., Kilpatrick, J., & Laborde, C. (Eds.). (1996). *International Handbook of Mathematics Education.* Dordrecht, the Netherlands: Kluwer.

Cobb, P., & Bauersfeld, H. (Eds.). (1995). *The emergence of mathematical meaning: Interaction in classroom cultures.* Hillsdale, NJ: Lawrence Erlbaum.

Cooney, T. (1994). Research and teacher education. In search of common ground developed. *Journal for Research in Mathematics Education, 25,* 608–636.

Cooney, T., & Krainer, K. (2000). The notion of networking: The fusion of the public and private components of teachers' professional development. In M. Fernández (Ed.), *Proceedings of the 22nd Annual Meeting of the North American Chapter of the International Group for the Psychology of Mathematics Education (PME-NA 22)* (Vol. 2, pp. 531–537). Columbus, OH: ERIC.

Cooney, T., & Shealy, B. (1994). Conceptualizing teacher education as field of inquiry: Theoretical and practical implications. In J. Ponte & F. Matos (Eds.), *Proceedings of the 18th PME International Conference* (Vol. 2, pp. 225–232). Lisbon, Portugal: University of Lisbon.

Dörfler, W. (1987). Die Genese mathematischer Objekte und Operationen aus Handlungen als kognitive Konstruktion [The genesis of mathematical objects and operations from actions as cognitive constructions]. In W. Dörfler & R. Fischer (Eds.), *Kognitive Aspekte mathematischer Begriffsentwicklung* (pp. 55–126). Vienna: Hölder-Pichler-Tempsky.

Fischer, R., Krainer, K., Malle, G., Posch, P., Zenkl, M. (Eds.). (1985). *Pädagogik und Fachdidaktik für Mathematiklehrer* [Pedagogy and subject-specific methodology for mathematics teachers]. Vienna: Hölder-Pichler-Tempsky.

Fischer, R., & Malle, G. (1985). *Mensch und Mathematik: Eine Einführung in didaktisches Denken und Handeln* [Man and mathematics: An introduction to didactical thinking and doing]. Mannheim, Germany: Bibliographisches Institut.

Freudenthal, H. (1983). *Didactical phenomenology of mathematical structures.* Dordrecht, the Netherlands: Reidel.

Fullan, M. (1993). *Change forces. Probing the depths of educational reform.* London: Falmer Press.

Grouws, D., & Schultz, K. (1966). Mathematics teacher education. In J. Sikula (Ed.), *Handbook of research on teacher education* (pp. 442–458). New York: Macmillan.

Jaworski, B. (1999). What does it mean to promote development in teaching. A response to Ron Tzur's paper: Becoming a mathematics teacher-educator: conceptualising the terrain through self-reflective analysis. In O. Zaslavsky (Ed.), *Proceedings of the 23rd PME International Conference* (Vol. 1, pp. 183–193). Haifa, Israel: Israel Institute of Technology.

Jaworski, B., Wood, T., & Dawson, S. (1999). (Eds.). *Mathematics teacher education. Critical international perspectives.* London: Falmer Press.

Krainer, K. (1993). Powerful tasks: A contribution to a high level of acting and reflecting in mathematics instruction. *Educational Studies in Mathematics, 24,* 65–93.

Krainer, K. (1994). PFL-Mathematics: A teacher in-service education course as a contribution to the improvement of professional practice in mathematics instruction. In P. Ponte & J. Matos (Eds.), *Proceedings of the 18th International Conference for the Psychology of Mathematics Education* (Vol. 3, pp. 104–111). Lisbon, Portugal: University of Lisbon.

Krainer, K. (1998). Some considerations on problems and perspectives of mathematics teacher in-service education. In C. Alsina, J. M. Alvarez, B. Hodgson, C. Laborde, & A. Perez (Eds.), *The 8th International Congress on Mathematical Education (ICME 8)* Selected Lectures (pp. 303–321). Sevilla, Spain: S.A.E.M. Thales.

Krainer, K. (1999). Promoting reflection and networking as an intervention strategy in professional development programs for mathematics teachers and mathematics teacher educators. In O. Zaslavsky (Ed.), *Proceedings of the 23rd PME Conference* (Vol. 1, pp. 159–168). Haifa, Israel: Technion – Israel Institute of Technology.

Krainer, K. (2003). Innovations in mathematics, science and technology teaching (IMST2): Initial outcome of a nation-wide initiative for upper secondary schools in Austria. *Mathematics Education Review, 16,* 49–60.

Krainer, K. (2005). Pupils, teachers and schools as mathematics learners. In C. Kynigos (Ed.), *Mathematics education as a field of research in the knowledge society. Proceedings of the First GARME Conference* (pp. 34–51). Athens, Greece: Hellenic Letters.

Krainer, K. (2006). How can schools put mathematics in their centre? Improvement = content + community + context. In J. Novotná, H. Moraová, M. Krátká, & N. Stehlíková (Eds.), *Proceedings of the 30th Conference of the International Group for the Psychology of Mathematics Education* (Vol. 1, pp. 84–89). Prague, Czech Republic: Charles University.

Krainer, K. (2007). Mathematics teacher education: A practical field and an emerging field of research. In Mathematisches Forschungsinstitut Oberwolfach (Ed.), *Professional development of mathematics teachers – Research and practice from an international perspective* (pp. 43-45). Proceedings of a conference held at the MFO, 11–17 Nov. 2007, Report No. 52/2007. See http://www.mfo.de/cgi-bin/path?cgi-bin/tagung_espe?type=21&tnr=0746

Krainer, K., Goffree, F., & Berger, P. (Eds.). (1999). *On research in mathematics teacher education. European research in mathematics education I.III*. Osnabrück, Germany: Forschungsinstitut für Mathematikdidaktik.

Lin, F.-L., & Cooney, T. (Eds.). (2001). *Making sense of mathematics teacher education*. Dordrecht, the Netherlands: Kluwer.

Llinares, S., & Krainer, K. (2006). Mathematics (student) teachers and teacher educators as learners. In A. Gutiérrez & P. Boero (Eds.), *Handbook of research on the psychology of mathematics education. Past, present and future* (pp. 429–459). Rotterdam, the Netherlands: Sense Publishers.

Loucks-Horsley, S., Hewson, P. W., Love, N., & Stiles, K. E. (1998). *Designing professional development for teachers of science and mathematics*. Thousand Oaks, CA: Corwin Press.

Mullis, I. V. S., Martin, M. O., Beaton, A. E., Gonzalez, E. J., Kelly, D. J., & Smith, T. A. (1998). *Mathematics and science achievement in the final year of secondary school: IEA's Third International Mathematics and Science Study*. Boston: Center for the Study of Testing, Evaluation, and Educational Policy, Boston College.

Peter-Koop, A., Begg, A., Breen, C., & Santos-Wagner, V. (Eds.). (2003). *Collaboration in teacher education. Examples from the context of mathematics education*. Dordrecht, the Netherlands: Kluwer.

Rauch, F., & Kreis, I. (Eds.). (2007). *Lernen durch fachbezogene Schulentwicklung*. Innsbruck, Austria: Studienverlag.

Schön, D. A. (1983). *The reflective practitioner*. London: Temple Smith.

Schuster, A. (2008). *Warum Lehrerinnen und Lehrer schreiben [Why teachers write]*. Doctoral thesis. Klagenfurt, Austria: University of Klagenfurt.

Strässer, R., Brandell, G., Grevholm, B., & Helenius, O. (Eds.). (2004). *Educating for the future. Proceedings of an International Research Symposium on Mathematics Teachers Education*. Goeteborg, Sweden: NCM, Goeteborg University.

Tzur, R. (1999). Becoming a mathematics teacher-educator: conceptualizing the terrain through self-reflective analysis. In O. Zaslavsky (Ed.), *Proceedings of the 23rd PME International Conference* (Vol. 1, pp. 169-182). Haifa, Israel: Israel Institute of Technology.

Tzur, R. (2001). Becoming a mathematics teacher-educator. Conceptualizing the terrain through self-reflective analysis. *Journal of Mathematics Teacher Education, 4*, 259–283.

Wagenschein, M. (1970). *Ursprüngliches Verstehen und exaktes Denken [Original understanding and precise thinking]*. Vol. 2. Stuttgart, Germany: Klett.

Willke, H. (1999). *Systemtheorie II: Interventionstheorie* [System theory II: Intervention theory]. Stuttgart, Germany: Lucius & Lucius UTB.

Wittmann, E. Ch. (1995). Mathematics education as a "Design Science". *Educational Studies in Mathematics, 29*, 355–374.

Zaslavsky, O., & Leikin, R. (1999). Interweaving the training of Mathematics teacher-educators and the professional development of mathematics teachers. In O. Zaslavsky (Ed.), *Proceedings of the 23rd PME International Conference* (Vol. 1, pp. 141–158). Haifa, Israel: Israel Institute of Technology.

Zaslavsky, O., & Leikin, R. (2004). Professional development of mathematics teachers educators: Growth through practice. *Journal of Mathematics Teacher Education, 7*, 5–32.

Konrad Krainer
Institut für Unterrichts- und Schulentwicklung
University of Klagenfurt
Austria

SIMON GOODCHILD

10. A QUEST FOR 'GOOD' RESEARCH

The Mathematics Teacher Educator as Practitioner Researcher in a Community of Inquiry

The chapter begins by outlining features of mathematics that contribute to the challenges of teaching and learning the subject. These challenges lead the author into practitioner research, first as mathematics teacher and later as mathematics teacher educator. The development of the author as researcher through different types of research activity is then described. This culminates in developmental research in which the author is presently engaged; it is a project, Learning Communities in Mathematics, which is intended to develop and research better teaching and learning of mathematics through the establishment of communities of inquiry and co-learning partnerships with teachers. Notions fundamental to the project are introduced and examined, such as 'inquiry as a way of being' and 'critical alignment'. Finally the author examines the project to substantiate his assertion that it is 'good research'.

INTRODUCTION

The professional 'quest' that I describe is driven by a desire to do things 'better': to enable my students to learn mathematics better; to enable better teaching within the mathematics department I led; to do better research. In the latter part of this chapter I will explore the notion of 'good' research but I have not sought after 'good' research as if this were an attainable goal. Throughout my quest, if the desire to do things better is the driving force then the inspirational energy is provided by the special nature of mathematics. The chapter starts by offering an explanation of the special nature and how it led me to research in mathematics education. The chapter then proceeds to explain how my approach to research evolved after putting on the mantle of 'mathematics educator-researcher'. I describe some approaches to research and briefly comment on how these contributed to the development of my own understanding of 'good' research. The major part of the chapter is then given over to a description of one major developmental research project in which I have most recently been engaged. I do not claim any responsibility for the conception or initial design of the project. However, it has offered the possibility for me to see, and experience, research in mathematics education that, I believe, can claim to be 'good'.

MATHEMATICS IS DIFFERENT!

> Mathematics is different, indeed, and we will take notice of this fact when we will turn to education. From the very beginning it has been different. (Freudenthal, 1991, p. 9)

Hans Freudenthal defends his bold assertion by drawing attention to how mathematics has developed in two ways: in its content (the substance of mathematics) and in its form (the way in which the substance is presented). I interpret Richard Skemp (1982) similarly to be referring to 'form' and 'content' when he writes about challenges to learning, understanding and teaching mathematics arising from, respectively, the 'surface' and 'deep structures' of mathematics. Skemp explains the challenges:

> What we are trying to communicate are the conceptual structures. How we communicate these, or try to, is by writing or speaking symbols. The first are what is most important. These form the deep structures of mathematics. But only the second can be transmitted and received. These form the surface structures. Even within our minds the surface structures are much more accessible, as the term implies. And to other people they are the only ones which are accessible at all. (1982, pp. 281–282)

Heinz Steinbring (1998) elaborates on this special characteristic of mathematics as a subject of study. He explains that the relationship between the symbol/sign and the idea it represents is not unambiguous and is 'subject to developments and changes' (p. 162) because the objects of mathematics are *ideas* existing within the person rather than objects external to the person, which can be shared and explored.

The mathematics teacher educator is challenged by this dual nature of mathematics; the challenges are both ontological and epistemological. If one asks where does mathematics 'exist'? One answer might be 'in texts and writing,' but in these we perceive the 'form' of mathematics. The 'content' of mathematics consists of concepts, patterns, relationships, generalisations and arguments, etc. The 'content' of mathematics lies in the meaning, which the text exists to communicate. The meaning is a product of the mind of the knower.[1] Mathematics is, indeed, 'a difficult subject both to teach and to learn' (Department for Education and Employment, 1982, p. 67).

BECOMING A MATHEMATICS TEACHER - PRACTITIONER-RESEARCHER

I cannot pinpoint in my own development when I became aware of this two-sided nature of mathematics; I do not think it emerged until after I had started teaching

---

[1] This reflects a constructivist account of mathematics, but the argument can be supported from socio-cultural theory. Vygotsky writes about spontaneous and scientific concepts (1986) and the process of internalisation (1978). There is no way of knowing the exact nature of another person's concepts except through the behaviour, i.e. expressions of 'form' which they display.

the subject. Nevertheless, I am sure it was the challenge of trying to support 'my' students in developing their own 'deep structures' which ignited and sustained a passion for teaching (and learning) mathematics, and later researching teaching and learning. After 9 years as a classroom teacher I became a head of a large mathematics department, and thereby responsible for the quality of mathematical education in the school. I led a team of 12 mathematics teachers and I began to be concerned that some highly competent and experienced teachers in my team were not, apparently, sensitised to the dual nature of mathematics. They appeared to concentrate on teaching the form of mathematics without providing explicit opportunities for students to develop their own concepts of the content of mathematics.[2] That is, they appeared to favour an approach to teaching and learning that promoted procedural (Hiebert, 1986) and instrumental (Skemp, 1976) knowing of the subject.

As a school head of department I tried to influence 'my' team of mathematics teachers through the scheme of work, departmental discussions and (agreed) curriculum developments. It was at this stage that research began to form part of my practice and I took the first steps as a practitioner-researcher.[3] One of the first pieces of research I undertook was to evaluate the implementation of new curriculum materials that I and my colleagues had introduced (Goodchild, 1987). This enticing experience of research and the wish to be able to make a greater influence in the practice of teaching mathematics led me to a job in teacher education.

THE MATHEMATICS TEACHER EDUCATOR AS RESEARCHER

One of the things that took me by surprise when I started as a mathematics teacher educator was the rich diversity of approaches that I believed could be characterised as successful mathematics teaching. I expressed this as follows:

> My model of the mathematics classroom had to be revised .... I observed student teachers working with pupils, apparently successfully, yet adopting approaches I would not have attempted in my own practice. More frequently I observed students using the approaches I believe to be 'tried and trusted' with, apparently, little success - or worse. (Goodchild, 2001, p. 9)

Teaching and learning are far more complex than I had imagined up to this point. Thus I was led to engage in classroom research. In Jon Wagner's terms I entered a data extraction agreement with a mathematics teacher, (Wagner, 1997). In a data extraction agreement 'school settings are regarded as a resource from which researchers extract knowledge for distribution to other communities and

---

[2] This begs a question about what 'explicit' opportunities for developing concepts of mathematics might look like. Briefly, I mean that students will be led to focus on reasons and relationships as well as rules of manipulating symbols.

[3] I must make it clear that I believe the reflective practice as realized in the cycle of lesson planning, implementation, evaluation and re-planning, which forms the part of every teacher's routine activity is research in every respect except that it is rarely published.

locales' (p. 15). I wanted to inquire into how students were engaging with both the content and form of mathematics by trying to expose the goals towards which they were working. I followed a class for every mathematics lesson for nearly one complete school year in an effort to explore what was going on in the students' experiences of mathematics. For this research the teacher and the class became for me an arena from which I gathered (extracted) data. My intention was to avoid as far as possible making any contribution to the activity of the class because that would 'contaminate' my data.

I did have conversations with students about the mathematics in which they were engaged. I believe this might have been beneficial for them but that was not my intention. Also, I shared the completed report with the class teacher, she may have gained something from reading it but her 'enlightenment' was not my main intention. I wanted to elicit her reaction and I hoped for her approval of what I had written about her class and practice. The research certainly had an impact on my own practice but I am not sure that it has had any impact beyond that, despite being reported in a published monograph (Goodchild, 2001).

The only negative critique I have experienced relating to this classroom research has been my own, perhaps this is an indicator of its lack of impact. However, as I reflect on what I did, I am led (as an 'educational' researcher) to question its ethics. Teachers and classes do not exist for my convenience; they do not exist to satisfy my curiosity or quest for knowledge. To treat them as such might be acceptable in terms of sociological or anthropological research but it is not educational research! As Wilfred Carr and Stephen Kemmis (1986) assert,

> An adequate view of educational science ... must develop theories of educational practice that are rooted in the concrete educational experiences and situations of practitioners and which enables them to confront the educational problems to which these experiences and situations give rise. (p. 215)[4]

It might be argued that data extraction research can be justified if it is likely to have a beneficial impact on teaching and learning in general, as in clinical trials of new drugs. Such justification is undermined by the widespread assertion that a great deal of mathematics (and other) education research does not make an impact. Lynn Arthur Steen (1999) for example reviews an ICMI study to which many prominent mathematics education researchers contributed and observes that those contributing to the study were largely agreed that: 'Research has had essentially no impact on the practice of mathematics education. On this broad indictment, researchers and their critics agree' (p. 240). Steen then quotes Alan Bishop from the ICMI study 'The lack of relationship between research and practice is well documented' (Bishop, 1998, p. 35; quoted by Steen, 1999, p. 240). The ICMI study

---

[4] I realise that I am on 'shaky' ground here. I do think there is a value in sociological and anthropological research into school classrooms. These would be data extraction agreements. But they are not 'educational research' even though the findings of such research may be of value to teachers and policy makers.

took place in 1994; I do not believe much has changed since then. Perhaps things have become worse as evidenced by 'the maths wars' in the US (Klein, 2007). Further, as Richard Pring (2004) observes referring especially to David Hargreaves (1996), the critique is applied to educational research in general.

My assertion that classroom research (my own field of specialism) on the basis of data extraction agreements might be ethically questionable also concerns the relationship between the researcher and teacher. It is likely, through the highly privileged position of observer, that the researcher might be able to facilitate the teacher's reflection on his or her own practice. Barbara Jaworski (1994, 1998) was able to do this effectively. In my research I did not want to do this and in any case I sensed that such intervention would not have been welcomed by the teacher. It was not part of the agreement into which we had entered. Nevertheless, a question that sometimes confronts the classroom researcher is: what should I do when confronted by examples of 'bad practice'?

Researchers do not gain access to classrooms with the intention of 'improving' the practice therein. The researcher enters the classroom to collect data in an objective and non-judgemental manner. However, many educational researchers, like me, have extensive experience of working in a variety of classrooms: as teacher and teacher educator; as informed practitioner and observer. The 'data extraction' researcher is in a difficult position if confronted by practice which is believed to be detrimental to students' learning and development. To engage the teacher in discussion might jeopardise the agreement. To report episodes that could reflect badly on the teacher could result in the researcher (and other researchers) losing access to classrooms. Teachers will not give their consent to researchers' presence if they believe it is possible their practice will be published and criticised. Most of all making judgements of practice assumes a normative dimension which is counter to the objectivity claimed by the researcher. Nevertheless, starting with myself, I believe that everyone has the potential to improve in their professional practice, and I really do want the research in which I engage to make things better.

Experience as a mathematics teacher educator has led me to some beliefs about learning. Students learn mathematics by engaging in mathematical activities, not merely being 'told about' mathematics. 'Engaging' entails activity with the content of mathematics. Merely 'being told' may leave students at the surface level, conscious only of the form of mathematics. Students learn to be teachers of mathematics by engaging in the practice of teaching mathematics, not merely being 'told about' educational theory. I do believe there is a place for good clear explanations of mathematics, and presentations of educational theory. Nevertheless, understanding of theory, and the competencies of the mathematician or teacher are developed through the practice, and informed reflection on practice. Therefore, I believe that if teachers are to develop, and again I include myself in this, it will be through systematic attempts to reform what we do. By this I mean engaging in some form of action research. Practitioner-research for me means working alongside teachers in action research.

ACTION RESEARCH

I have found action research effective in developing my own practice, and it avoids the ethical objections of data extraction. Action research, as a form of teacher-research has developed over several decades. It is a cyclical process of planning, acting and reflecting which teachers pursue to work on an element of their own practice. As an example I will describe briefly an episode from my own practice (Goodchild, 1996). I was dissatisfied with students' engagement in a course on elementary number theory that I taught for several years. I believed that a problem solving approach might lead students to reflect more deeply on the 'content' rather than the 'form', which they experienced from my lessons. So I planned an intervention, I gave them a rather difficult text on irrational numbers (Hardy & Wright, 1960) and asked them to read and try to understand this in sections prior to discussions in class. The 'problem' I believed the students would face was to understand the text which was deep in meaning and use of symbols. Bob Burn, a professional colleague and trusted friend, suggested that in this I was possibly 'adopting a sadistic and tyrannical posture which was intended to generate incomprehension and distress' (personal communication, March 7, 1996). The outcome was rather unexpected and surprising: the students told me, very courteously, that they thought they would be able to discuss and work on the text more freely if I were not present! They explained that they felt I was judging and evaluating everything they uttered and that they would feel more free to voice their misunderstandings and share their insights if they did not feel they were being 'assessed' by an expert. In that the implemented activity led students to work on the content rather than the form of mathematics it was a success. Nevertheless, the students' response and Burn's critique made a great impact on me. The episode led me to reflect on my practice, not just in this number theory course but more generally and the knowledge I gained fed back to reform my regular practice.

Often teachers are stimulated to engage in action research through university continuing professional development programmes (e.g. Elliott, & Sarland, 1995; Zack, Mousley, & Breen, 1997). Governmental initiatives have also created opportunities for teachers' action research, for example the 'Best Practice Research Scholarships' that were supported for a three year period by the Department for Education and Skills in England (Furlong & Salisbury, 2005). I believe my own experience of action research has been of value and effective within my own practice. I will contend that one ethical approach to educational research is to encourage teachers to conduct their own action research projects. Nevertheless, Burn had serious concerns that my 'problem-centred' approach described above was not ethical. Even with action research it is easy to cross the boundary of what is acceptable. However, action research can be a rather lonely occupation.

In Japan a form of collaborative action research, known as 'lesson study' has been a feature of teachers' professional development for many decades. Clea Fernandez and Makoto Yoshida (2004) describe lesson study as consisting 'of the study or examination of teaching practice

> ... [it is] a well-defined process that involves discussing lessons that they [teachers] have first planned and observed together. Study lessons are "studied" ... in an attempt to explore a research goal that the teachers have chosen to work on (e.g., understanding how to encourage students to be autonomous learners). (p. 7)

Stigler and Hiebert (1999) credit lesson study for the developments in mathematics teaching that have been sustained in Japan over the last 40 to 50 years. Lesson study involves a group of teachers in a school who meet first to collaboratively plan the 'study lesson'. One member of the group teaches the lesson whilst the others observe. The group then meets again to discuss the lesson. This might be followed by a revised lesson plan which is implemented by another member of the group (i.e., not the teacher who first implemented the lesson), again observed by the others. The group shares reflections and often produces a report, which might be made available to teachers in other schools through local bookshops. Lesson study is strongly recommended by the Japanese government and local education authorities but it is not compulsory. However Stigler and Hiebert observe:

> By all indications, lesson study is extremely popular and highly valued by Japanese teachers, especially at the elementary school level. It is the linchpin of the improvement process. One elementary school teacher interviewed ... remarked "You won't find a school without research lessons". (1999, p. 111)

According to Fernandez and Yoshida (2004) lesson study is found in 'the vast majority of elementary schools and many middle schools' (pp. 15–16). However, lesson study is not common in high schools; Fernandez and Yoshida suggest this is due in part to the focus on preparing students for college entrance examinations. At the elementary level, outsiders, such as 'curriculum or subject specialists, university professors' (a school principal quoted by Fernandez and Yoshida, p. 16) are sometimes invited to share in the process. I have not had the opportunity, as a teacher, to participate in a lesson study team but I can see it as combining collaborative inquiry with action research. The question then arises, for me, where does the mathematics teacher educator-researcher fit in, other than at the periphery as mentioned above?

A variation of lesson study that originates in the developmental research of a university team is 'learning study' reported by Mun Ling Lo and colleagues, who write:

> The learning study differs from the Japanese lesson study model in that our research lessons are based on a theoretical framework of learning ... and we wish to find out how well the theory has worked. In this way the learning study is a learning study not in two, but in three senses, as researchers are supposed to learn from it as well. (Lo, Marton, Pang, & Pong, 2004, p. 193)

Learning study is a form of developmental research in which the researcher is also concerned with the coordination of development and research into the developmental process. Thus it differs from action research, although action research cycles may be embedded within developmental research.

DEVELOPMENTAL RESEARCH

An early explanation of developmental research within mathematics education is provided by Koeno Gravemeijer (1994). Jan van den Akker (1999) provides a broader discussion of the rationale, basic principles and methods of developmental research but I will base most of my remarks here on Gravemeijer's paper. Gravemeijer's explanation of developmental research rests on a number of closely connected cyclical processes, or feedback loops. Fundamentally there is a cyclical process between development and research. Theory and evidence from prior research leads to an envisaging of development, this leads to actions which are evaluated and fed back into a new cycle of envisaging and action. I will refer to this as the 'developmental research cycle.' The developmental process is presented as a cycle between thought experiment and practical experiment; that is thinking through the consequences of some action, guided by theory, then implementing the action which leads to the adjustment of the theory that led the action, and so on. I will refer to this as the 'development cycle.' The research process, too, is presented as a cycle between global theories that are 'concretized in local theories,' which are tried out in practice, analysed and lead to reconstruction of the global theory. This I will refer to as the 'research cycle'. The research cycle guides the development cycle, which in turn nurtures the research cycle. I interpret and summarise Gravemeijer's discussion in Figure 1.

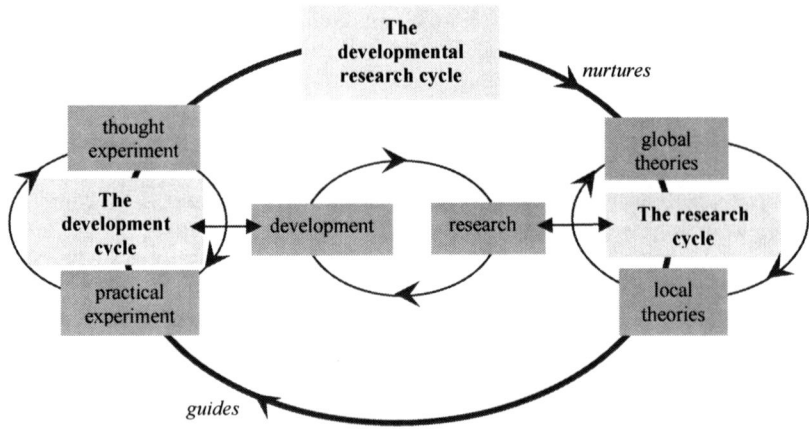

Figure 1. The Developmental Research Cycle

In Gravemeijer's explanation the detailed reporting of the cyclical processes is a key outcome of developmental research. He argues that knowledge of the developmental process is essential if the outcome of the process is to be understood, and refers to Freudenthal's assertion:

A QUEST FOR 'GOOD' RESEARCH

> ... developmental research means: experiencing the cyclic process of development and research so consciously, and reporting on it so candidly that it justifies itself, and that this experience can be transmitted to others to become like their own experience. (Freudenthal, 1991, p. 161)

An important point here is to note that the transmission of experience is not a simple matter. The 'candid reporting' is essential not merely for the benefit of informing a reader, but rather that readers of the report are sufficiently informed so that they can work on their own practice and thus create their own experiences. Inevitably, as similar developmental processes are pursued in different sites so experiences will vary. However, the candid reporting should also facilitate reflection on the variation that arises.

Jaworski and I have summarised developmental research as

> research which both studies the developmental process and, simultaneously, promotes development through engagement and questioning. ... Not only are research questions defined and explored ... but the whole research process is subject to question and exploration. We look critically at our research activity while engaging in and with it. (Jaworski & Goodchild, 2006, p. 353).

For the past three years I have been working with Jaworski and 10 other didacticians[5] in a developmental research project in Norway called 'Learning Communities in Mathematics' (LCM). This project has led me into research partnerships with teachers and other didacticians. It has provided a welcome opportunity to collaborate with teams of teachers in the style of Japanese lesson study and learning study as described above. The remainder of this chapter will be based upon experiences gained within this project. It will become clear, that the cycles of development and research described by Gravemeijer are central to the work of LCM. Both Gravemeijer and LCM are fundamentally concerned with educational research but whereas Gravemeijer focuses on curriculum development in terms of the development of resources for- and approaches to teaching, LCM takes teaching and learning mathematics as the focus.

Inquiry communities

The LCM project has opened up possibilities to develop the underlying theory of communities of inquiry and the brief outline given here is based on several published papers that describe this development (Jaworski, 2003, 2004a, 2004b, 2005, 2006a, 2007; Jaworski & Goodchild, 2006).[6] The present outline cannot, and

---

[5] In the project the university 'researchers' are referred to as 'didacticians' because teachers are also held to be researchers.
[6] The long list of references to Jaworski's writings is purposeful in this context. The outline of Gravemeijer's account of developmental research has drawn attention to the research cycle in which the research cycle is nurtured by the development cycle. The papers by Jaworski reveal a steady development of theory that has progressed alongside (symbiotically with) the LCM project.

is not intended to, provide a fully reasoned account of the theory and the interested reader is referred to the aforementioned references.

Developmental research is established on 'a global a priori theory ... (which) guides developmental work' (Gravemeijer, 1994, p. 449). In LCM inquiry is taken as 'a fundamental theoretical principle and position' (Jaworski, 2006a, p. 187). *Inquiry* is seen as both a developmental and learning 'tool' and a means to 'developing *inquiry as a way of being*' (ibid. p. 187 italics in original). The latter expression being derived, in part, from Marilyn Cochran Smith and Susan Lytle's notion of 'inquiry as stance' which they use 'to describe the positions teachers and others who work together in inquiry communities toward knowledge and its relationships to practice. ... the metaphor is intended to capture the ways we stand, the ways we see, and the lenses we see through' (Cochran Smith & Lytle, 1999, p. 288). In LCM there are three strands to inquiry: 'inquiry in mathematics', 'inquiry in teaching mathematics' and 'inquiry in developing the teaching of mathematics' (Jaworski, 2005, p. 103).

A second principle in LCM, explicit in the above, is that of 'community'. There is a substantial literature about 'inquiry communities' (e.g., Goos, 2004; Schoenfeld, 1996; Wells, 1999) and much of this refers to Etienne Wenger's development of the concept of a 'community of practice' (Wenger, 1998). Jaworski draws attention to Wenger's account of the formation of identity through a community of practice. 'Practice' according to Wenger 'is the source of coherence of a community' and he describes three 'dimensions of practice': 'mutual engagement', 'a joint enterprise' and 'a shared repertoire' (Wenger, 1998, pp. 72–73). Teachers for example will form their identity through engagement in the practice of teaching, didacticians through the practice of 'didacting' (Jaworski, 2007). Wenger also writes about alignment as a means through which 'we become part of something big because we do what it takes to play our part' (Wenger, 1998, p. 179). Jaworski then extends the notion to

> critical alignment in which it is possible for participants to align with aspects of practice while critically questioning roles and purposes as a part of their participation for ongoing regeneration of the practice. (2006a, p. 190)

In brief summary, the LCM project recognises the existence of communities of students, teachers in their schools, and the community of didacticians in the university. These communities gain their coherence through the practices within which each engage: learning mathematics, teaching mathematics, and developing the teaching of mathematics. It further seeks to form a community of teachers and didacticians (and students) in which members are critically aligned to the practices in which they are engaged. Critical alignment as operationalized in the process of 'critically questioning,' is a product of 'inquiry as a way of being,' this, as I have explained above, is taken as the fundamental principle of learning and development.

The activities of LCM, which function in both developmental and research roles include:
- workshops bringing teachers from 8 schools and didacticians together;

- didacticians, occasionally joined by teachers, planning workshops,
- didacticians meeting to discuss the progress of research and development within the project;
- teachers meeting within their own school teams occasionally joined by one or more didacticians. (It was a requirement for any 'school' to participate in the project that at least three teachers be involved and that the school principal support the activity);
- didacticians visiting mathematics classes and often making video films of lessons;
- teachers and didacticians discussing the video recordings of lessons;
- teachers engaged in action research.

These activities are underpinned by adherence to an 'inquiry cycle' of plan-act-observe-reflect-feedback-plan, and so on. In addition there are activities that appear more directly related to a research focus of the project, these include longitudinal mathematics tests, given to pupils in grades 4, 7, 9, and 11 on two occasions (May and September) each year, and focus group interviews with teacher teams in each school. As the focus group interviews inevitably lead teachers to reflect on their activity within the project it can reasonably be claimed that these interviews also have a developmental role. Similarly the students' responses to the longitudinal test questions are also used to stimulate inquiry and thus have a development role.

It is possible to relate LCM more explicitly to the foregoing account of developmental research. First, the research cycle is evident as the global principle of inquiry is realised in local theories. These include the use of inquiry style mathematics tasks in small group activity within workshops as a means of community building, and teachers engaging in the inquiry cycle as they plan and implement activities in their classes. The development cycle is evident as didacticians plan, implement, and reflect upon workshops and then feed back knowledge gained into fresh planning cycles. The inquiry cycle, which is a realisation of the global theory, is also a realisation of the development cycle described by Gravemeijer.

CO-LEARNING AGREEMENT

An immediate consequence of taking inquiry as a fundamental guiding principle is that all participants in LCM are engaged in inquiry processes, all participants are researchers. It is for this reason that the title 'didactician' is used for those attached to the university rather than the more usual 'researcher'. This draws attention to the fact that teacher participants and didacticians are partners in the research activity. Wagner (1997) refers to this as a 'co-learning agreement'. Both didacticians and teachers acknowledge that their intention is to learn about the process from their collaboration in development and research.

Thus in LCM three overarching research questions are addressed:

- How can teachers and pupils create a mathematical environment in their classrooms with suitable opportunity for pupils to learn mathematics with understanding and fluency?

- How can didacticians and teachers create a didactical environment in their interactive space (in schools and college) with suitable opportunity for teachers to develop mathematics teaching with understanding and fluency?

- How can ... didacticians create a (supra-didactical?) environment ... with suitable opportunity for didacticians to learn (didacting?) with understanding and fluency? (Jaworski, 2006b, p. 6)

Contingent on the co-learning agreement is that teachers should feel that they share the ownership of the project and have a contribution to make in the direction and development of the project. Inevitably there is a division of labour. Teachers are responsible for their classes and cannot hand this responsibility over to didacticians. Didacticians must accept responsibility for organising workshops because teachers do not have time to do this. Furthermore, teachers and didacticians bring different expertise and knowledge to the collaboration and thus are able to make different contributions. However, the differences are recognised to be differences in 'kind' and didacticians who steer the project are cautious to ensure that these differences are not perceived as (or develop into) differences in 'power' or 'authority'.

The preservation of the co-learning agreement has sometimes been pursued, consciously, in a way that appears contrary to the project's design. For example, the regular meeting of teachers within their school teams is seen as an important part of the development process, in terms of community development and inquiry development. At one stage it became apparent that teachers in some schools were not meeting as intended but the didacticians resisted suggestions to try to force the issue. It was probably recognised that teachers could not be 'forced' in any case. On the other hand, the didacticians' reluctance to impose demands on teachers has been the source of some criticism from the teachers. This is epitomised in the statement by one teacher, at the end of the first year of the project, 'In many ways we feel that we have done a lot for you, but maybe not got enough in return. We have had the workshops which have been interesting, but we would like you to come up with some suggestions.'[7]

The co-learning agreement and aim for teachers and didacticians to share ownership of the project are both consequences of the fundamental principle of developmental research in communities of inquiry. They are embedded in the developmental research cycle, as the research cycle guides, and is nurtured by, the development cycle. I want to argue here that developmental research in communities of inquiry is 'good' research.

"GOOD" EDUCATIONAL RESEARCH

Karl Hostetler (2005) draws attention to the possibility that the notion of 'good' research can be considered in two ways. Often research is evaluated as a 'methodological question', Hostetler argues that research should also be evaluated

---

[7] Extract transcribed from audio recording of the meeting.

in an ethical sense, he proposes: 'good research requires our careful, ongoing attention to questions of human well-being' (p. 16). Later he remarks, 'Good research is a matter not only of sound procedures but also of beneficial aims and results' (p. 17). This goes beyond the 'usual' ethical consideration of risk of harm to research participants, to considerations that the research will be positively for the 'good' of participants.

The fact that LCM has the goal of improving teaching and learning mathematics is not sufficient to provide an adequate response to the challenge. Any proposal for teachers to develop their practice involves a risk. Teachers may be challenged to work outside their own sphere of professional competence as they try out new and unfamiliar approaches. Pupils will be challenged to adjust to different ways of learning. For the teachers this poses a risk to their professional status as competent practitioners, and for their students it may pose a risk with regards to their performance in examinations that have a crucial bearing on their future. Theoretically the risks might be justified on the basis of decades of research evidence that demonstrates the value of inquiry approaches to teaching and learning. However, such justification is contested as for instance in the 'maths wars' in the United States, (e.g. Becker & Jacob, 1998; Schoenfeld, 2004; Klein, 2007). My claim that LCM is positively for the 'good' of participants is based on the principle of 'critical alignment' which has been argued to be fundamental to inquiry communities.

I see parallels between 'critical alignment' and Paulo Freire's notion of 'conscientization'. (I must emphasise that I do not wish to exaggerate the significance of 'inquiry communities' nor diminish the global challenge to confront oppression and poverty in Freire's work)

> The term 'concientization' refers to learning to perceive social, political, and economic contradictions, and to take action against the oppressive elements of reality. (Freire, 1972, p. 15, translator's note)

As I have admitted above, to relate 'critical alignment' to this seems first a failure of understanding of the human contexts (the oppressed poor on the one hand, and educators concerned with the development of mathematics teaching and learning on the other) within which the two terms are applied. It might also be argued that the concepts occur in different paradigms. Conscientization comes from a critical theory paradigm, and critical alignment from developmental research. My purpose is to try to demonstrate a mapping between the two concepts and contexts so that the concepts will be seen to be the same human attitude differing only with respect to the contexts within which each is used.

In the above definition there is reference to 'social, political and economic contradictions', in Freire's work these relate to the fundamental conditions in which people live. 'Critical alignment' is used in the context of 'practice', in particular the practices of teachers, didacticians and students. 'Practice' appears to be a local context in contrast to the national and global humanitarian context addressed by Freire. However, the practice of teachers, didacticians and students is subject to social, political and economic contradictions.

The culture of teaching is conservative (Stigler & Hiebert, 1999). Teachers tend to reproduce the practice they experienced themselves as students. The curriculum places pressures of content and pace which have been chosen with an 'ideal' pupil in mind. Classes do not comprise ideal pupils. The contradictions and tensions faced by teachers arising from 'external (political) control are cogently summarised by Wolff-Michael Roth and Yew-Jin Lee (2007) in their pen-picture of an imaginary teacher, Katherine. Michael Fullan (1993) also expresses the oppressive tensions and contradictions experienced by teachers: 'Productive educational change is full of paradoxes, and components that are often not seen as going together. Caring and competence, equity and excellence, social and economic development ...' But Fullan continues, (these components) '... are not mutually exclusive. On the contrary, these tensions must be reconciled into powerful new forces for growth and development' (1993, p. 4). Thus I argue that teachers are 'oppressed' in their practice by historic, economic, social and cultural contradictions over which they have little or no control. However, through 'critical alignment' they may become aware of their situation and explore the possibilities they have to make things better for their pupils. Critical alignment is the means, in Fullan's terms, of reconciling the tensions into 'powerful new forces for growth and development.'

Freire writes of conscientization:

> The correct method for a revolutionary leadership to employ in the task of liberation is, therefore, not 'libertarian propaganda.' Nor can the leadership merely 'implant' in the oppressed a belief in freedom, thus thinking to win their trust. The correct method lies in dialogue. The conviction of the oppressed that they must fight for their liberation is not a gift bestowed by the revolutionary leadership, but the result of its own conscientization. (1972, p. 42)

This can be reworked in the language of inquiry communities ...

The correct method for *didacticians* to employ in the task of *developing the teaching and learning mathematics in school* is, therefore, not *lectures on mathematics didactics*. Nor can *didacticians* merely *give teachers a set of tips, ideas and lesson plans*, thus thinking to win their trust. The correct method lies in dialogue. The conviction of *teachers* that they must *strive for better teaching and learning themselves* is not a gift bestowed by *didacticians*, but the result of their own *critical alignment*.

Freire also writes about the place of inquiry:

> ... apart from inquiry, apart from the praxis, men cannot be truly human. Knowledge emerges only through invention and re-invention, through the restless, impatient, continuing, hopeful inquiry men pursue in the world, with the world, and with each other. (p. 46, gendered language in original)

This seems, to me, to be in the same spirit as 'inquiry as a way of being'. Thus, my argument that developmental research in communities of inquiry is 'good' research in the ethical sense is based upon the claim that it empowers teachers, students, and didacticians, to begin to take control of their professional practices and empowers them to be, in their school teams, self sufficient in development. Jaworski (1998) makes the point similarly on the basis of research into teachers as researchers:

> In the research quoted here, teachers acknowledged that what they referred to as hard questions from the researcher were instrumental in enabling them to delve deeply into their own purposes and become more overtly aware of personal theories motivating their practice. Over time, teachers started to anticipate questions, and to ask their own questions. It was clear that this developing process of questioning led to explicit forms of enquiry on the part of teachers into their teaching. Teachers' increasing metacognitive awareness of classroom decisions and judgements promoted changes in and exploration of classroom activity aimed at students' better understanding of mathematics. (p. 4)

It is the transfer of agency from researcher to teacher, or the development of agency within the teacher-researcher which is the basis of the claim that developmental research in communities of inquiry is 'good' research. However, without evidence that the proposed development takes place it is very difficult to claim that developmental research is 'good'. It is to this theme with respect to Learning Communities in Mathematics that I turn very briefly.

THE IMPACT OF LEARNING COMMUNITIES IN MATHEMATICS.

The first thing to note is that the goal of LCM is ambitious: to develop inquiry communities between teachers and didacticians. Michael Huberman even argues that it is not possible; he asserts 'the literature also supports the view that attempting to create "learning communities" is institutionally naïve' (1999, p. 315). The aim of LCM is to achieve improved learning of mathematics in classrooms, develop the teaching of mathematics and research learning, teaching, and the development process. To achieve the goal of inquiry communities requires development of the practice culture, both in the way that teachers engage in their professional practice and the way that teachers and didacticians collaborate. Reaching such goals is a slow process; amongst other things it demands the growth of trust between all participants. The gains so far appear modest and it is too early to make hard claims about sustainability.

The project received funding for a period of four years, with workshops taking place in the course of three of these. Roughly there was a six month 'run up' to initiating work with teachers in the first workshop and a final six months in which didacticians continued to analyse and report data.[8] The first year of working with schools was seen principally as a period of community building. The second year was characterised by the development of inquiry within the community and the

---

[8] A huge data base was developed during the course of the project and it is likely that this will be a resource for research activity that will be sustained for some years.

third year a period of consolidation as teachers set and pursued their own goals within the established inquiry community.

This chapter is not the place for a comprehensive review of the research effort within LCM; more detail can be found in a book written by teachers and didacticians, which has been produced to disseminate information about the project (Jaworski et al., 2007).

I offer my own collaboration with a teacher of first to fourth grade children (Jørgensen & Goodchild, 2007) as an example of the 'co-learning agreement' to which the developmental-research project aspires. In one of the workshops it was proposed that teachers engage in a small piece of research to explore their pupils' understanding of pre-algebra and algebra concepts. A number of tasks that might be given to pupils for this purpose were proposed. One teacher (Kai Otto Jørgensen) responded to the challenge by adapting two of the proposed tasks for use with his first grade class. He invited me to observe and video record the lesson. Following the lesson Jørgensen and I, together with two other didacticians reviewed and discussed episodes from the video-recordings. This was the start of a collaboration that has now continued over many months. Jørgensen designs new activities for his class, I observe and video record the lessons and together we discuss the pupils' responses, stimulated by watching the video recording. The teacher admits that the collaboration has led him to explore deeper into his pupils' mathematical development. He acknowledges that he finds the combination of theoretical knowledge that I bring with his own rich experiential craft knowledge productive in the development of his own practice.[9]

This collaboration is the essence of a co-learning agreement. In the first instance, although the 'event' was initiated in a workshop by didacticians it was in response to the teacher's invitation that I visited the class. The didacticians had identified a broad theme, pupils' development of algebra, but the teacher interpreted this within the context of his own practice. I visited the class as a didactician, with several years of experience of teaching at the secondary level but I work with a teacher with much younger pupils. The teacher's own expert knowledge in this context is undisputed. The teacher acknowledges the collaboration as productive in the development of his own practice because our discussions facilitate his reflection on mathematical, epistemological and pedagogical issues. From my perspective I am able to pursue a research agenda that covers issues relating to pupils' understanding of number and pre-algebra, teaching and learning, and the developmental process. The teacher and I have separate but complementary goals in our collaboration, and we both acknowledge that we learn from our joint activity.

---

[9] This paraphrases the teacher's remarks to me in conversation as he reflects on the meaning of our collaboration for him. I want to add that the teacher also brings 'theoretical knowledge' to his practice, in particular through the application of an education in Gestalt psychology which he has had in the past.

SUMMARY

It is not enough for me to research in order to find out more about teaching and learning in classrooms for myself and to report to others (see Chapman, this volume). I want to engage in 'good' research. For me the practitioner researcher is one who wants to make 'things' better. It must be 'good' research in the sense set out by Hostetler (2005). I believe developmental research based on communities of inquiry is 'good' research and that it can make things better. It is ethical and does not exploit the participants in the research: it does not impose upon them anything other than to encourage a disposition that empowers them in their practice.

I began this chapter with Freudenthal's assertion that mathematics is different, his argument based on the special relationship between the content and form of the discipline. This leads me to reflect on how my personal account of practitioner research is coloured by being that of a *mathematics* teacher educator. To what extent has mathematics been influential? This is a difficult question to answer briefly and convincingly but I feel sure the answer is that the mathematics context is of crucial significance. It is the nature of the subject that makes it difficult to both teach and learn that leads mathematics educators to inquire into how to engage in their practice more effectively. Mathematics is the starting point and the special challenges in teaching and learning drive and inspire the research. It seems natural to accept inquiry as a developmental and problem solving tool because of the central place inquiry has in mathematical activity. In mathematics it is widely recognised that knowledge of the content of the subject is gained by personal engagement and reflection on the content, it is not passed on merely by oral or written 'telling'. Nor is understanding of the content gained by repetition of routine tasks that aid memory and recall of procedures in a behaviouristic fashion. It therefore seems natural to accept that knowledge of the deep and complex processes of teaching and learning must also be gained through personal experience of engagement in the knowledge creation processes. Thus perhaps mathematics teacher education researchers have a predisposition towards co-learning agreements. Moreover, perhaps there is a comradeship between mathematics educators, teachers and didacticians, which is the product of the common challenge they experience. All this is highly speculative. There are additional sources of motivation for research in mathematics education. For example, international studies of school pupils' mathematical performance convince us that there is need for improvement, which leads to the focus on teaching and learning. Developing mathematical competencies takes time; the development of teaching also takes time. Perhaps one essential characteristic of the mathematics teacher educator practitioner researcher is patience!

REFERENCES

Becker, J. P., & Jacob, B. (1998, June). Math war developments in the United States (California). *The International Commission on Mathematical Instruction Bulletin 44*. Retrieved May 11, 2007, from http://www.mathunion.org/ICMI/bulletin/44/MathWar.html

Bishop, A. J. (1998). Research, effectiveness, and the practitioners' world. In A. Sierpinska & J. Kilpatrick (Eds.), *Mathematics Education as a research domain: A search for identity.* Dordrecht, the Netherlands: Kluwer.

Carr, W., & Kemmis, S. (1986). *Becoming critical: Education, knowledge and action research.* London: RoutledgeFalmer.

Cochran Smith, M., & Lytle, S. L. (1999). Relationships of knowledge and practice: Teacher learning in communities. *Review of Research in Education, 24,* 249–305.

Department for Education and Employment (1982). *Mathematics counts: Report of the committee of Inquiry into the teaching and learning of mathematics in schools under the chairmanship of Dr. W. H. Cockcroft.* London: The Stationary Office.

Elliott, J., & Sarland, C. (1995). A Study of 'Teachers as researchers' in the context of award-bearing courses and research degrees. *British Educational Research Journal 21*(3), 371–386.

Fernandez, C., & Yoshida, M. (2004). *Lesson study: A Japanese approach to improving mathematics teaching and learning.* Mahwah, NJ: Lawrence Erlbaum.

Freire, P. (1972). *Pedagogy of the oppressed.* Harmondsworth UK: Penguin.

Freudenthal, H. (1991). *Revisiting Mathematics Education: China Lectures.* Hingham, MA: Kluwer Academic Publishers.

Fullan, M. (1993). *Change forces: Probing the depths of educational reform.* London: Falmer.

Furlong, J., & Salisbury, J. (2005). Best Practice Research Scholarships: an evaluation. *Research Papers in Education 20*(1), 45–83.

Goodchild, S. (1987). *An evaluation of a curriculum development project in mathematics.* Unpublished master's thesis, University of London.

Goodchild, S. (1996). Learner empowerment through a problem-centred course. *Mathematics Education Review 7,* 8–16.

Goodchild, S. (2001). *Students' Goals: A case study of activity in a mathematics classroom.* Bergen, Norway: Caspar Forlag.

Goos, M. (2004). Learning mathematics in a community of inquiry. *Journal for Research in Mathematics Education 35,* 258–291.

Gravemeijer, K. (1994). Educational development and developmental research in mathematics education. *Journal for Research in Mathematics Education 25,* 443–471.

Hardy, G. H. & Wright, E. M. (1960). *An introduction to the theory of numbers.* Oxford, UK: Oxford University Press.

Hargreaves, D. (1996). *Teaching as a research-based profession.* Teacher Training Agency Annual Lecture. London.

Hiebert, J. (Ed.). (1986). *Conceptual and procedural knowledge: The case of mathematics.* Hilsdale, NJ: Lawrence Erlbaum Associates.

Hostetler, K. (2005). What is "good" educational research? *Educational Researcher, 34*(6), 16–21.

Huberman, M. (1999). The mind is its own place: The influence of sustained interactivity with practitioners on educational researchers. *Harvard Educational Review, 69*(3), 289–319.

Jaworski, B. (1994). *Investigating Mathematics Teaching: A constructivist enquiry.* London: RoutledgeFalmer.

Jaworski, B. (1998). Mathematics teacher research: Process, practice and the development of teaching. *Journal of Mathematics Teacher Education 1,* 3–31.

Jaworski, B. (2003). Inquiry as a pervasive pedagogic process in mathematics education development. Paper presented at CERME3 Working Group 11, European Society for Research in Mathematics Education, Bellaria, Italy. Retrieved June 29, 2007, from http://ermeweb.free.fr/CERME3/Groups/TG11/TG11_Jaworski_cerme3.pdf

Jaworski, B. (2004a). Grappling with complexity: Co-learning in inquiry communities in mathematics teaching development. In M. Johnsen Høines & A. B. Fuglestad (Eds.), *Proceedings of the 28th Conference of the International Group for the Psychology of Mathematics Education: Vol. 1* (pp. 17–36). Bergen, Norway: Bergen University College.

Jaworski, B. (2004b). Insiders and outsiders in mathematics teaching development: The design and study of classroom activity. In O. Macnamara & R. Barwell (Eds.), *Research in Mathematics Education: Papers of the British Society for Research into Learning Mathematics, 6*, 3–22.

Jaworski, B. (2005). Learning communities in mathematics: Creating an inquiry community between teachers and didacticians. In R. Barwell and A. Noyes (Eds.), *Research in mathematics education: Papers of the British Society for Research into Learning Mathematics, 7*, 101–119.

Jaworski, B. (2006a). Theory and practice in mathematics teaching development: Critical inquiry as a mode of learning in teaching. *Journal of Mathematics Teacher Education 9*, 187–211.

Jaworski, B. (2006b). Developmental research in mathematics teaching and learning: Developing learning communities based on inquiry and design. In P. Liljedahl (Ed.), *Proceedings of the 2006 annual meeting of the Canadian Mathematics Education Study Group*. Calgary, Canada: University of Calgary.

Jaworski, B. (2007). Theory in developmental research in mathematics teaching and learning: Social practice theory and community of inquiry as analytical tools. Paper presented at CERME5 Working Group 11, European Society for Research in Mathematics Education, Larnaca, Cyprus.

Jaworski, B., & Goodchild, S. (2006). Inquiry community in an activity theory frame. In J. Navotná, H. Moraová, M. Krátká & N. Stehlíková (Eds.), *Proceedings of the 30th Conference of the International Group for the Psychology of Mathematics Education: Vol. 3* (pp. 353–360). Prague Czech Republic: Charles University in Prague.

Jaworski, B., Fuglestad, A. B., Bjuland, R., Breiteig, T., Goodchild, S., & Grevholm, B. (Eds.). (2007). *Læringsfellesskap i matematikk/Learning communities in mathematics*. Bergen, Norway: Caspar forlag.

Jørgensen, K. O., & Goodchild, S. (2007). Å utvikle barns forståelse av matematikk. [To develop children's understanding of mathematics] *Tangenten, 1/2007*. 35–40 & 49.

Klein, D. (2007). A quarter century of US 'math wars' and political partisanship. *BSHM Bulletin: Journal of the British Society for the History of Mathematics, 22*(1), 22–33.

Lo, M. L. Marton, F., Pang, M. F., & Pong, W. Y. (2004). Toward a pedagogy of learning. In F. Marton & A. B. M. Tsui, (Eds.), *Classroom discourse and the space of learning* (pp. 189–225). Mahwah, NJ: Lawrence Erlbaum.

Pring, R. (2004). *Philosophy of educational research*. (2nd ed.). London: Continuum.

Roth, W-M., & Lee, Y-J. (2007). "Vygotsky's neglected legacy": Cultural-historical activity theory. *Review of Educational Research, 77*(2), 186–232.

Schoenfeld, A. H. (1996). In fostering communities of inquiry, must it matter that the teacher knows "the answer"? *For the Learning of Mathematics, 16*(3), 11–16.

Schoenfeld, A. H. (2004). The math wars. *Educational Policy, 18*, 253–286.

Skemp, R. R. (1976). Relational understanding and instrumental understanding. *Mathematics Teaching 77*, 20–26.

Skemp, R. R. (1982). Communicating mathematics: Surface structures and deep structures. *Visible Language 16*(3), 281–288.

Steen, L. A. (1999). Theories that gyre and gimble in the wabe. *Journal for Research in Mathematics Education 30*(2), 235–242.

Steinbring, H. (1998). Elements of epistemological knowledge for mathematics teachers. *Journal for Mathematics Teacher Education 1*, 157–189.

Stigler, J. W., & Hiebert, J. (1999). *The teaching gap: Best ideas from the world's teachers for improving education in the classroom*. New York: The Free Press.

Van den Akker, J. (1999). Principles and methods of development research. In J. v. d. Akker, R. M. Branch, K. Gustafson, N. Nieveen, & T. Plomp (Eds.), *Design approaches and tools in education and training* (pp. 1–14). Dordrecht, the Netherlands: Kluwer.

Vygotsky, L. S. (1978). *Mind in society: The development of higher psychological processes*. Cambridge, MA: Harvard University Press.

Vygotsky, L. S. (1986). *Thought and Language*. Cambridge, MA: MIT Press.

Wagner, J. (1997). The unavoidable intervention of educational research: A framework for reconsidering researcher-practitioner cooperation. *Educational Researcher 26*(7), 13-22.

Wells, G. (1999). *Dialogic inquiry*. Cambridge, UK: Cambridge University Press.
Wenger, E. (1998). *Communities of practice*. Cambridge, UK: Cambridge University Press.
Zack, V., Mousley, J., & Breen, C. (1997). *Developing practice: Teachers' inquiry and educational change*. Geelong, Australia: Centre for Studies in Mathematics, Science and Environmental Education, Deakin University.

Simon Goodchild
Mathematics Education Research Group
University of Agder

SECTION 3

WORKING WITH PROSPECTIVE AND PRACTISING TEACHERS: WHAT WE LEARN; WHAT WE COME TO KNOW

CHRISTER BERGSTEN AND BARBRO GREVHOLM

11. KNOWLEDGEABLE TEACHER EDUCATORS AND LINKING PRACTICES

This chapter discusses knowledge of mathematics teacher educators in relation to their work on practices with student teachers and beginning teachers, in particular those aiming at linking theoretical course work and teaching practice in schools. The discussion builds on frameworks for mathematics teacher knowledge and reflections from research on mathematics teacher educators. An overview of practices used by mathematics teacher educators is given, and some more detailed accounts of documented practices that focus on links between theoretical course work and teaching practice. Such linking practices have the potential to bridge the didactic divide between disciplinary and pedagogical knowledge, often separated by the organisation of teacher education programmes. It is concluded that mathematics teacher educators not only need knowledge and experience in critical competence areas and activities involved in the work of mathematics teachers, but also to be knowledgeable persons in the activities they design, which comprises a critical awareness of the activity and of the others you are working with as well as a will to move forward to the best of the others.

INTRODUCTION

The research interest in mathematics education has expanded over time from a focus on the achievements of students to their activities in classrooms as cognizing subjects in a social context, and from the role and work of teachers as well as their beliefs and professional development, to the learning of student teachers in teacher education programmes and mathematics teacher educators. The societal-institutional impact on education adds to this list. One main outcome of these research efforts, and a source for further development, is the recognition of the complexity of institutionally organised mathematics education (Niss, 1999). This complexity is directly visible when analysing the work of mathematics teacher educators which spans over all the mentioned levels of activities (Krainer, this volume). While being locally autonomous, these levels of work relate to each other in a hierarchical dependence (Steinbring, 1998; Zaslavsky & Leikin, 2004). With this background in mind, this chapter investigates demands on mathematics teacher educators' knowledge for carrying out practices that make a difference for the development of knowledge to teach mathematics.

The professional development of mathematics teacher educators must thus be seen in relation to the objectives of teacher education, which in turn is related to the objectives of the teaching profession. This puts desired competencies of the

B. Jaworski and T. Wood (eds.), The Mathematics Teacher Educator as a Developing Professional, 223–246.
© 2008 Sense Publishers. All rights reserved.

mathematics teacher in the centre when working with beginning teachers in a teacher education programme. To give a structured view of existing strong practices that aim to develop such competencies in student teachers, there is a need to outline a framework of critical competencies for mathematics teachers and identify where and how within a teacher education programme such practices exist or can be developed. Of special interest is the potential, in this respect, of linking theoretical coursework with teaching practice, and how this is or could be treated in teacher education programmes. One aim of this chapter is to discuss what is required of mathematics teacher educators to accomplish this enterprise.

Teacher education programmes, organised within formal educational institutions, traditionally comprise three main strands – disciplinary studies, educational studies, and teaching practice (Comiti & Ball, 1996). The aim of these strands is to develop an integrated competence in student teachers often referred to as *teacher knowledge*. Winsløw and Durrand-Guerrier (2007) name the respective target knowledge components *content knowledge*, *pedagogical knowledge*, and *didactical knowledge*, noting that these components are viewed, in terms of weight and organisation, differently within different cultural traditions. The categories parallel the classic distinctions by Shulman (1986) between subject matter knowledge, pedagogical knowledge, and pedagogical content knowledge. In teaching practice, as an activity within a teacher education programme, all these components come into play in the contextual setting where they are supposed to be functional. This is where the student teachers can experience the viability of their own teacher knowledge. It is often witnessed by student teachers during teaching practice, that working along with an experienced practising teacher is when you really learn something about teaching (Bergsten & Grevholm, 2004). Personal experiences can here be related to the formal courses on theories of education or teaching methods. The relevance of teaching practice, especially when the student teachers are given the opportunity to try out didactic ideas which they have contributed to develop, has been shown to be very high in different national teacher education contexts, even when the differences are significant in terms of structure, organisation and curriculum (Favilli, 2006). The linking of teaching practice to theoretical courses offered in the education programme is therefore critical to the development of teacher knowledge in student teachers. The lack of such kinds of links within a teacher education programme has been termed a *didactic divide*, originally assigned to an observed gap between disciplinary and pedagogical knowledge (Bergsten & Grevholm, 2004). This divide may also exist between for example courses in general education and in mathematics education, or between mathematics content and teaching methods courses. To overcome this divide has been a driving force behind renewals of teacher education programmes in Sweden (ibid.), and may be seen as an incentive to the emergence of conceptualisations such as pedagogical content knowledge (Shulman, 1986) and mathematical knowledge for teaching (Ball & Bass, 2004; Adler, 2005).

Discussing the effect of teacher preparation programmes, Hiebert, Morris and Glass (2003, p. 201) note that teaching is a cultural practice, and changing one's view of teaching is therefore difficult, partly explaining why student teachers, when facing classroom challenges in real-time, often tend to teach as they were taught themselves. In a world with changing views of what kinds of knowledge is needed, changing school systems, and new generations of children entering school, learning to teach is a continuously ongoing process. The impact of teacher education programmes on how practicing teachers work has also been questioned:

> The evidence suggests that teachers' notions of good mathematics teaching are grounded more in their experiences as teachers than as participants in teacher education programs. (Wilson, Cooney, & Stinson, 2005, p. 105)

From that perspective it is even more critical to set up specific goals for the formal teacher preparation years. What are the critical competencies that form a necessary basis for this lifelong enterprise of learning to teach, and what practices can contribute to developing these critical competencies? To design and run such teacher education programmes, from the perspective of mathematics teaching, what is required of teacher educators?

This chapter discusses such knowledge of mathematics teacher educators in relation to their work on practices with student teachers and beginning teachers, in particular those aiming at linking theoretical course work and teaching practice in schools. The discussion builds on theoretical frameworks for mathematics teacher knowledge, leading into reflections from research on mathematics teacher educators. Then, after an introduction and overview of practices used by mathematics teacher educators, more detailed accounts of documented existing interesting *linking practices* are given, that is practices focused on links between theoretical course work and teaching practice. Finally, a discussion and concluding remarks are offered on the role and knowledge of mathematics teacher educators.

THEORETICAL FRAMEWORKS FOR MATHEMATICS TEACHER KNOWLEDGE

The goal of mathematics teaching is the students' learning of mathematics, that is to develop students' mathematical knowledge. In national educational systems, this target knowledge is set up in curricula in more or less general terms, often in a mix of descriptions of mathematical content and abilities. Such descriptions may be based on current educational, psychological, and didactical research. As an example, a sequence of Swedish national curricula for mathematics in compulsory school has had a strong influence from behaviouristic, cognitive, and socio-cultural perspectives (Wyndhamn, 1993). As mathematics education as a research discipline has evolved, more discipline-based descriptions of what it means to know mathematics have become available. Such descriptions form an important base for the identification of appropriate knowledge for teaching mathematics, which in turn sets up goals for mathematics teacher education, and as a consequence, desired competencies of mathematics teacher educators. In the following, some views on mathematical knowledge and teacher knowledge are briefly discussed, before we turn to the key issue of this chapter, the knowledge required by mathematics teacher educators and how this knowledge may be acquired by practices in teacher education.

CHRISTER BERGSTEN AND BARBRO GREVHOLM

Competency Approaches to Mathematics Teacher Knowledge

As an outcome of the Danish KOM-project, taking a *competency approach* to mathematical knowledge, the following definition is proposed:

> Possessing mathematical competence means having knowledge of, understanding, doing and using mathematics and having a well founded opinion about it, in a variety of situations and contexts where mathematics plays or can play a role. (Niss, 2004, p. 183)

In the project eight specific competencies were identified, forming two clusters included in mathematical competence: *The ability to ask and answer questions in and with mathematics*, and *The ability to deal with mathematical language and tools* (ibid., pp. 184–186). In addition, an emphasis on the application of mathematics, as well as its nature and historical development was made to build in learners a sound image of mathematics as a discipline. For a teacher to develop in students a mathematical competency of this kind, Niss (ibid.) also outlines a model for mathematics teacher competency. In this model, in addition to a general education component, four specific competencies are related to work with pupils: *Curriculum competency, teaching competency, uncovering of learning competency*, and *assessment competency*. Another two refer to the teacher's position in professional environments, that is *collaboration competency*, and *professional development competency*. The teacher's own mathematical competency is a prerequisite and highly involved in all these specific competencies.

A similar approach is taken in the proficiency concept in knowing and teaching mathematics: in the report *Adding it up* (Kilpatrick, Swafford, & Findell, 2001) *mathematical proficiency* is described to comprise five intertwined major strands. Based on this construct, five other but parallel strands are described to define *proficiency in teaching mathematics*:

- *Conceptual understanding* of the core knowledge required in the practice of teaching: knowledge of mathematics, knowledge of students, and knowledge of classroom practice
- *Fluency* in carrying out basic instructional routines
- *Strategic competence* in planning effective instruction and solving problems that arise during instruction
- *Adaptive reasoning* in justifying and explaining one's instructional practices and reflecting on those practices so as to improve them
- *Productive disposition* toward mathematics, teaching, learning, and improvement of teaching practice (Kilpatrick, 2004, p. 152).

The focus in this model is on the act of teaching rather than teacher knowledge in a more comprehensive sense, thus not emphasising dimensions such as assessment or curriculum. Adopting this approach, the goal of mathematics teacher education would then be to develop in student teachers each of these strands for proficiency in teaching mathematics.

Grevholm (2006) has summarised the content of mathematics teacher education in a concept map showing the professional development of a mathematics teacher (see also Grevholm, Even, Szendrei, & Carrillo, 2004). In this model the teacher's professional identity is seen to be built in five main distinct but interconnected areas:

– knowledge in mathematics in relation to teaching,
– competence to judge and diagnose pupils' learning in mathematics,
– knowledge about classroom management, methods and material,
– a personal view on and beliefs about knowledge and learning, and
– a professional language for mathematics teachers.

The concept map indicates the sources of all these competencies and skills that are based on practice, experiences, theoretical studies and research. Considering a mathematics teacher education programme, governed by societal demands, culture, and national identity, a critical issue is its potential to develop student teachers' *identity as a mathematics teacher* constituted through the five main areas (see also Lerman, 2001; Walshaw, 2004a), complemented by the identity as a person (see also Bednarz & Proulx, 2005).

In her overview of the mathematical education and development of teachers, Sowder (2007) identifies some major goals for professional development, including the development of mathematical and pedagogical content knowledge; an understanding of how students think about and learn mathematics; an understanding of the role of equity in school mathematics, and a sense of self as a teacher of mathematics. Building on Grossman's (1990) sub-division of pedagogical content knowledge into the four categories: conceptions of purposes for teaching subject matter (mathematics), knowledge of students' understanding, curriculum knowledge, and knowledge of instructional strategies for teaching particular topics, Sowder's educational and developmental goals broadly parallel the overall content of the three models for a mathematics teacher's competence described above.

Though it may sometimes be useful to refer to the broad categories of teacher knowledge components content knowledge, pedagogical knowledge, and didactical knowledge, the more fine grained analyses behind the four frameworks described seem more apt in discussions about the demands on mathematics teacher educators. They also question the traditional roles of different teacher educators and the separation of knowledge areas, often institutionally defined, that preserves existing didactic divides.

Even if these frameworks for mathematics teacher knowledge/competency all share a common core of elements constituting mathematics teacher knowledge, to make a comparison between them is problematic in many ways. However, all four descriptions presented above approach the issue of what is needed to function well as a mathematics teacher mainly by identifying a set of knowledge requirements, or competencies. Such approaches have been criticised, individual teachers' knowledge being a product of many types of knowledge created in quite diverse settings and rooted in 'local theories' [...] specific to their classroom situation. (Cooney & Krainer, 1996, p. 1170).

The Mathematics Teacher as a Knowledgeable Person

Some researchers instead put the main emphasis on the development of a deepened awareness or personal wisdom and sensitivity to those elements concerning mathematics, pedagogy, institutional constraints, beliefs, etc., that come into play in the classroom (Mason, 2002; Molander, 1998). This perspective puts more weight on the *ways of working* with prospective teachers during their education than on the more detailed content covered in course work. To *see* what is going on and be able to respond to it in a professional way may thus be a key element of a *mathematics teacher identity*, set up as a goal in Grevholm's model for the education of a mathematics teacher, and *a sense of self as a teacher of mathematics* in Sowder's description. It may also be seen as a part of the *professional development competency* in the KOM model, and *productive disposition* in the teaching proficiency model. As a consequence of this perspective, the traditional teacher education practices aiming at providing predefined knowledge and skills for teaching need to be replaced by more investigative approaches where student teachers engage in inquiry into "concrete phenomena" of teaching practice (Lampert & Ball, 1999).

According to Molander (1998), theory of knowledge is a special area of philosophy which can be studied without adding much to a person's success in moving forward in reality. It provides more or less ready-made parcels – empirism, positivism, critical rationalism, hermeneutics, evidence-based knowledge. Such a description can be useful but is treacherous if you forget to inquire critically into your own positioning – finding out about that may be much harder than buying an attractive parcel. More important than a ready made theory of knowledge is a continued reflection on knowledge. The *knowledgeable person* is the one who continues to learn, based on two main tasks: To find out where you stand, and try out a direction for where to go. Molander claims that knowledge exists only in the form of knowledgeable persons, which means persons who know how to proceed in the best way, and in so doing alert people so that thanks to their alertness and questions they can move further and learn better how tasks can be performed. Knowledge is, according to this view, what leads to the best for humans and thus also has an ethical stance. This is not knowledge *about something* but knowledge *in the activity*, where you live with the pupils, students, teachers, and material tools in a relation, which in principle is non-dualistic, involving a presence in the world together with others. It involves listening, noticing, and being aware (Mason, 2002). According to Watson and Mason (2007, p. 209),

> being knowledgeable about mathematics teaching influences classroom actions and knowing to act in the moment through having pertinent possibilities come to mind.

In terms of social practice theory, acquiring a professional identity and "becoming knowledgeably skilful" (Lave, 1988, p. 65) go in parallel supporting each other's development during apprenticeship learning in a community of practice. The process of becoming "knowledgeable in practice", as discussed by Walshaw (2004b) in a case study of a mathematics student teacher, requires "coming to terms with one's intentions and values, as well as one's knowing and being, in a setting of contradictory realities" (p. 563). For the work of mathematics teacher educators this means that information obtained from testing knowledge levels and characteristics is insufficient to inform on future practices. What is needed is rather

> to find ways to enable people to display evidence of identifying situations as instances of phenomena [...]; seeking fresh ways of acting in those situations whether mathematically, pedagogically or didactically; collecting evidence on which to reflect [..., and] trying out task-exercises with colleagues to test whether what has been noticed is accessible to others, and whether their future actions are also informed by being sensitised. (Mason, 2007, p. 6)

There is not much research available on how teacher education programmes and practices handle these different aspects of teacher knowledge or dispositions (Adler, Ball, Krainer, Lin, & Novotna, 2005; Even, 1999). Early work in this direction is found in Breen (1992) where a *reflection model* is adopted for course work with student teachers, with a focus on the strands social interaction, socialised fear, 'What is mathematics?', and personal conflict exercises. The model built on Gattegno's (1970) statement that "only awareness is educable" and the assumptions that the course should be about "learning about ourselves ... as human beings", make us engage with each other in social situations, seek out to increase energy, and "refine the ability to observe ourselves in different situations" (Breen, 1992, p. 35). In the next section we go on to discuss some research efforts concerning mathematics teacher education and the roles played by mathematics teacher educators.

RESEARCH ON MATHEMATICS TEACHER EDUCATION AND TEACHER EDUCATORS

At ICME10 a survey team reported from a study on research on mathematics teacher education (Adler et al., 2005). Some issues in the report are related to teacher educators' professional work. The authors claim that the differences between teacher educators and their 'learners' are increasing, that is between prospective and practicing teachers. We are dealing with different kinds of under-preparedness, a phenomenon that extends into the education of practicing teachers.

> Teachers need support if the goal of mathematical proficiency for all is to be reached. The demands this makes on teacher educators and the enterprise of teacher education are substantial and often under-appreciated. These, in turn, shape the context in which research on mathematics teacher education is developing. (ibid., p. 361)

The team also reports that ten years ago there was very little research on processes of mathematics teacher education (ibid., p. 362). They ask in the report what research in the field is contributing to the improvement of the education of teachers of mathematics. In trying to explain why many studies in the field are small scale studies and longitudinal studies following teachers over time are lacking, they claim that

> research in teacher education is often more complex since it deals not only with the beliefs, knowledge and practices of teachers but also students' beliefs and knowledge, as well as with the interaction between teachers and students, and the interaction between teacher educators and teachers. (ibid., p. 369)

In addition to the demands mentioned above, the report points to the fact that mathematics teacher educators' professional responsibilities include both research and teaching. Research is seen as one aspect of teacher educators' professional development. This kind of research is also an important part of teacher educators' learning to improve their practice (ibid., p. 371). Teacher educators have a double role of intervening and investigating, which means that they have to both improve and understand their own practice.

The complexity indicated above in studying mathematics teacher education is highlighted by Jaworski (2001), when she discerns three levels at which both teachers and teacher educators operate and reflect:

Level 1	Mathematics and provision of classroom mathematical activities for students' effective learning of mathematics;
Level 2	Mathematics teaching and ways in which teachers think about developing their approaches to teaching;
Level 3	The roles and activities of teacher educators in contributing to developments in (1) and (2). (Jaworski, 2001, p. 301)

Jaworski refers to Cooney's (1994) notions of *mathematical power* and *pedagogical power* as relating to Level 1 and Level 2, respectively, and suggests the term *educative power* to depict what comes into play at Level 3 (Jaworski, 2001, p. 302). This notion is interpreted by Zaslavsky, Chapman and Leikin (2003) as "the ability of teacher educators to draw on knowledge that is needed for facilitating teachers' mathematical and pedagogical problem solving" (p. 879). As a consequence of this model, the competency of a mathematics teacher educator must encompass all three kinds of *power*.

Jaworski also points to the fact that teacher educators' learning often takes the form of research. She wants to see teacher educators moving to be co-learners within a research culture. Drawing from one small study with three participants, she illustrates how the teacher educator is learning about pedagogy, about mathematics and about co-learning. Her overriding aim in doing so is to change a culture of dependency at so many levels of teaching and learning into one of mutual respect and responsibility (Jaworski, 2001, p. 316).

The professional development of mathematics teacher educators is presented as an ongoing life long process by Zaslavsky et al. (2003). This process occurs in various stages and contexts. They focus on trends in thinking about and practices within professional development programmes for mathematics educators. They start by offering a unifying conceptual framework, which takes into account the relations between different groups that can be labelled mathematics educators. The framework they present is based on the teaching triad (Jaworski, 1994), which contains the three corners *challenging content for students, management of students' learning* and *sensitivity to students*, and is further discussed in Zaslavsky and Leikin (2004). It constitutes a three layer hierarchical model of growth through practice of mathematics educators. The first layer contains students' mathematical knowledge and subjective interpretations of tasks, the second mathematics teachers' knowledge of the teaching triad and subjective interpretation of tasks and the third the same but for mathematics teacher educators. On top of the layers there is a space for the educators of mathematics teacher educators making 'learning offers' as planned activities in courses for the mathematics teachers and observing and varying the learning offers. The case reported in Tzur (2001) can be seen as an example of one person's travel through these four levels of learning mathematics, learning to teach mathematics, learning to teach teachers, and learning to mentor teacher educators.

After an extensive presentation of the changing nature of professional development programmes for mathematics teachers, Zaslavsky et al. (2003) discuss trends in the professional development of mathematics teacher educators. They claim that until recently mathematics teacher educators developed their own practices and procedures through their own practice. Similarly, Hiebert et al. (2003) state that "teacher educators mostly start anew, learning how to teach preparation courses more effectively" (p. 202). The professional development would demand growth in mathematical, pedagogical and educative powers of teachers and it is assumed that tasks should address these elements in a challenging and engaging way. Through examples of tasks the authors elaborate further on the connections between teaching, learning and tasks with respect to mathematics educators. They suggest further study of the relations between the communities of mathematics educators and the processes of their ongoing growth.

The models by Jaworski and by Zaslavsky and colleagues described above, were applied by Peled and Hershkovitz (2004) in a study involving students, student teachers, teachers and teacher educators solving problems on proportionality. The researchers report how they "experienced indirect learning of the challenge we presented to our students (the teachers)" (p. 306). For the benefit of their roles as teacher educators, as researchers they

> developed better understanding of the mathematical challenge associated with the proportional reasoning problem, a stronger awareness of the role of sensitivity to their learners (the teachers), and of the role of reflection. (ibid., p. 299)

CHRISTER BERGSTEN AND BARBRO GREVHOLM

PRACTICES OF MATHEMATICS TEACHER EDUCATORS

Mathematics teacher educators are not a homogenous group of professionals. At least three categories can be discerned among them. They may come from a background in mathematics, pedagogy or as experienced teachers in school. Teachers mentoring student teachers during practicum serve as mathematics teacher educators but do not have that as their main professional task. In some cases we find educators based in research of mathematics education, but not in all countries. It is not uncommon that among teacher educators the level of persons with a research degree is lower than among other groups of academic teachers. These three main groups of teacher educators have somewhat different tasks and thus different practices, illustrating that not one single person, or one single competence, holds teacher education together (Grevholm, 2002). They have also been valued differently over time. As an example, during the 1970s and 1980s in Sweden the practitioner, the good and experienced teacher was highly valued. Later, after all teacher education was reorganised within the university system, teacher educators with a research degree were praised. Such a shift from appreciating practice based knowledge to more theoretical knowledge and research can be observed also in other Nordic countries. The fact that the three groups exist beside each other within teacher education contributes to preserving the didactic divide in different ways (Bergsten & Grevholm, 2004, 2005).

In mathematics education contexts the main agents involved are students and teachers in school, which may be called their home arena, and student teachers, teacher educators and mathematicians at the university home arena (or other teacher education institutions). In teacher education practices, such as practicum and tutoring, some of these agents make 'visits' to the home arena of other agents. At the core in all of these activities are the learning and teaching of mathematics, which may happen directly or indirectly (cf. Zaslavsky et al., 2003, p. 878). For example, in a mathematics class in school the students learn mathematics directly since it is their goal of the activity, while the teacher is teaching mathematics directly but at the same time indirectly is learning both about mathematics and mathematics teaching by the experience he or she is engaged in, including the observation of the students' learning efforts. To make visible the complexity of the educational game of mathematics teacher preparation, Table 1 shows direct and indirect learning and teaching events on the part of its different main agents.

It may happen that the mathematics teacher educator and the mathematician is the same person, or that the mathematics teacher educator is teaching mathematics content courses to student teachers without being a mathematician in the sense of a research mathematician. While discussing with student teachers the teaching of mathematical concepts and methods, the mathematics teacher educator may also be teaching mathematics, indirectly. The extent to which the teacher as a mentor is visiting the university, for specific training as a mentor or in general professional development activities, is normally much smaller than the teacher educator visiting school for observing and tutoring student teachers or for running professional development activities with teachers.

Table 1. *Learning and teaching activities of main agents involved in mathematics teacher education. S = Student; ST = Student teacher; MT = Mathematics teacher; MTE = Mathematics teacher educator; M = Mathematician; D = Directly; I = Indirectly.*

	S	ST	MT	MTE	M
Home arena (H)	School	University	School	University	University
Visiting arena (V)	-	School	University	School	-
Learning mathematics	D (H)	D (H) I (V)	I (H) D (V)	I (HV)	D (H) I (H)
Teaching mathematics	-	D (V)	D (H)	D (H) I (HV)	D (H)
Learning how to teach mathematics	I (H)	D (HV) I (H)	I (H) D (V)	I (HV)	I (H)
Teaching how to teach mathematics	-	-	I (H)	D (HV)	-

Table 1 does not include the level of learning how to teach the teaching of mathematics, or even teaching how to teach the teaching of mathematics. This relates to the level of directly educating the teacher educators for their professional task, a level in need of development (Even, 1999; Grevholm, 2002; Hiebert et al., 2003; Sowder, 2007), and included in the model of mathematics educators' growth presented in Zaslavsky et al. (2003, p. 881). However, as discussed above, most mathematics teacher educators develop their own practices, either out of previous experiences as teachers in school or in university, or 'anew' based on research and experimentations. As expressed by Peled and Hershkovitz (2004) about their work as teacher educators, "it is a common practice for us to investigate an issue in one group and then to reflect on our action, design new tasks and to try them in another group" (pp. 302–303).

The educational game displayed in Table 1 takes place in a more or less autonomous world of mathematics for learning and teaching, isolated from the world of applied mathematics and use of mathematical models in science and society in general (even if it refers to these), as well as the world of mathematical research. This world may be called *educational mathematics*, and to cope with it teachers and teacher educators develop an *educational knowledge of mathematics* (Bergsten & Grevholm, 2005). In terms of the anthropological theory of didactics, the specific mathematical organisations taught in educational institutions (such as schools or universities) are shaped and constrained by didactical transpositions that may lead to inconsistencies and fragmentations which in turn cause learning problems (e.g. Barbé, Bosch, Espinoza, & Gascon, 2005). For mathematics teacher educators, this complexity adds to the complexity of the organisation of mathematics teacher education itself, one critical facet of which is the didactic divide (Bergsten & Grevholm, 2004, 2005). One stance within teacher education programmes where this divide may be merged is the link between academic course work and teaching practice.

What Are the Practices of Mathematics Teacher Educators?

Different areas of practices used within practising teacher development include, among others, action research, case inquiry, narrative inquiry, student thinking, reform classroom context, and model lessons/illustrative units (Zaslavsky et al., 2003). Especially case studies are also useful in educational practices for prospective teachers and have shown to enhance a number of effective payoffs on teacher development (see McGraw, Lynch, Koc, Budak, & Brown, 2007, p. 96), and in particular, "engaging in case discussions can involve interplay between theoretical and practical knowledge" (ibid., p. 117). The range of existing practices in teacher education programmes is vast and points to the diversity of competences involved on behalf of the teacher educator. Practices are linked either to the parts of the education that take place in the university or to the practicum, when the student teacher is working in a school with teaching practice and investigative tasks inside the school setting. For the work at university the teacher educator has to plan working sessions for students. This demands an overview of what is wanted and possible for choosing content and teach it, whether it is mathematics, didactics or pedagogy or an integration of these areas.

Several different forms for organising the teaching during education exist: teacher or student led lectures, lessons, seminars, group sessions, group work without the teacher, problem based learning in small groups, micro teaching for fellow students, problem-solving sessions, school visits, lesson observations, interviews, essays about practice, use of authentic material such as videos or tapes from classroom or pupils' work, tasks to do during practice, teaching practice in school, degree thesis work and writing, homework, and so on. We are beginning to know more about existing practices related to such different forms of organisation, which are used and in what combinations, and their viability in developing a deepened teacher knowledge (e.g. Sowder, 2007; Zaslavsky, 2007). Research papers by teacher educators often investigate one such practice, seen from the inside through the teacher educator as researcher perspective.

Another practice for the teacher educator is to create working tasks for different needs for the student teachers, theoretical, practical, or investigative, and offer response and assess the outcome of the work with tasks. Such tasks can be for use in school or at university during class or for individual work. Research interest in the use of tasks in mathematics education has turned also to tasks used in the education of both prospective and practicing teachers (Watson & Mason, 2007; Zaslavsky, 2007). While preparing such tasks the teacher educator needs to know about and be able to use and teach about all kinds of tools and supporting material for mathematics teaching such as concrete material, graphical calculator, computer, video camera, camera, TV-programs, DVD, OH-projector, tape recorder and so on. Mathematics textbooks and other books and printed material, text and pictures on internet are part of the artefacts used in tasks – and the teacher educator must be aware of how to choose, combine, use and evaluate the materials (see the special issue on tasks in issues 4–6 in volume 10 of *Journal of Mathematics Teacher Education*). Also ethics around video recorded cases to discuss in teaching needs to be considered. Doerr and Thompson (2004) claim that multimedia case studies of practice can serve as vehicles for revealing the knowledge and practice of teacher educators, as they engage in supporting the professional development of prospective teachers. Obviously such case studies of practice can help to create a link between practice and theory of teacher education.

For the part of the education that is called practicum other practices are in play. Teacher educators visit students in class during school practice, observe, discuss and assess. How systematic this work is and what practices exist for such activities may vary significantly between different national and cultural contexts. There are many critical issues around practicum that deserve but have not yet found much research interest, such as common practices used to organise the practicum, collaboration of institutions with schools, and what is done in order to overcome the didactic divide between the different strands of the education: disciplinary studies, educational studies, and teaching practice. Such organisational constraints may strongly influence the possible practices by the individual teacher educators, adding to the complexity of their work. This puts high demands on the teacher educator and may require flexibility for example in ways of interacting with mentors in school concerning issues such as decisions about the tasks student teachers do in school, or how their work during practicum is being evaluated.

Examples of Linking Practices

Mathematics teacher educators have developed practices where research based theoretical tools are integrated with different activities in the teacher education programme. Such tools may help the student teachers to analyse and reflect upon their experiences from teaching practice in a more focused and systematic way, relating it to theoretical perspectives highlighted in course work, and thus increase the awareness and deepen the understanding of critical aspects of the teacher's role in the mathematics classroom (Rye Ejersbo, 2007). We call *linking practices* such ways of working that aim at creating functional bridges between academic course work in the home arena of the student teachers and teaching practice in the visiting arena, so that each of these may profit from experiences and reflections in the other. A selection of practices from the literature is presented below, all having the explicit or implicit aim to connect student teachers' experiences from both these arenas by enhancing efforts of a reflecting and investigative stance toward the learning and teaching of mathematics (cf. Dewey, 1933; Ball & Cohen, 1999). As for teachers' practices, most practices of teacher educators are not documented and thus not shared widely. The practices discussed here can thus only be seen as examples of the critical role of linking practices for the work of mathematics teacher educators by their potential to contribute to bridge the didactic divide, and of what demands and learning of mathematics teacher educators they afford.

Teachers' professional language. Grevholm and Bergsten (2005) describe a practice aimed at developing a professional language in student teachers, using a model that includes natural study groups for work with group tasks. During practice student teachers notice that their ability to talk about mathematics with pupils needs to be improved. By linking this experience to work at university, where student teachers are offered learning opportunities in group work, to try to express themselves in a problem solving situation and afterwards about the situation, they become motivated to work towards an improved professional language. The group work was documented in written form and the sessions videotaped, so that teacher educators could follow up afterwards needs that were expressed in the group work (ibid.).

One aspect of the professional language is to be able to ask questions to pupils in meaningful ways. Moyer and Milewicz (2002) offer suggestions for the teaching of the skill of mathematics questioning in teacher education courses. They examined categories of questions used by prospective teachers in audiotaped interviews with children and had the teachers reflect on their own questioning. Professional language skills include both listening and responding in ways that pupils can understand. Wallach and Even (2005) have explored teachers' listening to students and discuss the complexity of what pupils are saying, showing and doing. The aspect of listening and hearing is often implicit in the theoretical part of teacher education but can be made much more explicit during practice and a link between the parts is necessary for the development of the professional language. The importance of listening, as emphasised by Cooney and Krainer (1996), also relates to the issue of mutual respect between the different agents in educational games, such as between teachers and students in classroom settings or between teacher educators and teachers in professional development activities, forming a basis for co-learning arrangements (see Jaworski, 2001).

Rosu and Arvold (2005) report on a study of questioning that took place in a secondary mathematics teacher education programme. After an initial sequence of investigations into questioning, the student teachers studied questioning in practice during their second semester practicum, including meetings and discussions. Whether the focus in these discussions was on learning *for* or *in* practice, these studies "generated a milieu of learning appropriate for the multiple meanings and contexts of teaching experiences" (ibid., p. 4). It was observed that an inquiry approach was supported by the study of questioning, which helped student teachers to develop knowledge on students' mathematical understanding. However, this focus on questioning also created a milieu in teacher education where questioning as practice and questioning as theory do not come into conflict. To develop and maintain an inquiry stance in teachers, the study reported supports a practice of questioning as a learning milieu.

Post-observation meetings. Important in the work to link programme coursework to practicum is the work of the mentor, as well as the student teacher's post-teaching discussions with the mentor and the teacher educator that take place within many programmes. Since such conversations traditionally tend to have an evaluative character in terms of normative statements, rather than focus on epistemological aspects of mathematical knowledge and learning, they risk hindering the student teacher's development of a teacher identity. To avoid this,

Johnsen Høines and Lode (2007) investigated didactical conditions for a *subject based discussion* to support a more productive reflective approach. This is also emphasised by Mewborn (2000), who suggests that "field experiences are a site for content-specific *instruction*" (p. 42). To learn, in the sense of developing a teacher identity, from imitation of the mentor as a model teacher, the student teacher also needs to understand the rationale behind the activities of the mentor (Nilssen, 2003).

Recognising that post-observation meetings during practicum between student teacher, mentor, and teacher educator often tend to focus on classroom management rather than on aspects of how mathematical knowledge *per se* has been handled during the lesson (Brown, McNamara, Hanley, & Jones, 1999; Mewborn, 2000), Rowland, Thwaites and Huckstep (2005) suggest an empirically based framework called the *knowledge quartet* aimed at giving a structure to such discussions of teachers' mathematical knowledge in the classroom. The first dimension of *foundation* refers to subject matter knowledge as well as beliefs and understandings related to the teaching and learning of mathematics developed during academic course work. During lesson planning and actual teaching, teachers' "knowledge-in-action" defines the second dimension, *transformation*. Of interest here is for example how examples are chosen and used to support student learning. How the teacher provides links and handles different cognitive demands of separate parts of mathematical content constitutes the third dimension of the quartet, *connection*. Finally, to account for the unexpected, for decisions impossible to plan for about developments of classroom activity, the dimension of *contingency* completes the quartet. Elements of mathematical knowledge in lesson episodes can be captured and understood, in discussions at post-observation meetings during practicum, when structured by the four dimensions of the knowledge quartet (ibid.).

Based on the two assumptions that a lesson in mathematics is "exactly what those involved see in it" and that classroom interaction is very complex, depending on the mathematical content under discussion, lines of arguments used, interaction patterns, and how students participate, Gellert and Krummheuer (2005) argue that a focus on a "collaborative interpretation of classroom interaction" (p. 2) may be productive for learning from teaching practice. To be able to "uncover" what was behind the development, or flow, of a lesson, a group of teachers and student teachers, along with the researchers, met to analyse a chosen videotape of a 15 minutes lesson sequence. To give structure to the interpretations produced, three techniques developed in mathematics education research were adopted, that is interaction analysis, argumentation analysis, and participation analysis. Seen as members of a "community of interpretation" (p. 3), this group was also involved in different communities of practice, and moved during the meetings from peripheral to full participation in this community as they became more competent in classroom interaction interpretation. By using a heterogeneous community of interpretation it was possible to make different interpretations of classroom interaction visible and as a consequence open up for change and development of teaching approaches. The rationale behind this outcome relates to differences and changes of perspective, contrasting the "centred stance of teaching practitioners" and "de-centred stance of observers" in the "re-centring stance of legitimate self-regulation of a community of interpretation" (p. 5).

Practices with a focus on writing. As a means for analysing practicum, and providing own lived experiences as cases for reflection in theoretical didactic courses, Chapman (2005) suggests an approach of using *stories of practice* in a teacher education programme. Rather than judgements about good or bad teaching, the focus is on sense making. As the first stage of a sequence of four, student teachers are asked to write one story of "good" teaching, one of "bad" teaching, and one "memorable", from their own teaching during practicum, or from their own observations of teaching. The story is to include a complete mathematics lesson with as much detail as possible, including what teachers and pupils have said, but should be descriptive rather than normative. The second stage is one of initial self-reflection, where the student teachers write journals on why they think their stories represent good or bad teaching, which they share and discuss with their peers, providing reasons for what stories they like or do not like. During the third stage the stories are used during the semester to interpret theory and for the analysis of actual practice during practicum with a focus on making sense of mathematical content and discourse in the classroom, as well as alternative approaches. The fourth and final stage, at the end of the course/semester, aims at a final self-reflection by rewriting the previous story "in the way he/she would want it to unfold", in order to "provide an alternative way of conducting it in term of engaging students in the content to facilitate deep understanding of it" (Chapman, 2005, p. 4). The student teachers then write journals to compare their two stories, to share and discuss with their peers. Data from an observed sequence with 26 student teachers showed that this approach of writing stories of practice provided a constructive means to articulate their thinking of mathematical teaching and learning in a holistic way, with self-reflections prompting conflicting beliefs and shifts of thinking. The analysis of practicum initiated an increased awareness of critical aspects of teaching not previously noted, leading to a more inquiry-oriented approach and recognition of the need of a deep understanding of mathematics to be able to support their students' learning.

Another kind of story writing was involved in a practice described and evaluated in Bergsten and Grevholm (2004, pp. 132-137). During the second semester of the mathematics studies in a teacher education programme for secondary school level, a sandwich model of mixed course work in mathematics education (C) and teaching practice (P) was designed for a 1C+2P+2C+3P weeks period, with a concluding summing up seminar at the university. In this way theory and practice could benefit from each other, emphasised by the particular tasks designed for the practicum periods. In particular, for the final practicum period, where the main work for the student teacher was taking full responsibility for the teaching in one mathematics class, supervised by the ordinary teacher of the class (the mentor), the examining task was to write an essay titled *This is how I am as a mathematics teacher*. As a basis for this essay the student teacher kept a diary of all the lessons he or she had taught during the three weeks, and made a video tape of one lesson, which was discussed with the mentor. In addition, reflections from the course work and literature were included. These essays proved to be of high value for the student teachers as well as for the teacher educators/researchers. For example, the authors note that

> the importance of being aware of your own views on the teaching and learning process and of mathematics, and how you act and develop as a teacher, is frequently touched upon in the essays. (ibid., p. 136)

The essay writing task was also

> seen as valuable to develop such self awareness, and by some also as the task that opened up the eyes to see that this is important. (ibid., p. 136)

By the reading of these essays and the discussions in the follow-up seminar the teacher educators learned much about their students' experiences of practicum and its relation to course work to build on in future educational activities.

There are also other aspects of writing in teacher education as a means for learning and for connecting theory and practice. Problems exist with how to examine different parts of the education in order for students to give value to all parts. One such issue concerns the research based aspects of the education (see Grevholm, 2004), and how to make those visible during practicum. Grevholm, Berg and Johnsen (2006) give an example of a task for practicum, where student teachers are acting as research assistants and doing data-collection within a larger research project. They interview pupils individually based on answers given in a written test. The students experience that interviewing is a valuable instrument for learning about pupils' level of achievement in mathematics. This task links theory and practicum in multiple ways, where the student teachers in the required written report need to make these links visible and reflect on the role of the professional language.

Seeing lessons as experiments. In science knowledge grows through experimentation. Similarly, by treating lessons as experiments knowledge about teaching develops and grows, and is suggested by Hiebert et al. (2003) as a viable method to learn to teach. It is necessary to work on experience to learn from it:

> teachers need to design lessons with clear goals in mind, monitor their implementation, collect feedback, and interpret the feedback in order to revise and improve future practice. (ibid., p. 206)

This type of practice, related to the lesson study construct (Stigler & Hiebert, 1999), is suggested also for mathematics teacher educators for them to learn from their practice, that is a model where treating "lessons (for prospective teachers) as experiments becomes the routine, on-going activity for course instructors" (ibid., p. 214). Such ways of working also set up a model of working for the prospective teachers taking part in these "experiments".

Common for all the linking practices presented above, with a focus on teachers' professional language, post-observation discussions, student teachers' writing, or treating lessons as experiments, is that they are initiated and designed by the mathematics teacher educators to connect student teachers' experiences during practicum to a theoretical research based didactical discourse. There is a deliberate purpose to engage student teachers, and often their mentors as well as themselves, in activities that require an inquiry stance of noticing and being aware, building on communication and respect between the different agents in collaborating and supporting environments, while maintaining a focus on mathematical knowledge. In the next section we will discuss the role, demands and learning of the mathematics educator in relation to these linking practices, and what we mean by a knowledgeable teacher educator.

MATHEMATICS TEACHER EDUCATORS AS KNOWLEDGEABLE PERSONS

The linking practices described above focus on different aspects of links between theoretical course work and teaching practice (practicum), that is between main strands of a teacher education programme often more or less separated by the didactic divide. Where this divide is visible in the overall organisation of the programme (i.e. model 2 in Jaworski & Gellert, 2003, p. 832, where four different models of organisation are discussed), the rationale is that theory studied in course work is being applied during practicum. However, due to placement factors, influence of teaching traditions and the beginning teacher's struggles with the practical management of teaching, what happens is often that practicum lives its own life independent of the academy within the autonomy of the local practice (cf. Steinbring, 1998). Most of the practices described require an organisation with a built in structure for facilitating feedback processes of communication and reflection (as in models 3 and the ideal model 4 in Jaworski & Gellert, 2003). They also have in common an emphasis on process aspects, such as a reflective stance and activities of communication within different kinds of groups of co-learners, which may be seen as a consequence of the developmental trend observed that "teacher education is moving toward a more process orientation in which teachers are encouraged to be reflective beings" (Cooney & Krainer, 1996, p. 1163).

In practices related to a professional language it seems necessary to exploit the link to practicum in order to *experience* the power of questioning and listening as well as its difficulties. Grevholm and Bergsten (2005) designed the practice to create an experiential basis to *motivate* student teachers to work with their language, that is to establish the relevance of the task. The task design also integrated direct work on mathematical knowledge, thus linking language and the object of teaching. The questioning practice (Rosu & Arvold, 2005) used a multiple linking process from a theoretical study to investigation and discussion in practicum, by *creating a learning milieu* within the teacher education programme that bridged the didactic divide and enhanced an inquiry stance. The design and implementation of these practices could happen only because the mathematics educators *found it important* for the student teachers to discover, experience and reflect upon the role of language, questioning and listening, in order to be able to act for the best of their future pupils. To make the practice functional to this end the educators need to possess an enactive knowledge of the phenomenon of professional languages and of the mathematical content under discussion, as well as an ability to notice and respond to the students' work on the task.

The practice reporting on stories (Chapman, 2005), based on a constructivist approach as it took its starting points in the student teachers' own stories from their own experiences, supported their development of an increasing awareness of mathematics teaching as well as an inquiry-oriented approach with a focus on mathematical knowledge. In common with the essay task in the practice reported in Bergsten and Grevholm (2004) was writing as a medium for analysis and reflection. Here linking course work and practicum was the very format of the task design, made possible by the structure of the programme. An increased critical awareness of the student teacher's own positioning was one of the benefits of this practice. By the documentations produced in these practices, and the group discussions taking place, the mathematics teacher educators could increase their own awareness of not only their students' awareness but also of their responses to issues discussed in the course work. These essays and stories also provided rich sources of critical classroom events that the student teachers noticed and found worthwhile to reflect on, some of which the teacher educators could use to support theoretical course work.

One kind of practice that explicitly addresses the link between the different home or visit arenas (i.e. the university and the school) as well as between courses and practicum, is post-teaching discussions between the student teacher (whose teaching is the object of discussion), the mathematics teacher, and the mathematics teacher educator (Johnsen Høines & Lode, 2007; Gellert & Krummheuer, 2005; Rowland et al., 2005). In one case the process from theory to practice served itself as the topic of discussion. By the persons represented in the "community of interpretation" their different perspectives are naturally displayed and reflected upon. That the practices presented explicitly set focus on subject matter issues rather than evaluative or class management matters support a co-learning setting with reduced power relations in play. Being present in all three milieus, that is course work at the home arena and a taught lesson and follow-up discussion at the visiting arena, the mathematics teacher educator gets involved in own teaching, learning, and reflecting activities as well as in those of the student teacher(s). Such shifts of attention necessitate an awareness of the phenomena in focus and how they are being noticed by the others in the mutual work. This puts demands on the educator to be knowledgeable in the activity, which also requires a knowledge or competence basis on which to build professional ways to take the participating agents' reflections and knowledge further.

Seeing the lesson as an experiment has a flavour around it as being the most natural way to link theoretical course work in a teacher education programme to a teaching practice that is reflected upon and discussed back at the university. Using it as a practice also for the teacher educators' teaching in their own courses for student teachers (Hiebert et al., 2003) would not only parallel their learning of their own practice to that of the student teachers when doing their practicum, but also offer an indirect learning of how the student teachers may experience learning of their own practice. This would contribute to the improvement of their practices for linking course work and practicum.

In all these practices a "twofold demand" is placed on mathematics educators, since it "requires attention to the state-of-art developments in the field such as considering mathematics learning as a social, constructive process" as well as seeing these developments as "relevant to teachers' practical concerns" (Cooney & Krainer, 1996, pp. 1168-1169). The linking practices aim at connecting these two dimensions as worked on in theoretical courses and teaching practice, respectively, and thereby bridging the didactic divide. The examples illustrate the need of mathematics teacher educators not only to have knowledge and experience in those competence areas involved in the work of mathematics teachers discussed above, but also to be knowledgeable both in the activities they design, such as the linking practices, and in such teaching and learning activities in which these practices are designed to make the student teachers knowledgeable. We have tried to provide a hint of the complexity of and requirements for this enterprise by displaying the wide range of practices and levels of analysis involved at different levels in the educational world of mathematics discussed in the preceding sections. In Table 2 a sketch of some of the aspects of what we mean by a knowledgeable teacher educator is outlined.

Table 2. The knowledgeable teacher educator

	Past	Present	Future
State	Learnt from theory and practice	Awareness and sensitivity	Stance toward continued learning
Action	Building on earlier experiences	Noticing and inquiring	Further inquiry
Ethics	Know where I stand	Care and decide for next step	Toward what is better for humans

For the mathematics teacher educator, as any person, there is a past, a present and a future. He or she has in the past learnt from theory and practice and is now using those earlier experiences in being aware *and* present *and* noticing *and* inquiring into mathematics learning and teaching at all levels. For the future he or she will continue to learn by being a curious person who inquires further into that area. The teacher educator has made it clear to him/herself his or her positioning (in knowledge theory) and is able to take the next step going further to a situation better for humans. Since what at one specific point on the timeline belonged to the future later will belong to the past, there is also in Table 2 a hidden cyclic process of learning and development of the *educative power* of the mathematics teacher educator.

Just as for students to learn mathematics or for teachers to learn mathematics teaching there is no ready made 'parcel' to buy from somebody else for a mathematics teacher educator to learn from about mathematics teaching and learning. Seeing this educational world of mathematics as a system, with autonomous subsystems linked together hierarchically, all its agents live there together and must learn and act together to move forward.

REFERENCES

Adler, J. (2005). Mathematics for teaching: What is it and why do we need to talk about it. *Pythagoras*, *62*, 2–11.

Adler, J., Ball, D., Krainer, K., Lin, F-L., & Novotna, J. (2005). Reflections on an emerging field: Researching mathematics teacher education. *Educational Studies in Mathematics*, *60*, 359–381.

Ball, D. L., & Cohen, D. K. (1999). Developing practice, developing practitioners: Toward a practice-based theory of professional education. In L. Darling-Hammond, & G. Sykes (Eds.), *Teaching as the learning profession: Handbook of policy and practice* (pp. 3–32). San Francisco, CA: Jossey-Bass.

Ball, D. L., & Bass, H. (2004). Knowing mathematics for teaching. In R. Stræsser, G. Brandell, B. Grevholm, & O. Helenius (Eds.), *Educating for the Future. Proceedings of an International Symposium on Mathematics Teacher Education* (pp. 159–178). Stockholm: The Royal Swedish Academy of Sciences.

Barbé, J., Bosch, M., Espinoza, L., & Gascon, J. (2005). Didactic restrictions on the teacher's practice: The case of limits of functions in Spanish high schools. *Educational Studies in Mathematics*, *59*, 235–268.

Bednarz, N., & Proulx, J. (2005). Practices in mathematics teacher education programs and classroom practices of future teachers: From the teacher educator's perspectives and rationales to the interpretation of them by the future teachers. In *The Fifteenth ICMI Study. The professional education and development of teachers of mathematics*. Águas de Lindóia, Brazil, 15–21 May 2005. (CD-ROM proceedings)

Bergsten, C., & Grevholm, B. (2004). The didactic divide and the education of teachers of mathematics in Sweden. *Nordic Studies in Mathematics Education*, *9*, 123–144.

Bergsten, C., & Grevholm, B. (2005). The didactic divide and educational change. In *The Fifteenth ICMI Study. The professional education and development of teachers of mathematics*. Águas de Lindóia, Brazil, 15–21 May 2005. (CD-ROM proceedings)

Breen, C. (1992). Teacher education and mathematics: Confronting preconceptions. *Perspectives in Education*, *13*, 33–44.

Brown, T., McNamara, O., Hanley, U., & Jones, L. (1999). Primary student teachers' understanding of mathematics and its teaching. *British Education Research Journal*, *25*, 299–322.

Chapman, O. (2005). Stories of practice: A tool in pre-service secondary mathematics teacher education. In *The Fifteenth ICMI Study. The professional education and development of teachers of mathematics*. Águas de Lindóia, Brazil, 15–21 May 2005. (CD-ROM proceedings)

Comiti, C., & Ball, D. (1996). Preparing teachers to teach mathematics: A comparative perspective. In A. Bishop et al. (Eds.), *International handbook of mathematics education, part 2* (pp. 1123–1153). Dordrecht: Kluwer Academic Publishers.

Cooney, T. (1994). Teacher education as an exercise in adaptation. In D. B. Aichele & A. F. Coxford (Eds.), *Professional development of teachers of mathematics: 1994 yearbook*. Reston, VA: National Council of Teachers of Mathematics.

Cooney, T., & Krainer, K. (1996). Inservice mathematics teacher education: The importance of listening. In A. Bishop et al. (Eds.), *International handbook of mathematics education, part 2* (pp. 1155–1185). Dordrecht: Kluwer Academic Publishers.

Dewey, J. (1933). *How we think: A statement of the relation of reflective thinking to the educative process*. Boston: D.C. Heath and Co.

Doerr, H. M., & Thompson, T. (2004). Understanding teacher educators and their pre-service teachers through multi-media case studies of practice. *Journal of Mathematics Teacher Education*, 7, 175–201.

Even, R. (1999). The development of teacher leaders and inservice teacher educators. *Journal of Mathematics Teacher Education*, 2, 3–24.

Favilli, F. (Ed.) (2006). *LOSSTT-IN-MATH - Lower secondary school teacher training in mathematics. Comparison and best practices*. Pisa: PLUS – Pisa University Press.

Gattegno, C. (1970). *What we owe children*. London: Routledge and Kegan Paul.

Gellert, U., & Krummheuer, G. (2005). Collaborative interpretation of classroom interaction: Stimulating practice by systematic analysis of videotaped classroom episodes. In *The Fifteenth ICMI Study. The professional education and development of teachers of mathematics*. Águas de Lindóia, Brazil, 15–21 May 2005. (CD-ROM proceedings)

Grevholm, B. (2002). Lärarutbildning – Utbud, utbildare och anordnare (Teacher education – Programmes, educators and organisations). NCM-rapport 2002:1. Göteborg: NCM, Göteborgs universitet.

Grevholm, B. (2004). What does it mean for mathematics teacher education to be research based? In R. Stræsser, G. Brandell, B. Grevholm, & O. Helenius (Eds.), *Educating for the Future. Proceedings of an International Symposium on Mathematics Teacher Education* (pp. 119–134). Stockholm: The Royal Swedish Academy of Sciences.

Grevholm, B. (2006). Matematikdidaktikens möjligheter i en forskningsbaserad utbildning (The opportunities of mathematics education in a research based teacher education). In S. Ongstad (Ed.), *Fag og didaktikk i lærerutdanning. Kunskap i grenseland* (pp. 183–206). Oslo: Universitetsforlaget.

Grevholm, B., Even, R., Szendrei, J. & Carillo, J. (2004). From a study of teaching practices to issues in teacher education. Thematic Working Group 12, CERME3. In M. A. Mariotti et al. (Eds.), *Proceedings of CERME3*. Electronic publication. Pisa: University of Pisa.

Grevholm, B., & Bergsten, C. (2005). The development of a professional language in mathematics teacher education. In *The Fifteenth ICMI Study. The professional education and development of teachers of mathematics*. Águas de Lindóia, Brazil, 15–21 May 2005 (CD-ROM proceedings).

Grevholm, B., Berg, C., & Johnsen, V. (2006). Student teachers' participation in a research project in mathematics education. In E. Abel, R. Kudzma, M. Lepik, T. Lepmann, J. Mencis, M. M. Ivanov. & M. Tamm (Eds.), *Teaching mathematics: Retrospective and perspectives*, 7th international conference May 12–13, 2006 (pp. 61–68). Tartu, Estonia: Tartu Ulikool.

Grossman, P. L. (1990). *The making of a teacher: Teacher knowledge and teacher education*. New York: Teachers College Press.

Hiebert, J., Morris, A. K., & Glass, B. (2003). Learning to learn to teach: An "experiment" model for teaching and teacher preparation in mathematics. *Journal of Mathematics Teacher Education*, 6, 201–222.

Jaworski, B. (1994). *Investigating mathematics teaching: A constructivist enquiry*. London: Falmer Press.

Jaworski, B. (2001). Developing mathematics teaching: Teachers, teacher educators, and researchers as co-learners. In F.-L. Lin, & T. J. Cooney (Eds.), *Making sense of mathematics teacher education* (pp. 295–320). Dordrecht: Kluwer Academic Publishers.

Jaworski, B., & Gellert, U. (2003). Educating new mathematics teachers: Integrating theory and practice, and the role of practising teachers. In A. J. Bishop, M. A. Clements, C. Keitel, J. Kilpatrick, & F. K. S. Leung (Eds.), *Second international handbook of mathematics education* (pp. 829–9875). Dordrecht: Kluwer Academic Publishers.

Johnsen Høines, M. & Lode, B. (2007). Meta-level mathematics discussions in practice teaching: An investigative approach. In C. Bergsten, B. Grevholm, H. Strømskag Måsøval, & F. Rønning (Eds.), *Relating practice and research in mathematics education, Proceedings of NORMA 05* (pp. 311–323). Trondheim: Tapir Academic Press.

Kilpatrick, J. (2004). Promoting the proficiency of U.S. Mathematics teachers through Centers for learning and teaching. In R. Stræsser, G. Brandell, B. Grevholm, O. Helenius (Eds.), *Educating for *the future. Proceedings of an international symposium on mathematics teacher education* (pp. 143–157). Stockholm: The Royal Swedish Academy of Sciences.

Kilpatrick, J., Swafford, J., & Findell, B. (Eds.) (2001). *Adding it up: Helping children learn mathematics.* Washington DC: National Academy Press.

Lampert, M., & Ball, D. L. (1999). Aligning teacher education with contemporary K-12 reform visions. In L. Darling-Hammond & G. Sykes (Eds.), *Teaching as the learning profession: Handbook of policy and practice* (pp. 33–53). San Francisco: Jossey-Bass.

Lave, J. (1988). *Cognition in practice: Mind, mathematics and culture in everyday life.* Cambridge, UK: Cambridge University Press.

Lerman, S. (2001). A review of research perspectives on mathematics teacher education. In F. Lin & T. Cooney (Eds.), *Making sense of mathematics teacher education* (pp. 33–52). Dordrecht: Kluwer Academic Publishers.

Mason, J. (2002). *Researching your own practice: The discipline of noticing*, London: Routledge Falmer.

Mason, J. (2007). ICMI Rome 2008. Notes towards WG2. Paper to be presented at the Symposium on the occasion of the 100th anniversary of ICMI, Rome 5–8 March 2008.

McGraw, R., Lynch, K., Koc, Y., Budak, A., & Brown, C. A. (2007). The multimedia case as a tool for professional development: An analysis of online and face-to-face interaction among mathematics pre-service teachers, in-service teachers, mathematicians, and mathematics teacher educators. *Journal of Mathematics Teacher Education, 10,* 95–121.

Mewborn, D. (2000). Learning to teach elementary mathematics: Ecological elements of a field experience. *Journal of Mathematics Teacher Education, 3,* 27–46.

Molander, B. (1998). Kunskapens former och beteckningar – och kunniga människor (Forms and representations of knowledge – and knowledgeable persons). *Utposten, Nr 5* [available 2008-01-02 at www.uib.no/isf/utposten/1998nr5/utp98506.htm]

Moyer, P. S., & Milewicz, E. (2002). Learning to question: Categories of questioning used by preservice teachers during diagnostic mathematics interviews. *Journal of Mathematics Teacher Education, 5,* 293–315.

Nilssen, V. (2003). Mentoring teaching of mathematics in teacher education. In N. A. Pateman, B. J. Doherty, & J. Zilliox (Eds.), *Proceedings of the 27th Conference of the International Group for the Psychology of Mathematics Education* (Vol. 3, pp. 381–389). Honolulu, HI: PME.

Niss, M. (1999). Aspects of the nature and state of research in mathematics education, *Educational Studies in Mathematics, 40,* 1–24.

Niss, M. (2004). The Danish KOM-project and possible consequences for teacher education. In R. Stræsser, G. Brandell, B. Grevholm, & O. Helenius (Eds.), *Educating for the future. Proceedings of an international symposium on mathematics teacher education* (pp. 179–190). Stockholm: The Royal Swedish Academy of Sciences.

Peled, I., & Hershkovitz, S. (2004). Evolving research of mathematics teacher educators: The case of non-standard issues in solving standard problems. *Journal of Mathematics Teacher Education, 7,* 299–327.

Rosu, L. M., & Arvold, B. (2005). Questioning as a learning milieu in mathematics teacher education. In *The Fifteenth ICMI Study. The professional education and development of teachers of mathematics.* Águas de Lindóia, Brazil, 15–21 May 2005 (CD-ROM proceedings).

Rowland, T., Thwaites, A., & Huckstep, P. (2005). Elementary teachers' mathematics subject knowledge: The knowledge quartet and the case of Naomi. *Journal of Mathematics Teacher Education, 8,* 255–281.

Rye Ejersbo, L. (2007). *Design and redesign of an in-service course: The interplay of theory and practice in learning to teach mathematics with open problems.* Doctoral thesis. Copenhagen: Danish Pedagogical University.

Shulman, L. S. (1986). Those who understand: Knowledge growth in teaching. *Educational Researcher, 15,* 4–14.

Sowder, J. (2007). The mathematical education and development of teachers. In F. K. Lester (Ed.), *Second handbook of research on mathematics teaching and learning* (pp. 157–223). Charlotte, NC: Information Age Publishing.

Steinbring, H. (1998). Elements of epistemological knowledge for mathematics teachers. *Journal of Mathematics Teacher Education, 1*, 157–189.

Stigler, J. W., & Hiebert, J. (1999). *The teaching gap: Best ideas from the world's teachers for improving education in the classroom.* New York: Free Press.

Tzur, R. (2001). Becoming a mathematics teacher educator: Conceptualising the terrain through self-reflective analysis. *Journal of Mathematics Teacher Education, 4*, 259–283.

Wallach, T., & Even, R. (2005). Hearing students: The complexity of understanding what they are saying, showing, and doing. *Journal of Mathematics Teacher Education, 8*, 393–417.

Walshaw, M. (2004a). Pre-service mathematics teaching in the context of schools: An exploration into the constitution of identity. *Journal of Mathematics Teacher Education, 7*, 63–86.

Walshaw, M. (2004b). Becoming knowledgeable in practice: The constitution of secondary teaching identity. In I. Putt, R. Faragher, & M. McLean (Eds.), *Mathematics education for the third millennium: Towards 2010. Proceedings of the 27th Annual Conference of the Mathematics Education Research group of Australasia* (pp. 557-564). Sydney: MERGA, Inc.

Watson, A., & Mason, J. (2007). Taken-as-shared: A review of common assumptions about mathematical tasks in teacher education. *Journal of Mathematics Teacher Education, 10*, 205–215.

Wilson, P. S., Cooney, T. J., & Stinson, D. W. (2005). What constitutes good mathematics teaching and how it develops: Nine high school teachers' perspectives. *Journal of Mathematics Teacher Education, 8*, 83–111.

Winsløw, C., & Durrand-Guerrier, V. (2007). Education of lower secondary mathematics teachers in Denmark and France. *Nordic Studies in Mathematics Education, 12*(2), 5–32.

Wyndhamn, J. (1993). Problem-solving revisited: On school mathematics as a situated practice. *Linköping Studies in Arts and Science, 98*. Linköping: Linköping University.

Zaslavsky, O. (2007). Mathematics-related tasks, teacher education, and teacher educators. The dynamics associated with tasks in mathematics teacher education. *Journal of Mathematics Teacher Education, 10*, 433–440.

Zaslavsky, O., Chapman, O., & Leikin, R. (2003). Professional development of mathematics teacher educators: Trends and tasks. In A. J. Bishop, M. A. Clements, C. Keitel, J. Kilpatrick, & F. K. S. Leung (Eds.), *Second international handbook of mathematics education* (pp. 877–917). Dordrecht: Kluwer Academic Publishers.

Zaslavsky, O., & Leikin, R. (2004). Professional development of mathematics teacher educators: Growth through practice. *Journal of Mathematics Teacher Education, 7*, 5–32.

Christer Bergsten
Linköping University

Barbro Grevholm
University of Agder

AMY ROTH MC DUFFIE, COREY DRAKE, AND
BETH HERBEL-EISENMANN

12. THE ELEMENTARY MATHEMATICS METHODS COURSE

Three Professors' Experiences, Foci, and Challenges

This chapter presents mathematics teacher educators' work with prospective elementary teachers (university students in their final stages of teacher preparation) in a "Methods of Teaching Elementary Mathematics" course. In the U.S. and other countries, this kind of course typically focuses on developing knowledge and skills for teaching in conjunction with understanding students' mathematics learning. We discuss our common course goals, key activities and resources that we believe are important for the development of these students as elementary school mathematics teachers. We then examine our roles in facilitating these activities and our own professional development through participation in practice. Our intent is to critically describe, analyse, and reflect on our practices as instructors of this course and to extend our personal discussion to a more public dialogue about key goals and activities for such a methods course.

INTRODUCTION

Over the past few years, mathematics teacher educators (MTEs) have recognised two critical needs in teacher preparation: (1) connecting research and theory to practice by situating learning in authentic experiences (e.g., Lave & Wenger, 1991; Putman & Borko, 2000) and, (2) developing mathematics knowledge for teaching as a specialised form of mathematics knowledge (e.g., Ball, Lubienski, & Mewborn, 2001). Researchers have studied specific strategies, approaches, and activities that support prospective teachers' development including: preparing lessons as a vehicle for anticipating and understanding teaching and learning (Hiebert, Morris, Berk, & Jansen, 2007; Peterson, 2005); using K-12 mathematics curriculum materials to develop teachers' understandings of teaching and learning (Lloyd, 2006; Papick, Beem, Reys, & Reys, 1999); selecting and using rich mathematics problems to promote teacher learning (Crespo & Featherstone, 2006; Flowers & Rubenstein, 2006); using case studies to situate teachers' learning in practice (Silver, Mills, Castro, & Ghousseini, 2006); and interviewing children to gain understanding of their mathematical thinking (Jacobs, Ambrose, Clement, & Brown, 2006; Moyer-Packenham, 2004; Philipp et al., 2007). Although the field of mathematics teacher education is developing a research base for specific strategies and activities to facilitate prospective teachers' learning, there still needs to be more discussion about our practice--how to draw on these various approaches and activities as we design and teach our mathematics "methods" courses.[1] Indeed, in a preliminary analysis of North American mathematics methods course syllabi, Taylor (2006) found wide variability in goals, objectives and assignments, underscoring the need for MTEs to focus on the purposeful design of mathematics methods courses. More concentrated attention to course goals and overall design could help MTEs make more informed decisions about which activities and strategies to use under particular circumstances. From a perspective of *teaching as learning in practice* (Lave, 1996; Jaworski, 2006), we describe and analyse our methods courses for prospective elementary teachers as a first step in engaging in discussions about our practices as MTEs and about the goals, limitations, key activities, and resources involved in the design of methods courses more generally.

COURSE GOALS AND CONTEXTS

This chapter represents a synthesis of many conversations and collegial experiences among the three authors. Although we completed graduate studies at different universities and independently developed our methods courses, several years ago we began conversations while attending professional meetings and found that we shared an interest in research on teachers' learning and development. In autumn, 2006 Beth and Corey organized and Amy attended a 3-day meeting with MTEs who were interested in researching teaching and learning practices in the elementary mathematics methods courses in North America. At this meeting the group of MTEs presented perspectives and activities that were found effective for prospective teachers' learning and discussed a research agenda for the field. Through these experiences we, the three authors, were struck by how similar our courses and assignments were, but we also realized that we could each improve our individual practices and our courses by sharing and clarifying what we do, how our courses evolved, and why we have made certain decisions to make changes in the our instruction. As we collaborated on this chapter and looked back on our practices in this course, the discussions and process of writing the chapter clarified further the common values, beliefs, and goals that were evident in our practices. Below, we briefly describe the process we used in writing this chapter.

Individually, we each wrote about our core beliefs and essential aspects and activities of our courses. We then shared, examined, and discussed our collection of statements. *Prospective teachers developing understanding of children's mathematical thinking and learning* quickly emerged as a central element in framing our courses and decision-making. In reflecting on why children's mathematical thinking seemed to be so important, we believed that this emphasis

---

[1] The *elementary mathematics methods* course typically is a required course in teacher education programs in North America, as it was in the programs in which we taught. The course usually is taught after the mathematics content courses are completed and prior to a full-time student teaching practicum. The course focuses on developing pedagogical knowledge for teaching and learning mathematics.

was a result of the influence of Cognitively Guided Instruction (CGI), research on elementary teaching and learning practices (Carpenter, Fennema, Franke, Levi, & Empson, 1999). From our perspective, this research project underscored the need for a paradigm shift in the field from thinking about teaching in terms of how to present material to thinking about teaching in terms of how children might understand and learn concepts. Although the field of mathematics education has been in the process of shifting views for many years (cf. National Council of Teachers of Mathematics [NCTM], 1989), we found that many of our prospective teachers continue to enter our classes with the former view on teaching mathematics and the need for many experiences in order to begin to view teaching and learning differently. Moreover, research had shown that the CGI framework has great potential for affecting prospective teachers' beliefs about teaching mathematics (Vacc & Bright, 1999). We believed that focusing on children's thinking and learning was a promising approach. Indeed, although we agreed that our courses extend well beyond the ideas presented in CGI, the underlying perspective of focusing on children's ways of thinking and doing mathematics permeated what we valued in our courses. Next, our discussion turned to articulating specific goals that represented this perspective.

Course goals. After continuing to share detailed information about our courses, we identified a unifying and common goal for our students across our courses: This goal was to prepare prospective teachers to be deliberate and reflective practitioners whose instructional decisions are grounded in *children's mathematical thinking and learning* (CMTL). This goal is central to our courses because we believe that attending to CMTL is foundational to effective and equitable teaching practices (i.e., planning, implementing, and assessing for students' learning) and consistent with current research and theory on teaching and learning (e.g., Donovan et al., 1999). After identifying this central goal, we discussed key components of our courses that we perceived as needed to support prospective teachers in using children's thinking as a foundation for their future practice. This discussion led us to the ideas shown in Figure 1.

This figure represents our three common (and overlapping) goals with CMTL at the centre for each of the goals. In the overlapping regions, we included specific teaching activities we used in our courses. In the process of writing this chapter, we discussed activities that we individually used in our courses, but writing the chapter created a new lens for examining these activities. Once again we found it interesting that we shared many common activities in our courses. In reflecting on how we had arrived at such similar instructional decisions, we came to the realization that this outcome reflected a widely held belief among MTEs that teachers need to develop skills for listening to children's ideas and thinking, for reflecting on their practice in light of student thinking, and for planning and adjusting instruction based on children's understandings and experiences (c.f. NCTM 1991). Although the activities described do not comprise all of the experiences that we might include in any given course, we selected them because for us, they represent some of the most essential experiences for prospective teachers in order for them to develop understandings and perspectives about

CMTL. Faced with the constraint of very limited instructional time - only one semester (typically 15 weeks) for the course at our respective universities - we discussed how as our courses evolved, we each (independently) moved toward emphasising a few key areas for prospective teachers' learning. Following the NCTM Standards (1989, 2000) the mathematical content area most emphasized was Number and Operation, paralleling the primary focus in North American elementary school mathematics curricula; the mathematical process area was problem solving and reasoning. Focusing on these selected topics reflected current research on learning that calls for learning deeply key ideas and processes and connecting those ideas to form stronger and more meaningful understandings (e.g., Donovan et al., 1999).

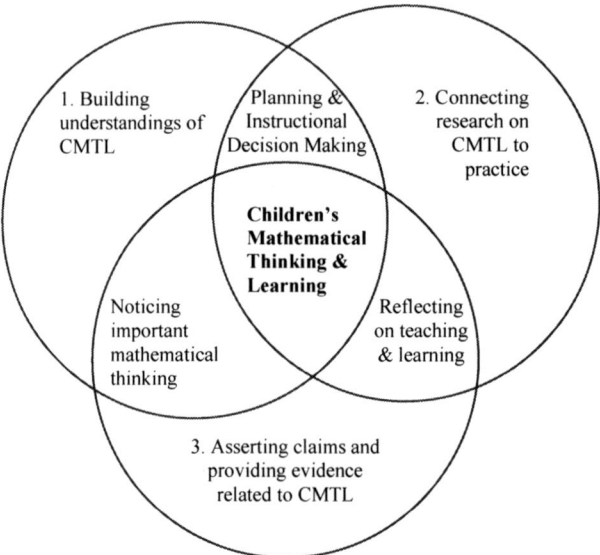

Figure 1. Course Goals Framework Focused on Children's Mathematical Thinking and Learning (CMTL)

Course contexts. As graduate students and later as faculty, we each taught courses enrolling approximately 25 students in a wide-range of situations that reflect the diverse ways in which mathematics methods courses are taught in North America.

For example of the variability of our collective experience, we taught this course over a semester to a year in duration and with practicum experiences that range from no practicum placement offered with the course to a few hours of practicum experiences concurrent with or at the end of the course to university courses held in schools where prospective teachers participate in intense school-based work. The practicum influenced whether and how field-based activities and assignments were used in the courses. For example, when a practicum experience connected closely with the course and ran throughout the semester, we could assume regular interaction between our prospective teachers and pupils, coordinate university work with school-based experiences, and require prospective teachers to report on observations, interactions and/or work with children from their practicum placement. Without an associated practicum, the logistics involved with arranging for prospective teachers to work with children were a challenge, and thus, we either had to be flexible about how activities were completed with children or we were unable simply to assign activities that required our prospective teachers to work with children.

ACTIVITIES DESIGNED TO MEET COURSE GOALS AND THEIR EVOLUTION

To illustrate how we attempted to meet the goals shown in Figure 1, we present activities that we found best exemplified our common goals and reflected our belief that prospective teachers' learning should be situated in practice (Lave, 1996; Jaworski, 2006). We categorize the activities and corresponding assignments into two types: (1) using videos and children's written work; and (2) using the CGI framework (Carpenter et al., 1999) and other research-based resources. In what follows, we describe the activities, how they addressed course goals, and how we enacted them with our students. Then we share our reflections on common issues and challenges.

Using Video and Written Artefacts to Illustrate CMTL

Throughout the semester, our prospective elementary teachers engaged in class activities that used video and print materials as artefacts for analysis and discussion of CMTL (see Appendix for a list of these artefacts). A variety of resources were selected so that prospective teachers could engage in a range of activities and topics to help prepare them for interacting with children. Although we varied somewhat in which resources we selected for given topics, we found that we agreed that all the resources listed in the Appendix had the potential to serve our goals. We also found that we used these resources to promote a similar developmental trajectory for prospective teachers' learning – that of how to use children's thinking for conjecturing, instructional planning and decision making. Recommendations from the authors of the resources stressed the importance of teachers engaging in the type of mathematics their students would encounter prior to participating in other activities around CMTL (e.g., see Appendix for Schifter, Bastable, & Russell, 1999) our developmental trajectory began with prospective teachers solving a problem they later saw in a video and/or children's written work. The prospective teachers solved the problem using their own strategies and/or as they anticipated children might solve it. This provided time for prospective teachers to engage in the mathematics involved in a problem and develop an awareness of the variety of strategies so that they could prepare to notice how children solved the same problem. For example, Amy asked her prospective teachers to calculate a product for 29 multiplied by 12, using as many approaches as they could imagine (both correct and incorrect). For a problem such as this one, initially many prospective teachers only thought about a standard algorithm, but after some time and discussion, they realized that children might see other ways of finding an answer. Next, the prospective teachers examined a video clip or written work of children solving this problem, and then paraphrased and summarized key aspects of what children said or did. Summarizing the video clip focused the prospective teachers' attention on observing for children's mathematical thinking rather than noticing less pertinent features (e.g., a teacher's management strategy) or waiting for their professor to tell them what just happened (as Amy did in her initial work using these video clips). Then, the teachers analysed CMTL, responding to prompts such as: "How did the child think about the problem?", "What approach did the child use?", "How might you question or prompt a child during a lesson if you observed this work?", "What do you conjecture about the child's mathematical understanding?", "What is your evidence for this conjecture?", and "Based on this evidence, what would you do next instructionally – in the next minute, day, week, or year? What problem would you pose next? Why would this be a potentially productive problem?"

After several experiences engaging in this developmental trajectory, we found that prospective teachers developed skills for answering these questions and, in particular, for generating specific claims based on detailed evidence related to children's mathematical thinking. Through these questions and discussions, we perceived that prospective teachers began to view CMTL as a resource for instructional decision-making and to understand that the more they know about children's thinking, the more detailed and deliberate they can be in their conjectures, instructional planning, and decision making.

Using the CGI Framework to Facilitate Skills in Planning, Making Instructional Decisions, and Reflecting on Instruction

Although research served as the basis for all activities, the role it served was more in the foreground for some activities. We believed that grounding our prospective teachers' work in research was important in order that they develop a knowledge base to anticipate CMTL, identify and understand CMTL, and employ effective approaches for supporting learning. Given that the courses focused on mathematics for elementary school, we endeavoured to facilitate prospective teachers' understanding of the specialized research on teaching and learning for this level. We present two research-based assignments in this section: an *interview project* from Amy's and Corey's courses (although Amy and Corey developed these assignments independently, their assignments were similar and thus described as one, with variations noted in the text) and a *reflective journal* from Beth's course.

The interview project. The interview project involved interviewing children and then designing a lesson based on information gained from the interviews. First, prospective teachers read research-based texts on the teaching and learning of a mathematics content area that was relevant to the children in their practicum classroom. Corey required prospective teachers to focus on CGI. In Amy's class, prospective teachers with primary grade (K-2, 5-7 year olds) placements selected articles related to the CGI project; however, students in upper elementary placements (4-6, 9-11 year olds) tended to select topics such as fractions, proportional reasoning, and computational methods with large whole numbers and used other research. Next, prospective teachers wrote an interview script for a 10 to 30 minute interview based on children's ages and practicum classroom teachers' advice. Prospective teachers explicitly selected or wrote interview items, possible student responses, and possible follow-up probes, connecting their ideas to research. The intent was for these teachers to engage in a thought experiment to anticipate how children think about and approach tasks. In class they discussed how to adapt scripts to adjust to a child's needs and background and how a carefully designed script helps to prepare the prospective teachers to make appropriate adjustments. The prospective teachers submitted drafts of interview scripts to Amy or Corey and their practicum teachers for feedback; in most cases, prospective teachers were asked to make substantial revisions. Common areas for revision included: scripts needed better connections to research (e.g., they included addition and subtraction problems without carefully identifying how the items aligned with the CGI framework); scripts included too many or too few items for the time allotted; and, the numbers selected for problems were not age appropriate.

After revising their interview scripts, prospective teachers administered their interviews to two or more children in their practicum classroom. By working with children in their practicum placements, prospective teachers had prior knowledge about children's strengths and limitations that informed the interview process, similar to a regular classroom teacher working with his or her class. Following the interviews, prospective teachers wrote a report in which they analysed and reflected on children's responses (connecting their analysis to research) and on the process. Figure 2 shows prompts guiding this report.

Although prospective teachers' depth of analysis and interpretations varied, the assignment compelled them to assert claims and provide evidence for CMTL with the children they encountered daily (as compared to children's solutions in a video or in children's written work where it can appear "staged" to sceptical prospective teachers). Prospective teachers often reflected on the importance of listening to and watching children as they did mathematics (noticing CMTL in real-time) and how this process revealed more than simply examining children's written work or viewing a video. Further, they experienced how children's thinking is often surprising and challenges our assumptions about what children can do. For example, one prospective teacher, Julie, reflected,

> I found the interview approach to be a useful tool because I had made some judgments about what I thought the students knew, or how smart they were, and I was surprised at how wrong I was […] I realised how each student needs to solve math problems differently, as they think about them in different ways which I wouldn't have otherwise seen.

In addition, prospective teachers commented that their analyses revealed patterns that aligned with research on the topic and that reading the research prepared them to notice and identify children's strategies. Julie continued her reflection and explained,

> I found out the research from Part I of my interview [project] stated some of the very methods they used. Counting up, making ten, and making doubles are all methods they used in math. These are all strategies I never used or heard of until this class, and it is really what they are all doing [...]!

I. *Analyse and interpret your data* (children's oral and written responses, actions and methods, expressions and time required to respond, etc.). Look for patterns both within one child's response and among the children's responses. Often by comparing responses you learn more about an individual child, and you begin to see common approaches, gaps, and/or misconceptions in children's mathematics.
 A. Describe in-depth children's understandings, developing conceptions, and/or misconceptions of the topic including: how each child responded to the interview items; what these responses indicate about the child's thinking, understanding, and/or conceptions (analyse and interpret!); and which questions were more difficult to respond to and why. Provide supporting evidence for your conclusions by giving specific examples of responses.
 B. Include information about each child's demeanour as well as previous academic performances (e.g., consider the child's attitudes, dispositions, and work style).
 C. Explain how these findings will inform your instruction (e.g., If you were the child's classroom teacher, what would you include in an instructional plan?).

II. *Reflect on the Interview Process*. Your reflections should include any insights you gained regarding assessing through interviews, your topic, etc.
 A. To what extent was this approach helpful in understanding students' thinking, knowledge, and/or approaches?
 B. Did the assessment focus on important mathematics and elicit the information you intended?
 C. Did you find it necessary to change the presentation of items (Why)? If you were to use the interview again, would you change your items or script?
 D. Did students' responses surprise you (Explain and consider research on the topic)?

Figure 2. Prompts guiding the interview report

The last part of the project involved prospective teachers designing a lesson to meet the needs of children they interviewed. Based on their interview findings, prospective teachers designed a lesson from existing, high-quality instructional materials. Throughout the course, they participated in activities and discussions to help them identify the characteristics of high-quality materials, including discussing ideas about tasks with high-level cognitive demand (Stein & Smith, 1998). Given that most practicing teachers use existing, published instructional materials in mathematics teaching, we required prospective teachers to start with existing materials (often the same materials used by their practicum teacher) and select tasks or entire lessons that seemed appropriate for the children interviewed.

In many cases, prospective teachers needed to adapt materials. In designing lessons, prospective teachers wrote about what they anticipated would happen with regard to children's questions, approaches, gaps in understandings, misconceptions, and challenges prior to teaching. These teachers indicated what they planned to do in response to anticipated actions or to elicit certain approaches and thinking. They included samples of written work they expected to see from children and discussed how the interview informed the lesson. In Julie's case, as she analysed children's interview responses, she found that although the children had studied addition and subtraction, they needed more experience working with a variety of addition and subtraction problems in meaningful contexts to build stronger understandings for the operations. Using the CGI framework of word problem types (Carpenter et al., 1999) and the reform curriculum, *Investigations in Number, Space and Data* materials (TERC, 1998), Julie developed and selected a range of types of problems that were situated in contexts relevant to her children.

Reflective journals. After reading the CGI book and taking part in some of the professional development activities described in *A Guide for Workshop Leaders* (Fennema et al., 1999), Beth's prospective teachers worked with a small group of heterogeneously grouped children for three to five sessions. They were required to:
– Develop or use a set of problem solving tasks that built on their children's thinking and ideas, drawing from the ideas in the CGI program,
– Engage children with the problems and audio tape the session,
– Listen to the audio-taped interview and write a reflective journal entry,
– Use the information they learned to develop the next set of tasks.

The reflective journal, thus, became a tool for prospective teachers to record children's progress as well as their own thinking about the teacher-learner interaction. The reflective journals consisted of four parts, each of which focused prospective teachers' attention on an important aspect of teaching/learning: 1) mathematical content, 2) children's thinking, 3) teacher/learner interaction, and 4) other aspects about which they chose to write. In Part 1 of the reflective journal (see Figure 3), prospective teachers reflected on the mathematical tasks. This allowed them to consider both the mathematical ideas in the problems they wrote as well as those they heard in children's responses. The last three questions solicited information about how they were building on CMTL. This was the area that was most challenging for the teachers. They often began the semester by justifying the choices they made based on affective reasons rather than mathematical rationale. For example, a common journal response at the beginning of the course was: "The children really seemed to have fun with the problems we did this week, so we will do similar problems next time."

As the semester progressed, however, many prospective teachers began to identify mathematical reasons for their choice of problems, in particular connecting to CGI research (Carpenter et al., 1999) and citing a progression of student difficulty as their reason for offering a particular type of problem. For example, prospective teachers identified regrouping problems as being more challenging to students than problems that did not require regrouping. In addition, problem types

1. Attach a copy of the problems you gave the children.
2. What mathematical ideas did the problems contain?
3. Based on what you've heard in the interview, give 2-4 routes you might take in the problems you pose next. How do those problems build on or relate to the mathematical ideas posed in the problems you did last week? How do those problems build on the mathematical thinking the children are doing?
4. Which problems do you think you'll use next? WHY would you use those over the other possibilities?
5. AFTER you choose which problem(s) you will pose, describe ways you might extend the problem for children who are getting the problems quickly and ways you might provide support for the children who are struggling.

Figure 3. Part 1 of the reflective journal: Thinking about the mathematical tasks posed

that did not have an action involved (i.e., Compare or Part-part-whole) were identified as being more difficult than problems in which students could model the action in the problem (i.e., Join and Separate), and multiplication and division problems that used the numbers 3, 6, 7, or 9 were more difficult than problems that used numbers such as 2, 5, and 10.

Part 2 of the reflective journal (see Figure 4) provided space for the prospective teachers to collect data related to each child's content, processes, and attitudes. In this section, prospective teachers were required to provide sufficient evidence for their claims about each child. If further detail was needed, Beth provided follow-up questions for them to answer. For example, one prospective teacher wrote, "She counted to get the answer," to which Beth responded, "What did she say? Did she count by ones, tens or something else? What do her counting strategies indicate about her mathematical understandings?"

Child's name:			
1. Information about this child's "processes" (Include CLAIMS as well as EVIDENCE)			
Problem solving strategies; Methods of reasoning or proof.	Communication/ Articulation of ideas to you and other children.	Connections made to other children's ideas, to other mathematical ideas or to real life.	Representations used (to solve the problem and the kind of language used).
2. Describe your conjectures about this child's understandings with respect to number and operation after the interview. Provide evidence for claims you make			
3. Information about this child's attitude:			

Figure 4. Part 2 of the reflective journal: Focusing on children's thinking

At the end of the semester, the prospective teachers used their data to write case studies of children with whom they worked; these case studies then were shared with the practicum classroom teachers. As Beth's prospective teachers analysed their data, they considered prompts that focused on CMTL with particular attention to processes and making connections to CGI research such as: "How did children generate conjectures, look for patterns, and/or solve the problem in more than one way?," "How did they make connections to other mathematical ideas, other student's ideas, or their life outside the classroom?," "What level of problem solving strategies did they use (refer to CGI – direct modelling, counting strategies, derived facts) for what kinds of problems?"

Part 3 of the reflective journal project asked the prospective teachers to reflect on their experience; this changed each week to encourage reflection about various aspects of teaching/learning. Part 3 also provides an example of a way we incorporated professional literature in the courses. A set of course readings was organized around a series of questions one might consider when teaching. These questions aimed to help prospective teachers connect research to practice with a focus on planning and instructional decision making during interactions with children and reflection on these interactions. For example, in regard to classroom environment, the question was, "What kind of environment do I want to foster?", and prospective teachers read articles by Kazemi (1998) and Manouchehri and Enderson (1999) on classroom discourse. In focusing on questioning, prospective teachers considered, "What kinds of questions am I asking? How are students responding?" and read articles from Vacc (1993) and Rowan and Robles (1998). These teachers also listened to audio tapes of their interactions with children and then answered questions about their own practice in relationship to the readings. For example, after reading Kazemi's article, they were asked to pay attention to when and how they interrupted students as a means to examine the sociomathematical norms they were establishing in their interaction with children.

In Part 4 of the reflective journal, prospective teachers described aspects of the interactions that they found interesting, perplexing, or frustrating. They also reflected on what they learned from working with the children, what they noticed, and new ideas or wonderings. The four parts of the reflective journal played an integral role in helping students work toward the course goals we described in Figure 1.

INFLUENCES ON THE EVOLUTION OF OUR COURSES AND OUR PROFESSIONAL DEVELOPMENT

Initial Influences on and Issues for Developing Our Practices and Courses

We each taught our first mathematics methods courses while in graduate school. Although we attended different universities, lessons learned in our first experiences teaching the courses were quite similar. Consistent with the perspective of *teaching as learning in practice* (Lave, 1996), we found that serving in an apprentice role was important in gaining understandings for the complexities of teaching the mathematics methods course. We were strongly influenced by our collaborations with more experienced colleagues who shared their perspectives on important aspects of the course. These mentors made children's ways of thinking about and doing mathematics central to the design of their courses. We each carried this perspective into our future teaching of methods courses.

A second initial influence was our recognition that CGI and other video materials helped prospective teachers to connect research they had been studying in the course to actual images of children's mathematical thinking and learning. The videos of children seemed to authenticate and bring to life the nature and depth of CMTL. Moreover, while using video clips, prospective teachers were visibly more engaged, physically leaning toward the monitor and becoming endeared to children's ways of doing mathematics. At the same time, we noticed that the materials alone did not ensure prospective teachers' learning. We identified three primary issues in using CGI and video materials to meet our course goals for developing understandings for CMTL (see Figure 1), as described below.

First, we found that prospective teachers needed to engage with 'real' students about mathematics during the course – not just read about children's mathematics or view it on a video clip. This finding is not surprising, however these opportunities for interaction are not always built into teacher education programs or mathematics methods courses, because they require access to children during the course and collaboration with mentor classroom teachers. Identifying this need motivated us to develop assignments in which prospective teachers engaged with children doing mathematics during their practicum experiences, such as the interview project and the reflective journals, discussed earlier.

A second issue emerged as we began to understand how video could be used as an instructional tool for prospective teachers' learning: we realized the need to develop strategies for providing adequate support for prospective teachers learning from video materials without resorting to 'teaching by telling'. We believed that if these teachers began to view children's thinking as a *resource*, or instructional tool, they might then be motivated to learn to notice, understand, and build on CMTL and leave the methods course more prepared to teach mathematics. Researchers have found that an important skill in teaching mathematics is noticing and interpreting important learning opportunities and interactions in the classroom (Jacobs et al., 2004, 2006; Van Es & Sherin, 2002). However, we questioned whether we were effectively facilitating and supporting our students in noticing of children's mathematical thinking and learning in video clips and children's written work. For instance, we began to see patterns in the questions the teachers asked and noticed that some questions focused on mathematical content (e.g., "How can I make this problem more complex for the students?"), some questions focused on students' developing mathematical processes (e.g., "How do I get someone who is getting the answer quickly to explain their thinking?"), and some questions focused on social issues (e.g., "How can I get students to listen to each other better?"). In an attempt to have our students' insights align with our instructional goals, we developed questions to serve as prompts such as those listed in the activities described in Figures 2, 3, 4, and 5.

While we realised the need for additional support for these teachers, we also recognised the risk of providing too much assistance in viewing and using children's work. For example, Amy noticed that during early efforts to facilitate video discussions she tended to point out what she thought prospective teachers should be seeing and how these observations related to CGI frameworks – essentially telling her students what to see and what it meant – rather than providing opportunities for them to engage in noticing and determining the child's meaning. While the prospective teachers seemed to be making meaning for the frameworks, it was not clear to Amy that her efforts were preparing them to learn to notice and interpret CMTL in their own practices. Our challenge to provide learning opportunities instead of telling or explaining was very similar to the experiences reported in research on teachers' struggles to resist telling in changing to more student-centered practices (c.f., Smith, 1996). These early experiences and the realization of the challenge to change in our own teaching contributed to the redesign of our courses.

Continuing Efforts to Develop Our Practices and Courses

As we gained experience teaching the course, we continued to learn and develop our practices as our field developed. As Jaworski (2006) argued,

> Teaching develops through a learning process in which teachers and others grow into the practices in which we engage [...]. We can use inquiry as a tool to enable ourselves and others to engage critically with key questions and issue in practice (p. 187).

In our efforts to continually improve our practice, we strove to inquire into our own practices and find ways to promote students' learning individually and collectively. We recognized that an essential component in order to accomplish the goals of our course was to support prospective teachers in developing an *inquiry stance* toward their future practices. In rethinking, we redesigned the course to increase our focus on the use of children's mathematics and their thinking and learning as the object of course work. That is, rather than focus on covering the teaching and learning of mathematics content, we continually worked to centre the 'content' of our courses more on children's mathematical thinking and learning. By focusing more on CMTL, prospective teachers would move toward viewing teaching as inquiry about learning. To achieve this goal, we found that our students did not necessarily enter our courses with skills to anticipate, notice, and reflect on CMTL. Consequently, we carefully designed and connected activities to enable prospective teachers' to build professional knowledge about CMTL and provided opportunities to apply these understandings in practice. As we illustrated above, we benefited from an increased number of available published professional development materials for U.S. practicing teachers and prospective teachers (See Appendix). This range of resources from which to select videotape recordings of children sharing their strategies for solving problems and children's written work that featured specific instructional strategies (e.g., noticing and preparing effective questions) or mathematical topics (e.g., place value and how children develop an understanding for tens and ones). In addition, materials often provided ideas for leading discussions to facilitate activities so that our prospective teachers could deeply examine CMTL and use more sophisticated skills for instructional planning and decision-making.

Another factor that influenced change was the increased adoption of the reform curriculum materials developed from funding from the National Science Foundation in the 1990's. For example, in the schools surrounding Amy's university, during the past five years almost all districts adopted *Investigations in Number, Space and Data* (TERC, 1998) as their instructional materials. Amy's students regularly observed classroom teachers using these materials in ways that aligned with course goals depicted in Figure 1. As prospective teachers came to class with more examples of successes and challenges from watching practicum classroom teachers using materials that focused on developing CMTL, Amy decided to provide time regularly in class for sharing these experiences. This sharing helped prospective teachers connect the course content and goals to observed practice of teaching. Prior to observing classrooms that had adopted these curriculum materials, prospective teachers consistently described a lack of opportunities to observe teaching that considered CMTL, and these discussions often degraded into complaints about the practicum classroom teachers.

In addition to the advent of new instructional resources, reading and conducting research in our field influenced what we included in our assignments and class activities. As evidenced throughout this chapter, research in mathematics education informed initial course development and continues to influence on-going redesign and revisions. We also changed our courses based on our own research on prospective teachers' development (e.g., Drake, 2005; Herbel-Eisenmann, 2002; Roth McDuffie, 2004). For instance, Amy conducted a study in which she examined prospective teachers' experiences during their student teaching practicum and found that being *deliberate* in practice was at least as important as reflecting on practice (Roth McDuffie, 2004). For the prospective teachers studied, whenever children's responses or ideas were not expected, the prospective teachers abandoned plans to implement lessons designed to focus on CMTL and instead went back to 'teaching by telling' and showing procedures to children, the form of teaching they experienced when students themselves in school. However, when they dedicated time to anticipating and deliberating over what might happen in a lesson, they were more successful in implementing lessons as planned and focused on developing CMTL. Engaging in this research resulted in Amy focusing more on encouraging prospective teachers to deliberately anticipate CMTL in planning lessons. In addition, we continually change the content of our classes based on the wisdom gained individually from practice and from our collaboration. Consistent with our course goals for prospective teachers' practices, we focus on our prospective teachers' learning and understandings and then use this information in planning instruction and decision making to improve our courses.

FUTURE DIRECTIONS AND PRESSING QUESTIONS

Teaching as learning in practice (Lave, 1996; Jaworski, 2006) describes our work on two levels. First, as MTEs who strive to continually improve our practices, we learned by describing, analysing, reflecting on, and adjusting our instruction. Second, a primary reason for focusing the course on CMTL was to support prospective teachers in developing an inquiry-oriented practice so that critical analysis and reflection becomes embedded in their work. In designing the elementary mathematics methods courses, we focused on identifying useful artefacts of CMTL and understanding strategies for effectively facilitating prospective teachers' discussions and learning that are prompted by particular video clips or examples of children's work. We also developed additional ways to connect prospective teachers' work in our course with their experiences in the field-based practicum in the classroom. In other words, we asked ourselves: How do we support prospective teachers in noticing, analysing, and building on CMTL as a resource in the context of the realities of daily classroom life? How do we create spaces for our students to share with their classmates what they have learned in this process? Finally, how do we provide opportunities for prospective teachers to move beyond understanding children's mathematical thinking as a resource (already a significant learning task for many of our students) to understanding children's mathematical funds of knowledge (e.g., Civil, 2006; Moll, 1992; Turner & Strawhun, 2007) as an even greater resource for instruction – a resource that connects to and builds on who children are as individuals and as members of families and communities? This shift is a particular challenge given that prospective teachers' experiences are increasingly different from those of the children, families, and communities with whom they are working.

As we collaborated to write this chapter, we identified many similarities in our beliefs, values, and approaches to the course which resulted in our common framework for course goals as shown in Figure 1. To enhance opportunities to learn from practice, MTEs (as an inquiry community) need more discussion and research about questions such as the following. What goals seem essential to prospective teachers' development? What approaches and types of activities seem to be most effective in attaining these goals? How do the experiences in mathematics methods courses support and influence prospective teachers as they begin and build their practices? What are common beliefs and values among MTEs and are these beliefs and values valid relative to how they learn? Our course goals and discussion of activities and their evolution might serve as a first step in a more global examination of the practice of preparing prospective teachers in mathematics teaching and learning.

APPENDIX

The following list contains resources we used in our courses for prospective teachers to observe, gain understandings for, and analyse CMTL.

Carpenter, T., Fennema, E., Franke, M., Levi, L., & Empson, S. (1999). *Children's mathematics: Cognitively guided instruction.* Portsmouth, NH: Heinemann. (Book, Facilitator's guide, videos).
Carpenter, T. P., Franke, M. L., & Levi, L. (2003). *Thinking mathematically: Integrating arithmetic and algebra in elementary school.* Portsmouth, NH: Heinemann. (Book, videos)
Empson, S. B. (2007). *Case study of four second-graders.* Retrieved from www.edb.utexas.edu/empson, July 10, 2007. (Case study, children's work)
Philipp, R., Cabral, C., & Schappelle, B. (2006). *IMAP CD ROM: Integrating mathematics and pedagogy to illustrate children's reasoning.* Upper Saddle River, NJ: Prentice Hall. (Facilitator's guide, video).
Russell, S. J., Schifter, D., & Bastable, V. (2002). *Working with data.* Parsippany, NJ: Dale Seymour Publications. (Case book, facilitator's guide, video).
Schifter, D., Bastable, V., & Russell, S. (1999). *Developing mathematical ideas, Number and operations, Part I: Building a system of tens.* Parsippany, NJ: Dale Seymour. (Case book, facilitator's guide, video).
Schifter, D., Bastable, V., & Russell, S. (1999). *Developing mathematical ideas, Number and operations, Part II: Making meaning for operation.* Parsippany, NJ: Dale Seymour. (Case book, facilitator's guide, video).
Schifter, D., Bastable, V., & Russell, S. J. (2002). *Measuring space in one, two, and three dimensions.* Parsippany, NJ: Dale Seymour Publications. (Case book, facilitator's guide, video).
Schifter, D., Bastable, V., & Russell, S. J. (2002). *Examining features of shape.* Parsippany, NJ: Dale Seymour Publications. (Case book and facilitator's guide, video).
Stein, M. K., Smith, M. S., Henningsen, M., & Silver, E. (2000). *Implementing standards-based mathematics instruction: A casebook for professional development.* New York: Teachers College Press. (Case book and background reading).
Van de Walle, J. (2007). *Elementary and middle school mathematics: Teaching developmentally* (6th ed.). New York: Longman. (Methods course textbook with many examples of children's work).
WGBH (1995). *Teaching math: A video library, K-4.* Burlington, VT: Annenberg Foundation. (Facilitator's guide, videos).

REFERENCES

Ball, D. L., Lubienski, S. & Mewborn, D. (2001). Research on teaching mathematics: The unsolved problem of teachers' mathematical knowledge. In V. Richardson (Ed.), *Handbook of research on teaching* (4th Ed,) (pp. 433–456). New York: Macmillan.
Carpenter, T. P., Fennema, E., Franke, M. L., Levi, L., & Empson, S. (1999). *Children's mathematics: Cognitively guided instruction.* Portsmouth, NH: Heinemann.
Civil, M. (2006). Building on community knowledge: An avenue to equity in mathematics education. In N. Nasir & P. Cobb (Eds.), *Improving access to mathematics: Diversity and equity in the classroom* (pp. 105–117). New York: Teachers College Press.
Crespo, S., & Featherstone, H. (2006). Teacher learning in mathematics teacher groups: One math problem at a time. In K. Lynch-Davis & R. Rider (Eds.), *The work of mathematics teacher educators: Continuing the conversation. (AMTE Monograph 3)* (pp. 97–116). San Diego, CA: Association of Mathematics Teacher Educators.
Donovan, M., Bransford, J., & Pellegrino, J. (1999). *How people learn: Bridging research and practice.* Washington, DC: National Academy Press.

Drake, C. (2005). Community mathematics education: A framework for teaching elementary mathematics methods. In G. Lloyd, M. Wilson, J. Wilkins, & S. Behm (Eds.), *Proceedings of the 27th Annual Meeting of the North American Chapter of the International Group for the Psychology of Mathematics Education*, Roanoke, VA: Virginia Polytechnic Institute and State University.

Fennema, E., Carpenter, T. P., Levi, L., Franke, M. L., & Empson, S. B. (1999). *A guide for workshop leaders: Children's mathematics cognitively guided instruction*. Portsmouth, NH: Heinemann.

Flowers, J., & Rubenstein, R. (2006). A rich problem and its potential for developing mathematical knowledge for teaching. In K. Lynch-Davis & R. Rider (Eds.), *The work of mathematics teacher educators: Continuing the conversation) (AMTE Monograph 3)* (pp. 29–44). San Diego, CA: Association of Mathematics Teacher Educators.

Herbel-Eisenmann, B. (2002). Combining CGI with working with children: Examining reflections of preservice teachers. In D. Mewborn, P. Sztajn, D. White, H. Wiegel, R. Bryant, & K. Nooney (Eds.), *Proceedings of the 24th annual meeting of the North American Chapter of the International Group for the Psychology of Mathematics Education* (Vol. 3, pp. 1218–1221). Columbus, OH: ERIC Clearinghouse for Science, Mathematics, and Environmental Education.

Hiebert, J., Morris, A .K., Berk, D., & Jansen, A. (2007). Preparing teachers to learn from teaching. *Journal of Teacher Education, 58*, 47–61.

Jacobs, V., Ambrose, R., Clement, L., & Brown, D. (2006). Using teacher-produced videotapes of student interviews as discussion catalyst. *Teaching Children Mathematics, 13*, 276–281.

Jacobs, V. & Philipp, R. (2004). Mathematical thinking: Helping prospective and practicing teachers focus. *Teaching Children Mathematics, 11*, 194–201.

Jaworski, B. (2006). Theory and practice in mathematics teaching development: Critical inquiry as a mode of learning in teaching. *Journal of Mathematics Teacher Education, 9*, 187–211.

Kazemi, E. (1998). Discourse that promotes conceptual understanding. *Teaching Children Mathematics, 4*, 410–414.

Lave, J. (1996). Teaching as learning, in practice. *Mind, Culture and Activity, 3* (3), 149–164.

Lave, J., & Wenger, E. (1991). *Situated learning: Legitimate peripheral participation*. Cambridge, MA: Cambridge University Press.

Lloyd, G. (2006). Using K-12 mathematic curriculum materials in teacher education: Rationale, strategies, and preservice teachers' experiences. In K. Lynch-Davis & R. Rider (Eds.), *The work of mathematics teacher educators: Continuing the conversation) (AMTE Monograph 3)* (pp. 11–28). San Diego, CA: Association of Mathematics Teacher Educators.

Manouchehri, A., & Enderson, M. (1999). Promoting mathematical discourse: Learning from classroom examples. *Mathematics Teaching in the Middle School, 4*, 216–222.

Moll, L. C. (1992). Bilingual classrooms and community analysis: Some recent trends. *Educational Researcher, 2*, 20–24.

Moyer-Packenham, P. (2004). The interview assignment: Evaluating a teacher candidate's knowledge of mathematics content, questioning, and assessment. In T. Watanabe & D. Thompson (Eds.), *The work of mathematics teacher educators: Exchanging ideas for effective practice (AMTE Monograph 1)* (pp. 169–188). San Diego, CA: Association of Mathematics Teacher Educators.

National Council of Teachers of Mathematics. (1989). *Curriculum and evaluation standards for school mathematics*. Reston, VA: Author.

National Council of Teachers of Mathematics. (1991). *Professional teaching standards for school mathematics*. Reston, VA: Author.

Papick, I., Beem, J., Reys, B., & Reys, R. (1999). Impact of the Missouri Middle Mathematics Project on the preparation of prospective middle school teachers. *Journal of Mathematics Teacher Education, 2*(3), 301–310.

Peterson, B. (2005). Student teaching in Japan: The lesson. *Journal of Mathematics Teacher Education, 8*, 61–74.

Philipp, R., Ambrose, R., Lamb, L. C., Sowder, J. T., Schappelle, B. P., Sowder, L., Thanheiser, E., & Chauvot, J., (2007). Effects of early field experiences on the mathematical content knowledge and beliefs of prospective elementary teachers: An experimental study. *Journal for Research in Mathematics Education, 38*, 438–476.

Putnam, R., & Borko, H. (2000). What do new views of knowledge and thinking have to say about research on teacher learning? *Educational Researcher, 29*, 4–15.

Rosaen, C., Schram, P., & Herbel-Eisenmann, B. (2002). Using technology to explore connections among mathematics, language, and literacy in teacher education. *Contemporary Issues in Technology and Teacher Education,* [Online serial] *2*(3). Available: http://www.citejournal.org/vol2/iss3/mathematics/article1.cfm.

Roth McDuffie, A. (2004). Mathematics teaching as a deliberate practice: An investigation of elementary preservice teachers' reflective thinking during student teaching. *Journal of Mathematics Teacher Education, 7*, 33–61.

Rowan, T. E., & Robles, J. (1998). Using questions to help children build mathematical power. *Teaching Children Mathematics, 4*, 504–509.

Schifter, D. (1996). Facilitating students' construction of their own mathematical understandings. In D. Schifter (Ed.), *What's happening in math class, Volume 1: Envisioning new practices through teacher narratives* (pp. 9–43). New York: Teachers College Press.

Silver, E., Mills, V., Castro, A., & Ghousseini, H. (2006). Blending elements of lesson study with case analysis and discussion: A promising professional development strategy. In K. Lynch-Davis & R. Rider (Eds.), *The work of mathematics teacher educators: Continuing the conversation) (AMTE Monograph 3)* (pp. 117–132). San Diego, CA: Association of Mathematics Teacher Educators.

Smith, J. (1996). Efficacy and teaching mathematics by telling: A challenge for reform. *Journal for Research in Mathematics Education, 27*, 387–402.

Stein, M. K., & Smith, M. S. (1998). Mathematical tasks as a framework for reflection: From research to practice. *Mathematics Teaching in the Middle School, 3*(4), 268–275.

Taylor, M. (2006). Syllabus study: A structured look at mathematics methods courses. *AMTE Connections, 16*(1), 12–15.

TERC. (1998). *Investigations in number, data, and space.* Menlo Park, CA: Dale Seymour.

Thompson, P. (1994). Concrete materials and teaching for mathematical understanding. *Arithmetic Teacher, 41*, 556–557.

Turner, E. E., & Strawhun, B. (2007). Problem posing that makes a difference: Students posing and investigating mathematical problems related to overcrowding at their school. *Teaching Children Mathematics,* 13, 457–463.

Vacc, N. N. (1993). Questioning in the mathematics classroom. *Arithmetic Teacher, 41*, 88–91.

Vacc, N. N., & Bright, G. (1999). Elementary preservice teachers' changing beliefs and instructional use of children's mathematical thinking. *Journal for Research in Mathematics Education, 30*, 89–110.

Van Es, E., & Sherin, M. (2002). Learning to notice: Scaffolding new teachers' interpretations. *Journal of Technology and Teacher Education, 10*, 571–596.

Amy Roth McDuffie
Department of Teaching and Learning
Washington State University Tri-Cities

Corey Drake
Department of Curriculum and Instruction
Iowa State University

Beth Herbel-Eisenmann
Teacher Education
Michigan State University

PAT PERKS AND STEPHANIE PRESTAGE

13. TOOLS FOR LEARNING ABOUT TEACHING AND LEARNING

This chapter describes how we develop tools for our work in initial teacher education. We use the language of 'tool' in the widest sense of Vygotsky's use of the word. We discuss our philosophy of teaching, our style as teachers and the context in which we work, our activity system, all of which influence how we create experiences for prospective teachers. The different tools for sessions arise from our practical wisdom and professional traditions, which we define and describe, and, through reflection, discussion and professional and academic writing we show how these develop into more explicit tools for teacher development. By reviewing the types of tools we offer an insight into how these develop and further inform our practice.

1. INTRODUCTION

In Vygotskian theory learning can be mediated through the use of tools and artefacts. Vygotsky (1986) describes two modes of the development of concepts – these he labels as spontaneous and scientific with the two processes being closely connected.

> One might say that the developments of the child's spontaneous concepts proceeds upward and the development of his scientific concepts downwards … This is a consequence of the different ways in which the two kinds of concepts emerge. The inception of a spontaneous concept can be usually traced to a face-to-face meeting with a concrete situation, while a scientific concept involves from the first a 'mediated' attitude towards its object. (Vygotsky, 1986, pp. 193–194)

In discussing the professional learning of prospective teachers we find it useful to use Vygotsky's language, equating the development of spontaneous concepts about teaching coming from the prospective teachers' own experiences of schooling as well as their experience from engaging in teaching during the course. Scientific concepts Vygotsky says are developed through the context of instruction (ibid. p.158), mediated by the use of tools and leading to a conscious understanding of the subject. We take this process of learning to be developed through deliberate pedagogic acts (such as our teaching), deliberate interventions during the course of the teaching year. We see our role to be supporting a conscious understanding of teaching.

Scientific concepts, in turn, supply structures for the upward development of spontaneous concepts toward consciousness and deliberate use. (Ibid. p. 194)

In this paper we explore the nature of tools and the way in which we define and use *them to support development of scientific concepts of teaching*. We consider the tools we use to support others learning about the teaching of mathematics and the tools we use to develop our own learning about being teacher educators. Firstly we share the assumptions about learning that ground our approaches to working as teacher educators and the contexts that impact on our working. Secondly, we describe some of the tools that we use in our courses and the ways in which these impact on our own development as teachers.

2. ASSUMPTIONS AND BELIEFS ABOUT PROFESSIONAL LEARNING

2.1 Professional Learning

For us, teaching about teaching is a complex process mainly because teaching, in its many forms, is itself a complex profession. The school teacher is required to have knowledge about pupils, systems and structures; knowledge about styles of teaching and learning; knowledge about management, resources and assessment as well as knowledge about the subject. Research in the area of teachers' knowledge (for example, Brown & McIntyre, 1993; Cooper & McIntyre, 1996; Desforges & McNamara 1979; Ernest, 1989; Marks, 1990; Calderhead & Shorrock, 1997; Banks et al., 1999) offers definitions of professional knowledge as well as explanations for the different forms of knowledge that a teacher holds and the complexity of such knowledge,

It is clear that learning to teach involves more than a mastery of a limited set of competencies. It is a complex process. It is also a lengthy process, extending for most teachers well after their initial training. (Calderhead & Shorrock, 1997, p. 194)

Our beliefs about teaching lead us to agree with Buchmann (1984) that teachers need a rich and deep understanding of their subject in order to respond to all aspects of pupils' needs: 'Content knowledge of this kind encourages the mobility of teacher conceptions and yields knowledge in the form of multiple and fluid conceptions' (p. 46). Indeed some researchers see learning a professional practice as developing a capacity to interpret aspects of the field/context of action in increasing complex ways and to respond to those interpretations (Eraut, 1994; Sternberg & Horvath, 1995).

A sociocultural approach to learning echoes these interpretations. In sociocultural terms professional learning is evident in the capacity to interpret the 'object of our activities' (i.e., what it is we are focusing our energies and attention on) so that its complexity is increasingly revealed. For example, a prospective teacher might see a pupil as disruptive and badly behaved, but after conversations with a mentor or tutor might learn to interpret that behaviour as evidence that the pupil is disturbed and revise any responses. In sociocultural analyses this process can be considered as 'expanding the object' (Engeström, 1999), the identifying of more of the potential meanings in an event. Prospective teachers for example see teaching as a 'simple' process and work hard to avoid the unexpected while teaching (Desforges, 1995; Edwards, 1998). They avoid expanding the object and close down on complexity, looking for a simpler solution, a 'right' answer, and thus limit their learning.

Teacher education, then, involves helping prospective teachers to create richer interpretations about teaching, helping them to expand the object in order to respond intelligently to those interpretations – developing expertise. For this 'helping' we refer back to Vygotsky's idea of mediation to aid the development of scientific concepts about teaching. Later in the chapter we describe ways in which we construct tools for this mediation.

2.2 The Social Nature of Knowledge and Learning

A sociocultural approach also adds to understandings of such expertise by seeing it as

> (an) ongoing collaborative and discursive construction of tasks, solutions, visions, breakdowns and innovations. (Engeström & Middleton, 1996, p. 4)

That is, expertise is not located within one individual but is distributed across systems in the forms of other people and the artefacts that they have produced (Hutchins, 1991; Pea, 1993), or as what Bruner has called the 'extended intelligence' of settings (Bruner, 1996). The sociocultural line suggests that we are both shaped by and shape the environments in which we participate (Cole, 1996). Knowledgeable teachers who can interpret complexity and act on those interpretations are therefore likely to be found in organisations which allow this to happen. The community of practice in which we work is English in its context and international in its peer group. The metaphor for learning the practices of such communities is clearly participation. Feiman-Nemser (2001) suggests that a valuable approach to teacher learning and professional development lies in creating professional communities in which teachers learn through sustained, thoughtful discourse with other teachers. By participating in professional communities, teachers can share their expertise and learn from the expertise of others, bringing professional development closer to the work of teaching. As Feiman-Nemser (2001) describes it,

> The kind of conversation that promotes teacher learning differs from usual modes of teacher talk which feature personal anecdotes and opinions and are governed by norms of politeness and consensus. Professional discourse involves rich descriptions of practice, attention to evidence, examination of alternative interpretations and possibilities .... [Teachers] create new understandings and build a new professional culture. (p. 1043)

It is not easy to develop teacher communities of practice where the individual can engage in thoughtful challenge and debate about the ideas and issues of teaching practice. As tutors we have the advantage of belonging to the small community of mathematics educators in our school, the wider community nationally, the community of mathematics teachers in our partnership and each year the developing community of our prospective teachers. The creation of a new community depends on moving beyond polite discussion. Others recognise the issues, for example, Grossman and colleagues (Grossman, Wineburg, & Woolworth, 2001; Wineburg & Grossman, 1998) found that when trying to foster a community for teacher professional development, teachers were constrained by their failure to recognize difference as an opportunity for learning rather than a sign of interpersonal conflict. These difficulties resulted in part from their lack of experience in having discussions about teaching and learning that involve intellectual challenges. By its nature, a developing community needs to be built on trust and dialogue if learning about teaching is to move beyond the intuitive. In creating the tools to challenge and provoke, one of our roles is to offer a safe environment. But we also believe that creating some conflict and then negotiating different understandings from that conflict allows us to move beyond polite response into learning. Reflection upon our practice reveals the importance of tools which forge professional communities of teachers and the fostering of meaningful dialogue within those communities.

3. THE CONTEXT OF OUR WORK

In expanding the idea of learning communities mentioned above, we draw on the work of Engeström. His unit of analysis is what he calls the 'activity system' (Engeström, 1999). His analysis asks us to look at that system as a learning zone (i.e. a system that allows for the learning of participants to contribute to the development of the system). The framework (a heuristic aid to understanding what is going on) allows us to trace connections between the acts that make up the action of learning about teaching in the activity systems of our current practices (Cole & Engeström 1993; Engeström, 1999). Three aspects of the framework are described here, each of which impact on practice; they are the community, the division of labour and the rules of the activity system (see Figures 1 and 2).

Our main community is the University which provides the professional traditions for our role as teacher educators – the structure of the teaching commitment; that teacher educators for initial teacher education are employed for their teaching expertise; the expectation that this know-how will be transformed into knowledge about teaching education and the joining of a research community. In our particular context we also teach mathematics each year as we have a two-year course, where the first year is to help prospective teachers without a mathematics degree improve their mathematics.

Division of labour, in a micro sense, describes our team teaching; the course we teach has sufficient numbers of prospective teachers to allow two tutors to work on the course. For us this collaboration requires regular accounting for the decisions we take in the teaching moment as well as the analysis of planning and evaluation. In a macro sense, the division of labour is shared and underpinned by the partnership with the schools where our prospective teachers have their placements. We share beliefs about teaching with the teachers who will work in this partnership to create environments where prospective teachers will develop.

The rules of the course we teach are external to University; they are prescribed by government through the Training and Development Agency (for example, TDA, 2006) Secondary prospective teachers have to follow a 36 week course, 24 weeks of which have to be school-based. This means that the university sessions take place within a 12 week duration. So not only do we as tutors have to develop as teacher educators, we have a limited time frame to work with prospective teachers on developing, in Vygotsky's sense, scientific concepts about teaching. Thus a major part of our development lies in looking for efficient and effective experiences to promote the learning of our prospective teachers.

A rule of the university activity system identifies that we research and publish as well as teach. As we describe below, this rule acts as a tool for our development as teacher educators – helping to expand the object of our activities. One choice has been to research our practice, to work in the practitioner research paradigm, using our own teaching to explore teaching.

4. THE NATURE OF TEACHER-KNOWLEDGE: PERSONAL LEARNING

As we have already observed, a major part of our work lies in initial teacher education (ITE) with the education of prospective teachers of mathematics. So the majority of our development as teacher educators comes within this role. One of the major tools in education over recent years has been that of reflection (Schön, 1983). In considering the place of reflection in our prospective teachers' learning and as a way of describing the transformation of mathematics subject knowledge that our prospective teachers hold as graduate mathematicians into the knowledge of mathematics necessary for teaching the subject, we developed a model (see Figure 1). This model offers a structure for what we believe is important to teacher transformation in all aspects (for details of this model see Prestage & Perks, 1999). With roots in Aristotle we took three aspects of knowledge: (1) practical wisdom – knowledge from being in the classroom; (2) professional traditions - knowledge from existing school curriculum and practices and research; and (3) learner-knowledge – the prospective teachers own knowledge. All three aspects, we suggest, need to combine in decisions about classroom events.

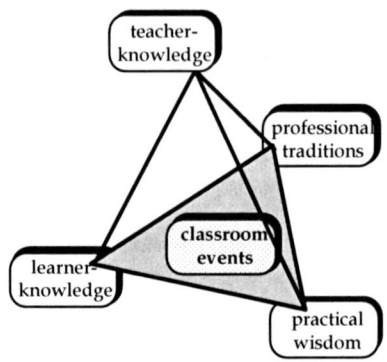

Figure 1. Teacher-knowledge tetrahedron

Our prospective teachers arrive with the learner-knowledge gained by being a pupil in school. Learner-knowledge (for example, the content knowledge of the topic, the experience of being assessed) requires transforming through deliberate reflection. Teaching requires a synthesis of reflections on this and the other elements, practical wisdom and professional traditions. In the diagram the struts of the tetrahedron represent the reflective/analytic process (Figure 1). Decisions about teaching are influenced by the prospective teachers' knowledge in these three areas to create classroom events. Further reflection on these builds to create expertise and multi-faceted teacher-knowledge. We would argue that it is in this way that our 'good' teachers come to own a better personal knowledge of all aspects mathematics teaching.

What then are the implications of this for thinking about a pedagogy for teacher education? Is there an equivalent teacher-educator-knowledge to which we aspire? We believe that just as teaching mathematics needs fluid and connected knowledge of mathematics (teacher-knowledge) so too mathematics educators need a fluid and connected understanding of teaching mathematics education – the teacher-knowledge of mathematics education – and that we need to articulate this.

In describing a model to classify the different elements that might come together to transform our learning about teaching into learning about teacher education, we use the same categories: (1) professional traditions (for us, the existing teacher education course, ways of working, research on mathematics teaching), (2) practical wisdom (the activities chosen for sessions) and (3) learner-knowledge (the knowledge we accrued from being teachers). These three come together through reflection to inform the implementation of our mathematics education sessions (Prestage & Perks, 2001a).

The learner-knowledge of the teacher-educator in our context (where we are appointed, at least in part, because of our mathematics teaching expertise) has particular facets. The learner-knowledge of mathematical subject knowledge for teaching about teaching mathematics is based upon our own learner-knowledge of mathematics, the practical wisdom of the experienced mathematics teacher and the professional traditions of curriculum, examination syllabuses and so forth. The teacher educator needs this teacher subject knowledge at the teacher-knowledge level. This mathematical learner-knowledge for a teacher educator then forms the basis, along with initial teacher education (ITE) professional traditions and the practical wisdom of teaching about teaching to create sessions for ITE students (prospective teachers) to work on their subject knowledge for teaching. Figure 1 lies at the learner-knowledge vertex in the tetrahedron in Figure 2. Professional traditions emerge from personal experiences, education and training, the current government teacher training policies, the mathematics and technology national curricula for teacher education (for example, TDA 2006), inspection criteria against which judgements are made and research. Practical wisdom is gained from considering what the prospective teachers need to know and how sessions might be constructed for them so that they engage in the ideas. The mathematics education sessions are built around learning opportunities, dependent upon the chosen tools. The continuing reflective process allows for us to develop not only as better teacher educators, creating more useful tools, but also better mathematics teachers.

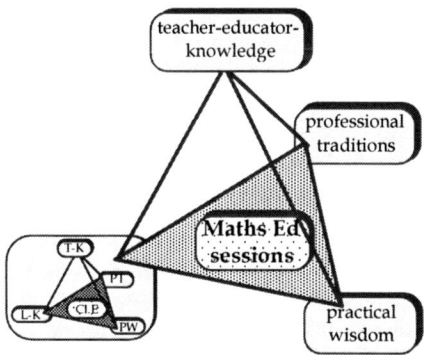

Figure 2. Teacher-educator-knowledge tetrahedron

For example, teaching our prospective teachers about planning lessons began initially with our own learner-knowledge of how we planned lessons, the professional traditions of our own training and school expectations and the practical wisdom of how to begin to set up expectations for others. This may have stayed at the level of unmediated or spontaneous knowing, with the sessions getting somewhat better over time but we would not necessarily have been creating our own further development. By sharing processes, research on their use (Perks, 1997) and constant reflection, our teacher-knowledge has adapted and the techniques used with our prospective teachers become more conducive to the multiple and fluid conception necessary to explore the complexities of lesson planning.

5. SURROGATE SITUATIONS

So what is our role? Our current system defines two contexts for our role: essentially inside school classrooms observing prospective teachers and in sessions in the university. It is the latter that we reflect on here. Given the complexities of developing as a teacher we can isolate/control some of the variables about teaching and learning and create sessions and tasks outside of the classroom in order to learn more about the classroom. We do this to help to expand the object, to reveal increasing complexity and work with tools to mediate Vygotsky's suggestion of developing 'scientific' concepts about teaching.

Clark (2005) uses the idea about surrogate situations to create internal and external props to map and solve complicated problems. The idea of surrogate situations summarises many our university sessions. In these sessions we work on particular aspects of teaching in order to create a variety of internal and external props for the prospective teachers to use when teaching.

> These mock-ups serve no primary purpose other than that of allowing human reason to get a grip on what might otherwise prove elusive or impossible to hold in the mind. Any given project will often rely on the use of multiple kinds of surrogate situations each of which highlights or makes available some specific dimension of what Gedenryd calls the 'future situation of use'. In this way, surrogate situations are not simple miniature versions of the real thing. Rather they are selected so as to allow us to engage specific and often quite abstract aspects of the future situation of use. (Clark, 2005, p. 237)

In fact our time in the university is vital to raise to a 'conscious level' (Vygotsky 1986) an understanding about teaching. We can reduce some of the complexity by controlling the variables about teaching and focus on particular aspects which we think will be useful for the 'future situation of use' Indeed Clark suggests that developing a fluency with surrogate situations 'depends to a certain extent on keeping the level of non essential detail quite low' (p. 237). We add to these surrogate situations the idea of labels to help create access to a shared experience/stored knowledge. A set of surrogate situations, easily brought to mind will allow, we believe, the prospective teacher to distribute attention in the classroom in new ways, as Clark suggests 'to parse the scene more quickly into its salient components' (p. 240).

Below we offer three different surrogate situations which have become tools for mediating learning about teaching – to offer ways of developing scientific understandings about teaching. Our *teacher knowledge* gives us many ideas that form the basis of sessions with the prospective teachers which in turn become defined and developed over time (or indeed rejected because of their limited value). *Professional traditions* of literature and of our research also inform and act as starting points for sessions. Writing and reflection moves the ideas to become more efficient tools and provoke our own learning. As with all teaching, serendipity plays its part, bringing together ideas which have more power collectively than is ever possible from the separate parts.

5.1 Paddington

An example of serendipity lies with our Paddington session. Paddington is a bear from darkest Peru and the hero of a series of well-known children's books. We read to the prospective teachers a fragment of the bear's experiences (Bond, 1971) answering questions on mathematics on a television quiz show. Why? Our t*eacher-knowledge* which acts as the *learner-knowledge for teacher educators*, reminds us of a good idea. We had used the story with our pupils and had the feedback that this hooked them into a lesson and helped their image of mathematics. Our *teacher-knowledge* also reminds us that pupils can react in different and unexpected ways to mathematics questions which is reflected in Paddington's responses to the quiz questions. The prospective teachers enjoy the story and in our original planning, the story and the ensuing discussion provoked some teacher learning. However the juxtaposition of the second part of the session - working on some text book material - brought about unexpected connections

The text book is a source of much material for pupils and our *teacher-knowledge* reminds us that pupils do not always understand the meaning of a question or even read the text question. During the session the prospective teachers did an exercise and noted how they were working and ready for the discussion phase. But shortly there were mutterings. The discussion that followed these mutterings revealed two styles of working and two lively labels. 'Doing a parrot' became shorthand for doing division by using the numbers and ignoring the text because 'Division' was the exercise heading. 'Doing a Paddington' emerged for answering a 'real' question where the responses to "How many nines in 171?" ranged from "none" through "one" to "19". The discussion of the other questions raised issues about the richness of tasks and the place of context.

The phrases 'Doing Paddington' and 'Doing a Parrot' were used by the prospective teachers during the year; a rich and powerful label had been created. This was a shorthand phrase that allowed them to reconnect to the activity, a memory prompt. Indeed it was a shorthand phrase that worked over a length of time. There was evidence that we had hooked them into considering the issues and, in the process, developed a language which enabled us to share ideas about pupils' mathematical behaviour quickly and share this with understanding. The story, a 'good idea', was identified as a hook to learning which gave rise to the label – a label for the difference between how pupils might read a question and how mathematics teachers expect them to read a question. The various Paddington style answers to the text book questions (answers not predicted by the answer scheme) also freed creativity for some of the prospective teachers. Their answers became more and more creative, as they exploited the context of the questions and some began to see ways of creating different tasks, connecting other areas of the curriculum

The analysis of the text book exercise as planned had now moved into a different dimension. The two parts of the session provoked in us a reflection on why the elements had come together in this way and what may be generalised to other sessions. For our learning as teacher-educators, the reflection on the session linked together hooks, labels and creative tasks. Writing about the session (Prestage & Perks, 1992) deepened our ideas as we noted our reactions to the session and the power of the labels. The act of writing extended our learning as we accounted for what happened in the session. The act of writing, a professional tradition and rule of our activity system acted also as a tool for learning as teacher-educators (see Figure 3 and 4).

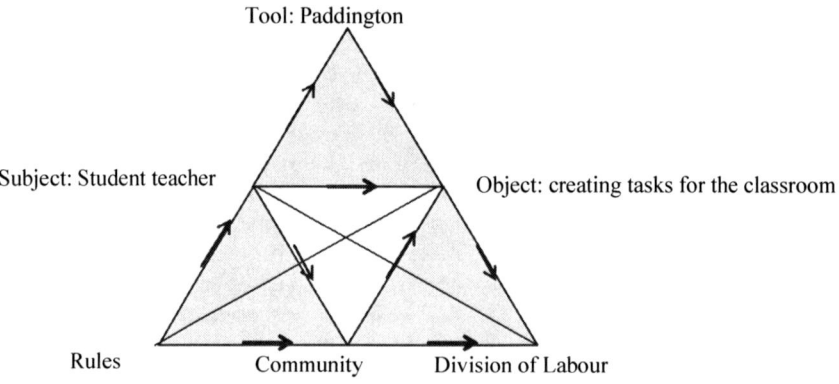

Figure 3. Activity system with 'Paddington' as a tool

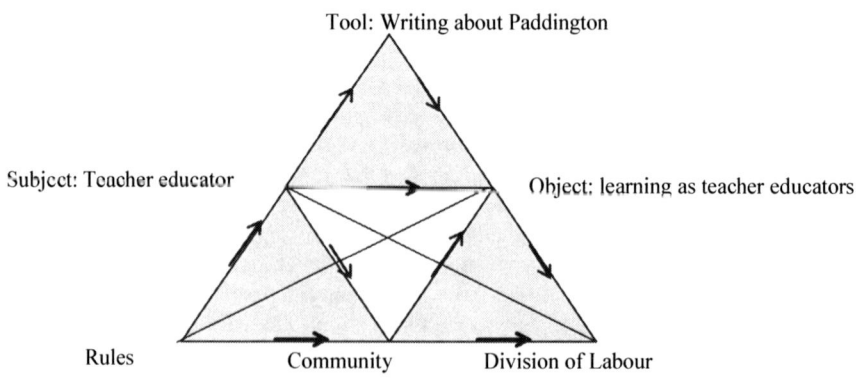

Figure 4. Activity system with professional writing as a tool

There appear to be different levels of tools in both the learning of our prospective teachers and our own level. The initial tasks, perhaps rather mundane activities in isolation, were tools for some mediated learning, but in themselves offered a context for only one issue – the first, a nice story to hook interest in learning, the second, giving rise to interesting analyses of a text book exercise. Together the tasks merged into a tool which we use yearly, a tool which has the capacity to help prospective teachers to create richer interpretations about teaching, to 'see' more in the event. We believe this tool offers the prospective teachers a vehicle for learning about:
– idiosyncratic approaches to the 'understanding mathematical tasks: pupils can see the world differently;
– styles of approaching mathematics tasks from a routine or from context: parrot versus Paddington;
– the tension between context and algorithmic routines linked to how you read a question;
– the role of context in expanding the variables of a problem: rich tasks can be created from exploiting the context.

Our learning lies in the rehearsing and refining and writing about our teaching – learning more about teaching and more about becoming a teacher-educator. Our knowledge about the power of hooks and labels are reinforced as tools for this context as well as the context of teaching mathematics.

5.2. Square Roots

Another powerful label in our repertoire is the surrogate situation "square roots" (Perks & Prestage, 1999) During their first week with us and in her sternest manner, Stephanie Prestage delivers to our prospective teachers a 'revision lesson' on the algorithm for calculating square roots. Our *teacher-knowledge* suggests the need to discuss: (1) the place of the algorithm, (2) the emotional effects of failure, and (3) the power of the teacher. The emotional response to this session is important and we deliberately set out to create those emotions. Breen (1991) suggests that harnessing emotions helps us to learn to assess and appreciate our own behaviour and attitudes towards people.

> I came to believe that the teaching of math, like the teaching of any other subject in schools is a 'political' activity. It either helps to create attitudes and intellectual models that will in their turn help students grow, develop, be critical, more aware and more involved, and thus more confident and able to go beyond the existing structures; or it produces students who are passive, rigid, timid and alienated. (p. 31)

The session begins promptly. Latecomers are met by a steely stare and angry comment or questions. The algorithm is presented quickly. The worked example disappears in seconds of its completion to be replaced by the exercise. For those who are clever enough, there are even some starred harder questions. A request for more explanation is greeted by a faster repetition of the algorithm in the same terminology. The prospective teachers work in silence, the tension in the room increases. Someone is reprimanded for using a calculator. There is a sigh of relief, when they are told to stop, but this is followed by confusion as the answers are read out. Prospective teachers are asked to put up their hands if they got more than ten correct, nine, eight down to zero; some confess later that they hid at this point: the public humiliation of getting the most or least correct spills into anger. The prospective teachers are asked to reflect and write down how they feel. A number are very angry. This does, however, seem to be one of the most important aspects of the session. Our acceptance of their expressions of anger is a major way in which a relationship is developed. Trust and the professional relationship it embodies is one of the major differences between our work with prospective and practicing teachers. Most of our work with practicing teachers is short term and we do not develop the same relationship. Only when working on long term projects (for example, Watson, de Geeste & Prestage, 2003) do similar relationships emerge.

The writing about this session (Perks & Prestage, 1999) offered a different dimension from Paddington. At the planning stage, the purposes and message of the session were clear; the writing was a re-confirming of the power of the activity for teacher learning. For our learning, the writing enabled us to stay with the session and not to change it for something less appropriate but also to refine and articulate the detail. The learning from this session reinforces the power of labels which prompt the memory. Indeed the words 'square roots' conjures for the prospective teachers feelings about the session that help them to reflect at a later date on their own algorithmic teaching moments

The surrogate situation creates empathy with many and less successful learners of mathematics for whom failure inhibits learning. Also using their emotions is powerful way to provoke prospective teachers into examining their own learning. The anger helped us to move this community of learners beyond polite response and surface engagement.

5.3. Givens

'What-if-not' is a label and a tool taken from our professional traditions (Brown & Walter, 1990). It contributed to our teaching of mathematics by creating new tasks for the classroom as well as impacting on our pedagogy. Successful sessions with prospective teachers from six to eighteen, confirmed the value of the tool for pupils' mathematical learning.. With our prospective teachers, the sessions always went well and the label got used by some. But there was little evidence that this tool had any influence on the prospective teachers' practice. What was explicit to us, the opportunity to work differently on the mathematics, the opportunities for differentiation, the stretching and connecting of our mathematics learner-knowledge, etc., did not become useful knowledge to our prospective teachers. We had to think again about sharing our *teacher-knowledge*. The lack of success was also a useful reminder that telling does not work and that sharing experience is not sufficient to provoke learning. 'What-if-not' could be used for personal creativity but for it to influence others we needed to take a different approach (Perks & Prestage, 1992).

We gave the prospective teachers a set of already changed tasks and asked them to identify what mathematics the pupils were doing and what decisions they would have to make. The discussion led to them identifying the problems pupils might have, for example in constructing a given diagram, deciding which bit to draw first, how much of the diagram was needed before the shape was fixed or how accurate they needed to be. The analysis before the creativity seemed to provoke a better approach to the changing of tasks. Importantly during the initial sessions new labels emerged - that of givens, so with practice prospective teachers were soon 'remove a given', 'change, a given' and 'add a given'.

The professional writing was extended from the articles into a book (Prestage & Perks, 2001b). For us this was the discipline we needed to return and review our thinking in more depth than in the post-session reflection. The routine of adding/removing/changing givens needed more examples and this extended our repertoire when working with our prospective teachers. The routine was clear in our heads and the repetition over many sessions means that we use the technique with a fluency envied by many of our prospective teachers. In having to be explicit about the routine and its use and developing a wider topic range in our examples, the writing provided another stage in our development. The tool had been honed and in later sessions helped the prospective teachers with their learning. It also deepened our understanding and expertise as teachers of mathematics.

When writing about Paddington again for the book a different issue arose. In the session we and our prospective teachers had been working intuitively on the ideas. In the moment and with the added energy of the group discussion we were all able to change the tasks and create new ideas to exploit the context of questions. The writing for the book provided impetus for analysing further what had been happening to create a routine to work on creating new tasks when alone. The new writing extended our learning as
– teachers, about changing tasks and
– teacher-educators, that some labels and heuristics help the prospective teachers to reconnect to the activity to access useful skills for the classroom.

6. CONCLUSION

We have described here three sessions that we use for teaching about teaching mathematics with prospective secondary mathematics teachers. We identified the potential learning for the prospective teachers as well as aspects which have deepened our understanding as teacher educators. These tools we believe help to encourage prospective teachers to work in a particular and focused way on aspects of teaching. The deliberate use of tools is the nature of teaching – to extend and provoke learning beyond the spontaneous to the scientific (using Vygotsky's labels). Every session has good aspects which will help most prospective teachers to think about some of the variables when in the classroom. Of course some will not. Prospective teachers might be seeing something different from that intended by the session. They may, not see the generality and the connections to other aspects of teaching but only learn the example or see the issue as relating to something else. For example, some prospective teachers may see the label 'square roots' as demonstrating the use of an algorithm that was not taught well, so 'square roots' stands as a reminder to explain properly, rather than encompassing the many other aspects we intended the label to stand for. Even those prospective teachers who remember 'square roots' with its many implications when commenting on their own teaching will tell you that this style of teaching is necessary if they are to 'deliver the curriculum', 'do what the pupils need', or 'do what the class teacher said/does'.

There are many tools that help to mediate our learning as teacher educators but writing is significant. Using reflection and writing to mediate for scientific understanding about aspects of these sessions helps us to improve the sessions, to extract the generality and to improve our understanding and learning about being teacher educators. Our ambitions (like many) are to hold deeper understandings and more fluid connections about the role and purpose of our teaching and so enhance that teaching. The recognition of those aspects of the tools which can be generalized over many session have helped in the creations of our "multiple and fluid conceptions" as teacher educators

The activity systems in which we work also help to expand our knowledge. The different communities offer different conversations, reflections and evaluations and research. The rule of our university activity system, that tutors should write and publish, has acted as a significant tool which mediates our continued learning as teacher educators. Articulating ideas for other to read, ideas which are often peer reviewed, leads to a conscious understanding of our subject. Writing has clarified ideas and purposes of the tools we use which have in turn become more explicitly used in sessions. In Vygotsky's words we have developed our understanding of scientific concepts for teacher education. As we work with prospective teachers to develop their capacity to expand their learning, we become more able to recognise our knowledge and more able to respond intelligently to the teaching of teachers and the teaching of mathematics.

REFERENCES

Banks, F., Leach, J., & Moon, B. (1999). New understandings of teachers' pedagogic knowledge. In J. Leach & B. Moon (Eds.), *Learners and pedagogy*, London: PCP.
Bond, M. (1971). *Mr Paddington at large*. London: Collins.
Breen, C. (1991). Concerning Smith and his (very brief?) reign of terror. *Pythagoras*, 25, 31–37.
Brown, S., & McIntyre, D. (1993). *Making sense of teaching*. Buckingham: Open University Press.
Brown, S.I., & Walter, M.I. (1990). *The art of problem posing* (2nd ed.).. London: Lawrence Erlbaum Associates.
Bruner, J. (1996). *The culture of education*. Cambridge: Harvard University Press.
Buchmann, M. (1984). The priority of knowledge and understanding in teaching. In L. Katz & J. Roth (Eds.), *Advances in teacher education* (Vol. 1, pp. 29–48), New Jersey: Ablex.
Calderhead, J., & Shorrock, S. B. (1997). *Understanding teacher education*. London: Routledge.
Clark, A. (2005). Beyond the flesh: Some lessons from a mole cricket. *Artificial Life*, 11, 233–244.

Cole, M.: (1996), *Cultural psychology*, Cambridge: Harvard University Press.
Cole, M., & Engeström, Y. (1993). A cultural-historical approach to distributed cognition. In G. Salomon (Ed.), *Distributed cognitions: Psychological and educational considerations* (pp. 1–46). Cambridge: Cambridge University Press.
Cooper, P., & McIntyre, D. (1996). *Effective teaching and learning: Teachers' and students' perspective.* Buckingham: Open University Press.
Desforges, C. (1995). How does experience affect theoretical knowledge for teaching? *Learning and Instruction*, 5(4), 385–400.
Desforges, C., & McNamara, D. (1979). Theory and practice: methodological procedures for the objectification of craft knowledge. *Journal of Education for teaching*, 5(2), 145–152.
Edwards, A. (1998). Mentoring student teachers in primary schools: Assisting student teachers to become learners. *European Journal of Teacher Education*, 21(1), 47–62.
Engeström, Y. (1999). Activity theory and individual and social transformation. In Y. Engeström et al, (Eds.), *Perspectives on activity theory* (pp 19–38). Cambridge: Cambridge University Press.
Engeström, Y. & Middleton, D. (1996). *Cognition and communication at work.* Cambridge: Cambridge University Press.
Eraut, M. (1994). *Developing professional knowledge and competence.* Lewes: Falmer.
Ernest, P. (1989). The knowledge, beliefs and attitudes of the mathematics teacher: A model. *Journal of Education for Teaching*, 15(1), 13–33.
Feiman-Nemser, S. (2001). From preparation to practice: Designing a continuum to strengthen and sustain teaching. *Teacher College Record*, 103(6), 1013–1055.
Grossman, P. L., Wineburg, S., & Woolworth, S. (2001). Towards a theory of teacher community. *Teachers College Record*, 103(6), 942–1012.
Hutchins, E. (1991). The social organisation of distributed cognition. In L. Resnick et al: (Eds.), *Perspectives on socially shared cognition* (pp. 283–307). Washington: American Psychological Association.
Marks, P. R. (1990). Pedagogical content knowledge: From mathematical case to a modified conception. *Journal of Teacher Education*, 41(4), 3–11.
Pea, R. (1993). Practices of distributed intelligence and designs for education. In G. Salomon (Ed.), *Distributed cognitions: Psychological and educational considerations* (pp. 47–87). Cambridge: Cambridge University Press.
Perks, P. (1997). *Lesson planning in mathematics: A study of PGCE students.* Unpublished PhD Thesis, University of Birmingham
Perks, P. & Prestage, S, (1992). Making choices (1): Making choices explicit. *Mathematics in School*, 21(3), 46–48.
Perks, P. & Prestage, S., (1999). Square roots – An algorithm to challenge. *Mathematics Education Review*, 10, 31–40.
Prestage, S., & Perks, P. (1992). Making Choices (2): '... Not if you're a bear. *Mathematics in School*, 21(4), 10–11.
Prestage, S., & Perks, P. (1999). Towards a pedagogy of initial teacher education. In L. Bills (Ed.), *Proceedings of the BSRLM Day Conference, November 1999* (pp. 91–96). University of Warwick: BSRLM.
Prestage, S. & Perks, P. (2001a). Models and super-models: ways of thinking about professional knowledge. In D. K. Jones, & C. Morgan, (Eds.), *Research in mathematics education* (Vol. 3, pp. 101–114). London: BSRLM.
Prestage, S. & Perks, P. (2001b). *Adapting and extending secondary mathematics activities: New tasks for old.* London: David Fulton
Sternberg, R. & Horvath, J. (1995). A prototype view of expert teaching. *Educational Researcher*, 24(Aug-Sept), 9–17.
T.D.A. (2006). *Handbook of guidance* (2006 edition). http://www.tda.gov.uk/upload/resources/doc/h/handbook_of_guidance_2006.doc Accessed 13/07/07
Vygotsky, L. (1986). *Thought and mind.* Cambridge, MA: MIT.

Wineburg, S., & Grossman, P. (1998). Creating a community of learners among high school teachers, *Phi Delta Kappan*, 79(5), 350–353.

Watson, A., de Geest, E., & Prestage, S. (2003). *Deep progress in mathematics: The improving attainment in mathematics project.*, Oxford: Oxford University

Pat Perks
School of Education
University of Birmingham

Stephanie Prestage
School of Education
University of Birmingham

VICTORIA SÁNCHEZ AND MERCEDES GARCÍA

14. WHAT TO TEACH AND HOW TO TEACH IT

Dilemmas in Primary Mathematics Teacher Education

In this chapter, we focus on how two mathematics teacher educators' learning is generated through dilemmas arising in the decision-making process on what to teach and how to teach it in a mathematics teacher education programme. These dilemmas were related to different alternatives among which we had to choose, among them, deciding between content and procedures provided by the institution or chosen by ourselves; between adapting them from other teacher education programmes or built on research; between considering the problems of practice as problems to be handled or problems for research. Dilemmas have allowed us to visualize the progress of two mathematics teacher educators, which permits us to make some contributions to the study of professional development, and identify/reflect on different ways of approaching this study.

INTRODUCTION

To begin with, this chapter requires us to take a position on how we understand the work of a mathematics teacher educator. For us, it consists of educating teachers to be self-sufficient professionals who are able to develop the tasks of their practice, be responsible for their actions, and reflect on these. This understanding is the result of our development as teacher educators and researchers. During the last decades, we have established a complex relationship between our professional practice, the information coming from advances in mathematics education research and our evolution as researchers. This relationship is the background on which we 'see' that development, focusing on our work with beginning teachers, in particular, on two key elements in a primary teachers' mathematics programme, what to teach and how to teach it; variables about which, as teacher educators, we can make some decisions within the institutional context of our university.

Of course, what to teach and how to teach it is a 'shared question' among mathematics teacher educators and researchers. Different authors have been interested in issues such as what mathematics teachers need to know and know how to provide constant quality instruction under diverse conditions (Adler, Ball, Krainer, Lin & Novotna, 2005), the learning of prospective mathematics' teachers from different methodological perspectives (Boero, Dapueto & Parenti, 1996), or what it means to prepare teachers of mathematics (Sowder, 2007), among other questions.

For us, the concept of dilemma, understood as a situation in which a complicated choice has to be made between two different things you could do, has been very useful for reflection on the issues stated above. For Tillema and Kremer-Hayon (2002), 'Dilemmas usually denote an argument presenting two or more equal alternatives of action, and a tension between alternatives of equally perceived values' (p. 595). Dilemmas of teaching have been used in different educational contexts and situations (Ball, 1993, Edwards & Mercer, 1987; Jaworski, 1994; Lampert, 1985; Tillema & Kremer-Hayon, 2002, 2005; Windschitl, 2002). Dilemmas of teacher educators can directly represent different professional views and perspectives and they provide a powerful construct for identifying different steps in their professional development.

In this chapter, we focus on how our learning is generated through the dilemmas arising in the decision-making process on what to teach and how to teach it in a mathematics teacher education programme. In particular, the dilemmas were related to different alternatives among which we had to choose. For us these dilemmas were: deciding between content and procedures provided by the institution or chosen by ourselves; between adapting them from other teacher education programmes or built on research; between considering the problems of the practice as problems to be handled or problems for research. They were choices and decisions that implied a tension between some alternatives that affected both our work and our way of considering it with respect to its philosophy and goals. These dilemmas are the 'thread' connecting the sections that follow.

IDENTIFYING DILEMMAS OF MATHEMATICS TEACHER EDUCATORS

In this section, we take a close look at an account of our practice as teacher educators by focusing on some particular situations that arose. The identification of dilemmas in these situations offers to others a perspective on how own professional development was generated.

'What and How': 'Given' by the Institution versus 'Chosen' by the Teacher Educator

Primary school teacher preparation, and the traditional debate on how those teachers should be educated, can be considered from different points of view. Focusing on primary education from a general point of view, Ishler, Edens and Berry (1996) emphasized the importance of considering primary teacher education programmes from a broader and comprehensive perspective, beyond surface information about specific courses or programmes that does not adequately characterize their complexity. These authors considered that social, political and cultural forces pressure both teacher education programmes and primary schools. In particular, teacher educators encounter pressure regarding the content and structure of the curriculum from the university where they work and from the demands of the schools where their students will develop their professional work.

These pressures, amongst others, might have led to the variety of teacher education systems in different countries. Comiti and Ball (1996) used three elements of teacher education to frame the descriptions of different countries' approach to the preparation of primary and secondary teachers for teaching mathematics: Student teachers, teacher education curriculum and teacher educators themselves. Focusing on the two last elements, in our country (Spain) in the 1980s, the curricular guidelines and the hours of the primary teacher study plan were set by national legislation stipulating the official degree in primary teacher education and specializations. This legislation also specified other features related to topics common to all the specializations, specific subjects, short description of contents, credits, knowledge areas, and so forth. In each Autonomous Community (Spain is organized in several Autonomous Communities), these features were further specified and developed and, finally, each university decided the particular details of organization of their own study plan and, in particular, the number of hours of mathematics education, number of topics included in those hours, and so forth. Without going into the specific details of each specialization, the study plan from the University of Seville had a general course in Mathematics during the first year and a Mathematics Methods course in the second or third year (depending on the specialization). The corresponding Department (Didactics of Mathematics) designed the programmes for these subjects, and each mathematics teacher educator was responsible for development and implementation of the programme.

We started our trajectory and development as mathematics teacher educators in this context. Our preparation consisted of a degree in mathematics or other related scientific subject. We had not received any education in pedagogical or psychological issues nor in mathematics education. One of us had some years of experience as a secondary mathematics teacher, and the other had always worked as primary mathematics teacher educator at the University of Seville. Beginning in 1990 and continuing to today, we are working together in the Department of Didactics of Mathematics, sharing ideas, doubts and knowledge.

At the beginning, the curricular orientations of the official documents, and some textbooks provided us with what to teach. How to teach was related to our own personal experience as students or mathematics teachers. Nevertheless, in the light of our experiences as professional teachers in our classrooms with student teachers, we felt the need to restructure the conditions of our own practice. This situation posed a conflict between our beliefs about what should be the work of a teacher educator (for us, autonomous professionals able to take their own decisions) and what we were really doing. Recalling this situation now, we recognize in it the idea of a dilemma as defined by Tillema and Kremer-Hayon (2005). They state that, "A dilemma denotes a potential action and opting for a practical strategy to manage inconsistencies between beliefs and practice" (p. 204). In this sense, we can identify a dilemma in our search of 'what to teach and how to teach it' that we can express in the following terms: 'given' by the institution versus 'chosen' by the teacher educator. Now, we can say that we tried to solve the dilemma looking for a personal alternative.

VICTORIA SÁNCHEZ AND MERCEDES GARCÍA

'What and How': Adapted from Other Teacher Education Programmes versus Built on Research

We became aware of the importance of knowing what other mathematics teacher educators were doing in other countries. The work of researchers, such as Simon (1994), Wittmann (1995), Lampert and Ball (1998), Goffree and Oonk (1999, 2001) enabled us to identify three important elements in the search for a framework in which we could think about the content and organization of a primary teacher education programme related to mathematics: a theoretical background, content, and how student teachers could access that content. The identification of these elements, and what different authors included in them, led us to propose either adopting some of these programs, and setting ourselves the task of trying to adapt them to our own context, or looking for referents in the field of mathematics education research that would help us to reflect on them and construct our own option.

The election of the second alternative led us to a two-stage process. In the first stage, the search for referents enabled us to generate our own view of the organization and content of the subjects related to mathematics education in the primary teachers' programme, based on what other mathematics teacher educators had done. In the second stage, identification of some problems related to the implementation in our classroom of that programme led us to resolve them using those referents to look for an answer. We describe these different stages below.

Stage 1. The organization and content of the mathematics methods course was our first concern. We started to look at different studies related to research on mathematics teacher education (Brown & Borko, 1992; Brown, Cooney & Jones, 1990; Grouws & Schultz, 1996) and from here we made some decisions. Firstly, we approached 'what' from the perspective of what a primary mathematics teachers must know. Authors such as Shulman (1986, 1987), Ball (1991), Cooney (1994), Bromme (1994), and many others, were referents for later studies such as Llinares and Krainer, (2006) and Ponte and Chapman, (2006). All of these studies started to characterize a body of scientific knowledge in the mathematics education field. The analysis of the contributions mentioned above, and others like those of Llinares (1991), made possible our initial approximation to the different domains of knowledge for teaching mathematics, and from which we began to infer components of the content of a primary teacher education programme as: knowledge of and about mathematics, knowledge of the learners and learning processes, and knowledge of instructive processes.

But this identification was not enough for us, because we need to operationalize the components in our classrooms. We started to look for 'teaching/learning spaces' that allowed these components to be developed in our classrooms and, in this context, the following question arose: What dimensions could these spaces be organised around? This was not an isolated question rather it was part of the process of our professional development, in which distinct perspectives played a key role, such as situated learning (Brown, Collins & Duguid, 1989; Lave & Wenger, 1991). As a result, we started to distinguish two different aspects in our program: mathematics as a science to be taught and learned, and a specific professional work involved in the teaching of mathematics.

In considering mathematics as a science that must be taught and learned at this specific level, Bishop (1988), Sowder (1992) and NCTM (1989, 2000) took us deeply into the particularities of the different systems that underlie mathematical thinking, understanding them as sets of concepts, procedures, interrelationships and different ways of reasoning. We considered these systems with respect to mathematical content at the primary level, and decided to include: algebraic, numerical, geometrical, statistical, and probabilistic thinking.

With respect to the professional tasks (e.g., planning a lesson, assessment, choosing textbooks and other instructional materials), we had to identify the systems of activity that enable teachers to learn those tasks. In particular, following Llinares (see 2004 for a recent version), we consider such systems of activity to be: (1) the organization of the mathematical content to be taught; (2) the management of the mathematical content and classroom discourse; (3) analysis and interpretation of the mathematical thinking of students.

Both '*mathematical thinking*' and '*systems of mathematics teachers' activity*' organized our ideas in the identification of 'what to teach' in our Mathematics Methods course. The different components of the program's content were developed in the teaching/learning spaces defined at the intersection of both dimensions (see Figure 1).

Mathematical Thinking \ Systems of mathematics teachers' activity	To organize the mathematical content for teaching	To manage of mathematical content and discourse in the classroom	To analyse and interpret what student teachers think/know
Numerical Thinking			
Algebraic Thinking			
Geometrical Thinking			
Statistical/Probabilistic Thinking			

Figure 1. Dimensions related to 'what to teach' in a mathematics methods course

Our second concern was the mathematical knowledge of prospective primary teachers. A problem that we had detected in our professional practice was the lack of adequate mathematical background of our student teachers. Their mathematical education was a blur between what they had to teach and what they had to know. Developing a specific mathematical component that was the basis of specific mathematics knowledge for primary teachers became something fundamental for us, because of the characteristics of our students.

In a process analogous to the one we had followed to generate the teaching/learning spaces, we tried to create other spaces into which that mathematical component could fit. On one hand, we considered the '*activities of mathematical practices*' (see Rasmussen, Zandieh, King & Teppo, 2005), which underlie any mathematical content and, sometimes, have specific characteristics, such as defining, justifying, and modelling, among others, and which are a part of *'doing mathematics'*. On the other hand, we considered mathematical content organized by major subject areas: Analysis, Geometry, Algebra, Statistics and Probability. We thus defined for ourselves two new dimensions, the intersections of which could be considered 'mathematical spaces'. Those spaces were the key to organizing the other part of our programme, the general course in mathematics (see Figure 2).

Mathematical content areas / Mathematical practices' activity	Defining	Justifying	Modelling	....
Analysis				
Geometry				
Algebra				
Statistic/Probability				

Figure 2. Dimensions related to what to teach in a course of mathematics for primary teachers

When we had characterized what we intended to teach, we approached 'how to teach' from the way in which we understood that knowledge can be constructed by student teachers. The theoretical ideas of many researchers that took part in the discussion of teacher learning and education some years ago have been summarized by authors such as Putnam and Borko (1997). These theoretical ideas as theoretical frameworks allowed us to understand the characteristics of cognition and the purposes of education discussed by researchers such as Resnick (1991), Brown, Collins and Duguid (1989), Cobb (1994), among others. Our approach was rooted in the ideas that brought us into contact with different views about the constructed nature of knowledge and beliefs and the social and situated nature of cognition. Moreover, we incorporated into our theoretical background the ideas of 'community of practice' and learning based in participation, drawn from Lave and Wenger (1991) regarding learning as increased participation in the practice of a community.

We decided that 'becoming a primary teacher' might be understood as the process of prospective teachers being introduced into the community of practice of primary teachers and acquiring an understanding of mathematics teaching. In this sense, student teacher learning could be considered as progressive participation in the community of practice through the use of tools, which enable student teachers to understand and undertake professional tasks, and learning to teach is seen as the identification and use of tools in solving professional tasks (García, Sánchez, Escudero & Llinares, 2003; Lave & Wenger, 1991; Llinares, 2002;). For us, the term 'tool' is used in a broad sense (Vygotsky, 1986). Tools are not only denoted as a physical object, but also extended to consider concepts and reasoning, and so forth, which enable and influence interaction within a community. Such tools may be classified as either technical or conceptual tools. Technical tools were those tools such as teaching materials and software, techniques for managing discussion of procedures, answers to problems, and so on. Conceptual tools were understood as those concepts and theoretical constructs that have been generated from research in teaching and learning in general, and in mathematics education in particular, leading to understanding and conducting situations in which mathematics is taught and learned. In this context, teacher education may be seen as the process of introducing prospective primary school teachers into the community of practice of primary teachers.

All these ideas led us to design 'hypothetical teaching/learning trajectories' (García, 2000; Simon, 1995; Simon & Tzur, 2004). In the mathematics methods course, the starting points were tasks very close to the professional tasks of primary teachers and, therefore, 'authentic activities' (Brown, Collins & Duguid, 1989). For instance, a task corresponding to the teaching/learning space situated in the intersection of 'Geometrical thinking/To organise the mathematical content' was the student teachers' interpretation of pupils' productions related to plane figures (Sánchez, García & Escudero, 2006). In the primary teachers' mathematics course, the tasks were mathematical situations. In both cases, the tasks were situated in the intersections of the different dimensions considered. Therefore, an active participation of the community members in the process of solving the tasks was generated through the work in small groups, in which the reasoning processes were made explicit, taking into account previous knowledge and beliefs (García, Sánchez, Escudero & Llinares, 2006). From our perspective, through these teaching / learning trajectories the generation of learning environments should be favoured.

Stage 2. In trying to advance in our choice, it could be said that we were characterising the programme from our point of view. However, the development of our work as teacher educators required another step: the programme had to become operative in our Mathematics Methods course and Mathematics course. This is when we began to identify some issues that we had to handle and resolve.

Didactic-mathematical reasoning process: Our reflection on the analysis of the videos and narratives of teaching situations, developed in primary classrooms by teachers that collaborated with us, and what our student teachers observed during their student teaching period (Sánchez, 2003) led us to value the importance of the ways of reasoning and managing the situations involved in teaching mathematics. We considered those aspects from the perspective provided by Wilson's model of pedagogical reasoning (Wilson, Shulman & Richert, 1987); it became another possible component of the primary teacher's specific education that, from our point of view, should be added to the components previously mentioned in our programme (García & Sánchez, 2002).

The content for this new component would include identification of the characteristics of how children learn mathematics in an interactive context that can lead to instructional modification, the assessment and analysis of the role of representations in texts to approach the problem of visualization in geometry, the analysis and comparison of different problem-solving processes, the mathematical content implied in the answers of the pupils in a task organizing the class to include both cognitive and mathematical diversity (avoiding a sole idea in the problem-solving tasks), among other things.

Returning to how to teach: Moreover, we became aware that student teachers do not yet belong to the mathematics teachers' community of practice, although they will become members of it. One of our aims of the primary teacher education programme was to provide a way for prospective teachers to become integrated in the community of primary teachers by creating *communities of learners*. These communities of learners were characterized by an activity we called 'learning to teach'. This activity should enable student teachers to get involved in the culture of their chosen profession. In our case, the activity was performed in learning environments generated in the teaching/learning trajectories (García, 2003), and the use of the tools would help the prospective teachers to solve the 'authentic activities' proposed in such a way that, while developing the community of learning, they fit into the mathematics teachers' community of practice (García et al., 2003).

Recalling our decision-making at the beginning of this section, we identify another important dilemma in our search for 'what and how'. Our choice had to be made between adopting some of the programmes provided by other researchers or looking for ideas in the field of mathematics education research that would help us to reflect on them and construct our own option. This dilemma can be framed as 'adapt' from other teacher education programmes versus 'build' on research. The second option was our choice, and the actions that were generated by our decision led us to a new situation.

'What and How': Problems of the Practice as Problems to Be Handled versus Problems of the Practice as Problems for Research

Up to this point, as teacher educators, we had been advancing in our search of what to teach and how to teach it. The identification of spaces in which to locate the components was an advancement in our approach to what to teach. In addition, our theoretical perspective had allowed us to consider the teaching/learning trajectories with new particularities, and to appreciate learning through the use of different types of tools, in the interactions among the different agents in the learning/teaching process.

Nevertheless, we were aware that there were still 'things to be solved'. In particular, we asked ourselves how we could assess our appreciation of students' learning and access the social processes that take place in learning environments. These questions raised a new situation: Should we limit ourselves to solving the problems created in our practice or should we transform these problems as a research object seeking solutions that could be validated by the international scientific community? We chose the second option and it was not an easy decision. This second option had implications in our own professional development because in considering theory and practice we had to manage elements of different nature.

Two ideas were very important in the development of this option. The first was our view of learning to teach as the identification and use of tools. We understand 'identification and use' as student teachers' simultaneously setting in motion the different tools, interaction, and communication of information among them that leads to reasonable decisions in the context of a learning environment. Secondly, according to Collins, Brown and Newman (1989), we considered the different dimensions that "constitute any learning environments: content, methods, sequence and sociology" (p. 477). These ideas were the background that allowed us to be aware of the importance of delving deeply both into the use of conceptual tools in the 'hypothetical teaching/learning trajectories' and the characterization of the social dimension of those environments.

Research Question 1. The first problem was formulated as a research question in the following terms: How do student teachers use conceptual tools to solve professional teaching tasks in the development of a teaching/learning trajectory? (García et al., 2006). The theoretical object of our study was 'the use of conceptual tools in the development of a professional task'. A detailed description of the research is not within the scope of this chapter; we only wish to point out that, from the results of our research, we were able to identify and characterize levels in the use of the conceptual tools. This allowed us to appreciate learning levels, as outlined and described elsewhere (García et al., 2003, 2006; García, Sánchez & Escudero, 2007). The characterization of these levels as a result of our research allowed us, as teacher educators, to assess the learning of students in our programme. But in addition, it made a small contribution to the knowledge-base of the field that can be validated (or not) in other studies and on which new work could be based.

Research Question 2. We are trying to examine the potential of using cognition/discourse/interaction (Kumpulainen & Mutanen, 2000; Schrire, 2006) to approach the dimension "sociology", considered by Collins, Brown and Newman (1989) as one of dimensions that intervene in the learning environments. These authors abstracted five critical characteristics affecting this dimension: situated learning, culture of expert practice, intrinsic motivation, exploiting cooperation and exploiting competition. From our point of view, the cognitive processes developed, the discourse generated, and the interactions that take place among the members of a student teachers' group were variables that could assist us in characterizing the role of that dimension in the construction of knowledge. In this context, we have asked the following research question: Would cognition, discourse and interaction together enable a researcher to approach the sociological dimension?

We have assessed both individual and social aspects of cognition through the learning levels of the individuals in the groups and the relationship between these levels and the level of corresponding group. Our previous research (Sánchez et al., 2006) had enabled us to characterize student teacher and group learning levels and identify the categories this study was based on. The discourse is analysed from the perspective of mathematics teacher domains of knowledge, adapting the work of authors such as Ben-Yehuda, Lavy, Linchevski and Sfard (2005) on arithmetical discourse. Interaction in the mathematics classroom has been analyzed from different perspectives: individual-psychological (Cobb, Wood & Yackel, 1991), radical constructivist (Von Glasersfeld, 1991), sociocultural (Kieran, Forman & Sfard, 2001) and epistemological (Steinbring, 2005). According to our theoretical perspective (García, Sánchez & Escudero, 2007), interaction alluded to the fact that independent individuals were affected by others. In particular, we considered didactic mathematical interaction in a group of student teachers as they tried to solve the task posed. We took into account the relationships amongst individuals in the group and the contributions of each of them to the group discussion.

As in the above research, we focus only on the results, which are showing that the sociological dimension could be reflected in the multiple uses of knowledge identified in the exchange of information when a task is performed. The three variables (cognition/discourse/interaction) considered together made it possible for us to observe a series of points along a path with which we could reconstruct a dynamic process of construction of knowledge, which progresses from an initial state to a final state. We think that this dynamic process is, in some way, related to the sociological characteristics of the construction of knowledge.

Looking at the situation presented at the beginning of this section from the theoretical perspective of the dilemmas, we can identify yet a third dilemma. Our choice had to be made between solving the problems created in our practice or transforming these problems as a research object. This dilemma can be stated as 'problems of the practice as problems to be handled versus problems of the practice as problems for research'. The option chosen and the actions developed have shown new characteristics of our professional development. Goals of research in mathematics education have now been incorporated into our professional domain as mathematics teacher educators.

In this chapter so far, we have used three dilemmas as a 'thread' connecting what has allowed us to describe a personal evolution viewed from an internal reflection of the 'actors' in the events. This description will be now a context for an external reflection, in which the trajectory is reified to be considered as an object of reflection.

DELVING DEEPER

There are many ways of presenting the information that is collected in books like this. Sometimes different studies and research are revised and, from here, questions, problems, contributions and future lines of research are identified. We have opted for extracting ideas about some of those aspects from reflection on our practice. Throughout this chapter, we presented an account that, firstly, described some teaching problems from the point of view of mathematics teacher educators and, secondly, identified dilemmas from a researcher's point of view.

A global glance at the account allows appreciation of the professional evolution of two mathematics teacher educators, shown in the development of a programme for primary teacher education with respect to mathematics. This evolution has been seen throughout some situations and choices for action identified as dilemmas. Along with the specific aspect of the practice the teacher educators considered (what to teach and how to teach it), these dilemmas can be classified as professionally self-oriented and have an internal attribution (Tillema & Kremer-Hayon, 2005). We now attempt to go from recognition of dilemmas to identification of some characteristics of a process of professional development and some lines of theoretical reflection, which allow us to advance our knowledge in the field of teacher education. We situate ourselves in another perspective, observing the account as a context in which that process and these lines can be identified.

Dilemmas have allowed us to recognize different ways of considering practice and theory, different goals, and different tools, which mathematics teacher educators identify and use to reach those goals, etc. These elements have enabled us to: (1) visualize the progress of two teacher educators, which allows us to make some contributions to the study of professional development; and (2) identify/reflect on different ways of approaching the study of professional development of mathematics teacher educators. Each of these aspects is considered in the following sections.

The Professional Development of Teacher Educators

In this chapter, we have used dilemmas as a theoretical element for 'looking at' how professionals (mathematics teacher educators) have considered some practical problems and have solved them in a certain way in the development of their work. Furthermore, we have recognized those situations as dilemmas, and we have seen how two teacher educators have looked for tools (specific of teacher educators) and applied them in solving the chosen option. We understand that, in identification and use, learning has taken place, and, therefore, is constructed knowledge. Consequently, we can consider the dilemmas as indicators of a learning process, being this process a part of the professional development of teacher educators.

In this context of professional development of mathematics teacher educators, identifying the processes that lead to dilemmas is a focus of interest. It is important then to investigate their origins. If we analyze the account presented here, we can recognize different sorts of awareness linked to the generation of each one of the dilemmas.

In the generation of the first dilemma, we see how the teacher educators became aware of problems in their practice in a local context. In some way, we could say there are not differences from the normal practice of teacher educators. In the second dilemma, teacher educators became aware that their particular situations are shared on two levels: other colleagues working on the same issues and other researchers offering results that they could be made use of in their own practice. Finally, the third dilemma, there is an awareness that these shared problems could be considered an object for research, their study being a suitable contribution to a scientific field.

Considering that professional development is a personal goal of many teacher educators, a social goal from an institutional point of view and a research goal in mathematics education field, it is important to show some 'indicators' that allow us to accede to this development, beyond a generic declaration of its importance. Dilemmas, as part of a trajectory, and the type of awareness that generates them are a step in that direction. But dilemmas can also be seen as a research element; a theoretical construct used to study the practice. In this sense, identification, study and analysis, and use of the information found through dilemmas can open new routes to studying the professional practice.

Different Ways of Approaching to the Study of Professional Development of Mathematics Teacher Educators

We now use the account above as a context in which we can approach different lines of inquiry in the learning and professional development of teacher educators. Their ways of considering the practice and the theory, their objectives and the tools that they identify and use to obtain those objectives in the dilemmas, are elements used now to establish those lines of study. The theory/practice relationship and the various emergent communities are issues that arise from their study, and provide lines of future research.

Theory/practice relationship: Throughout the process of professional development identified, we have been able to observe how two teacher educators establish different ways of relating theory and practice.

In this chapter, we have understood theory in a wide sense as a formal statement of ideas suggested to explain a fact or event or, more generally, an opinion or explanation. The term teachers' practice also includes professional tasks that achieve a purpose, making use of tools, and justifying and reflecting on their use. We would like to point out that our background (teacher educators and researchers) is the origin of a flexible view of theory and practice, without aligning them in a strict theoretical perspective.

The different relationships between theory and practice take place as distinct dilemmas and correspond to the different teacher educators' learning moments. Thus, in the first dilemma, a theoretical background that underlies the practice either in the situation presented or in the strategies generated is not realized. In the second dilemma, theory provides useful tools for decision-making related to the chosen alternative. We could say that the relationship is 'use of theory in practice'. Finally, in the third dilemma, the transformation of problems encountered in practice into objects of research allows us to observe a dialectical relationship between practice and theory. In this sense, the theory/practice relationship, which we have used elsewhere to describe our development as reflective members of a community (García, Sánchez & Escudero, 2007), is now considered as a theoretical tool for approaching to that development.

The community of practice: Considered as a theoretical element, the idea of the community of practice provides us with another view of the account. For us, that idea includes people who engage in a process of collective learning in a shared domain of human endeavour, portraying a social group in which its members share in giving (goals, purpose, ends, means, etc.) (Lave & Wenger, 1991; Wenger, 1998, 2007). Among these communities, we consider the communities of inquiry as characterized by Jaworski (2005, p.79):

- A community (of practice) in which one of the norms is an attitude of inquiry (e.g., involving critical questioning).
- Inquiry is *not* the practice, but a way of approaching practice.
- Inquiry can be used as a *tool* to develop inquiry as *a way of being*

In dilemmas, we can visualize how the issues faced by teacher educators change: their goals, the tools they use and how they make use of them in solving problems, and their relationships with other colleagues. These aspects, among others, characterize a community. Recalling the situations described in the account, we see in the first dilemma a lack of awareness of the existence of a group with common goals and tools, an 'absence of community'. In this case, the teacher educators could be identified as isolated professionals. In the second dilemma, the identification of goals, becoming aware of them as something shared, and the use of common tools, leads to their consideration as members of a community of practice, building relationships that enable them to learn from each other. In the third dilemma, new goals and tools identify features of a community of inquiry, involving critical questioning and in which questions can be asked and hypotheses tested (Jaworski, 2005). To summarize, consideration of the elements that characterize a community of practice can be useful for identifying dilemmas and, if nothing else, approaching the learning and professional development of mathematics teacher educators.

CONCLUDING COMMENTS

What has emerged in this chapter is the importance of the mathematics teacher educator's dilemmas, both in his/her own professional development and in the identification of future lines of research in the field. Furthermore, dilemmas can explain why an educational programme has been developed in a certain way. Without diminishing in any way the importance of the contents and orientation provided by experts, policy makers and institutions, other very different elements also take part in that development. In particular, in this chapter we have tried to show how the evolution of teacher educators has affected the development of an educational programme for prospective primary teachers. In this sense, dilemmas provide insights about what is really happening in that programme.

Finally, if we consider that primary teacher education programmes are composed of several parts, mathematics education being one of them, we conclude this chapter with a question: Can different matters and different ways of considering 'what' and 'how' be woven into a coherent whole that fulfils the overall goal of those programs? Obviously, these questions are different in a context of Secondary Mathematics Teacher Education, exclusively related (at least in Spain at present) to the teaching/learning of one subject matter. Nevertheless, in both cases, as mathematics teacher educators, we need to be more aware of the hidden messages that influence our way of understanding what teach and how to teach it, and how it can affect the students that we teach.

REFERENCES

Adler J., Ball D., Krainer K., Lin FL, & Novotna, J. (2005). Reflections on an emerging field: Researching mathematics teacher education. *Educational Studies in Mathematics, 60*(3), 359–381.
Ball, D. (1991). Research on teaching mathematics: Making subject-matter knowledge part of the equation. In J. Brophy (Ed.), *Advances in research on teaching: Teachers' knowledge of subject matter as it relates to their teaching practice* (Vol. 2, pp. 1–48). Greenwich, CT: JAI Press.
Ball, D. L. (1993). With an eye on the mathematical horizon: Dilemmas of teaching primary school mathematics. *The Primary School Journal, 93*(4), 373–397.
Ben-Yehuda, M., Lavy, I., Linchevski, L., & Sfard, A. (2005). Doing wrong with words: What bars students' access to arithmetical discourses. *Journal for Research in Mathematics Education, 36*(3), 176–247.
Bishop, A. (1988). *Mathematical enculturation*. Dordrecht: Kluwer Academic Publishers.
Boero, P., Dapueto, C., & Parenti, L. (1996). Didactics of mathematics and the professional knowledge of teachers. In A. Bishop, K. Clements, C. Keitel, J. Kilpatrick, & C. Laborde (Eds.), *International handbook on mathematics education* (pp. 1097–1121). Dordrecht: Kluwer Academic Publishers.
Bromme, R. (1994). Beyond subject matter: A psychological topology of teachers' professional knowledge. In R. Biehler, R. Scholz, R. SträBer, & B. Winkelman (Eds.), *Didactics of mathematics as a scientific discipline* (pp. 73–78). Dordrecht: Kluwer Academic Publishers.
Brown, J. S, Collins, A., & Duguid, P. (1989). Situated cognition and the culture of learning. *Educational Researcher, 18* (January–February), 32–42.
Brown, C. A., & Borko, H. (1992). Becoming and mathematics teacher. In D. A. Grouws (Ed.), *Handbook of research on mathematics teaching and learning* (pp. 209–239). New York: Macmillan.
Brown, S. I., Cooney, T. J., & Jones, D. (1990). Mathematics teacher education. In R. Houston (Ed.), *Handbook of research on teacher education* (pp. 639–656). New York: Macmillan.
Cobb, P., Wood, T. & Yackel, E., (1991). Learning through problem solving: A constructivist approach to second grade mathematics. In E. von Glasersfeld (Ed.), *Radical constructivism in mathematics education* (pp. 157–176). Dordrecht: Kluwer Academic Publishers.
Cobb, P. (1994). Constructivism in mathematics and science education. *Educational Researcher, 23*(7), 4.
Collins, A., Brown, J., & Newman, S. (1989). Cognitive apprenticeship: Teaching the crafts of reading, writing, and mathematics. In L. Resnick (Ed.), *Knowing, learning, and instruction. Essays in honour of Robert Glaser* (pp. 453–494). Hillsdale, NJ: LEA. Pb.
Comiti, C., & Ball, D.L. (1996). Preparing teachers to teach mathematics: A comparative perspective. In A. Bishop, K. Clements, C. Keitel, J. Kilpatrick, & C. Laborde (Eds.), *International handbook on mathematics education* (pp. 1123–1153). Dordrecht: Kluwer Academic Publishers.
Conney, T. J. (1994). Research and teacher education: In search of common ground. *Journal for Research in Mathematics Education, 25*(6), 608–636.
Edwards, D., & Mercer, N. (1987). *Common knowledge*. London: Methuen.
García, M. (2000). El aprendizaje del estudiante para profesor de matemáticas desde la naturaleza situada de la cognición: Implicaciones para la formación inicial de maestros. In C. Corral, & E. Zurbano (Eds.), *Propuestas metodológicas y de evaluación en la formación inicial de los profesores del área de didáctica de las matemáticas* (pp. 55–81). Oviedo: Universidad de Oviedo.
García, M., & Sánchez, V. (2002). Una propuesta de formación de maestros desde la Educación Matemática: adoptando una perspectiva situada. In L.C. Contreras, & L. Blanco (Eds.), *Aportaciones a la formación inicial de maestros en el Área de Matemáticas: Una mirada a la práctica docente* (pp. 59–88). Badajoz: Servicio de Publicaciones, Universidad de Extremadura.
García, M. (2003). La formación inicial de profesores de matemáticas: Fundamentos para la definición de un currículo. In D. Fiorentini (Ed.), *Formaçao de profesores de matemática. Explorando novos caminhos com outros olhares* (pp. 51–86). Campinas, Brasil: Mercado de Letras.

García, M., Sánchez, V., Escudero, I., & Llinares, S. (2003). The dialectic relationship between theory and practice in mathematics teacher education. In *CERME3 (Group 11) Bellaria*, Italy: http://www.dm.unipi.it/~didattica/CERME3.

García, M., Sánchez, V., Escudero, I., & Llinares, S. (2006). The dialectic relationship between research and practice in mathematics teacher education. *Journal of Mathematics Teacher Education, 9*(2), 109–128.

García, M., Sánchez, V., & Escudero, I. (2007). Learning through reflection in mathematics teacher education. *Educational Studies in Mathematics, 64*(1), 1–17.

Goffree, F., & Oonk, W. (1999). Educating primary school mathematics teachers in the Netherlands: Back to the classroom. *Journal of Mathematics Teacher Education, 2*(2), 207–214.

Goffree, F., & Oonk, W. (2001). Digitalizing real teaching practice for teacher education programs. The MILE approach. In F. Lin, & T. Cooney (Eds.), *Making sense of mathematics teacher education* (pp. 51–86). Dordrecht: Kluwer Academic Publishers.

Grouws, D., & Schultz, K. (1996). Primary education. In J. Sikula, T. Buttery, & E. Guyton (Eds.), *Handbook of research on teacher education* (pp. 442–458). New York: Macmillan.

Ishler, R. E., Edens, K. M., & Berry, B. W. (1996). Primary education. In J. Sikula, T. Buttery, & E. Guyton (Eds.), *Handbook of research on teacher education* (pp. 348–377). New York: Macmillan.

Jaworski, B. (1994). *Investigating mathematics teaching*. London: Falmer.

Jaworski, B. (2005). Learning communities in mathematics: Research and development in mathematics teaching and learning. In *Proceedings of NORMA 05, Fourth Nordic Conference on Mathematics Education* (pp. 71–96). Trondheim, Norway: Tapir Press.

Kieran, C., Forman, E., & Sfard, A. (Eds.) (2001). Bridging the individual and the social: Discursive approaches to research in mathematics education. A PME Special Issue. *Educational Studies in Mathematics, 46*(1–3).

Kumpulainen, K., & Mutanen, M. (2000). Mapping the dynamics of peer group interaction: A method of analysis of socially shared learning processes. In H. Cowie, & G. van der Aalsvoort (Eds.), *Social interaction in learning and instruction: The meaning of discourse for the construction of knowledge* (pp. 141–161). Amsterdam: Pergamon Press.

Lampert, M. (1985). How do teachers manage to teach? Perspectives on problems in practice. *Harvard Educational Review, 55*, 178–194.

Lampert, M., & Ball, D.L. (1998). *Teaching, multimedia and mathematics: Investigations of real practice*. New York: Teachers College.

Lave, J., & Wenger, E. (1991). *Situated learning. Legitimate peripheral participation*. New York: Cambridge University Press.

Llinares, S. (1991). *La formación de profesores de matemáticas*. GID Universidad de Sevilla.

Llinares, S. (2002). Participation and reification in learning to teach: the role of knowledge and beliefs. In G. C. Leder, E. Pehkonen, & G. Torner (Eds.), *Beliefs: A hidden variable in mathematics education?* (pp. 195–210). Dordrecht: Kluwer Academic Publishers.

Llinares, S. (2004). Building virtual learning communities and the learning of mathematics teacher student. Invited Regular Lecture at ICME-2004. Copenhague, Denmark.

Llinares S., & Krainer K. (2006). Mathematics (student) teachers and teacher educators as learners. In A. Gutierrez, & P. Boero (Eds.), *Handbook of research on the psychology of mathematics education* (pp. 426–459). Rotterdam: Sense Publishers.

NCTM (1989). *Curriculum and evaluation standards for school mathematics*. Reston, Virginia: NCTM.

NCTM (2000). *Principles and standards for school mathematics*. Reston, Virginia: NCTM.

Ponte, J., & Chapman, O. (2006). Mathematics teachers' knowledge and practices. In A. Gutierrez, & P. Boero (Eds.), *Handbook of research on the psychology of mathematics education* (pp. 461–494). Rotterdam: Sense Publishers.

Putnam, R., & Borko, H. (1997). Teacher Learning: Implications of New Views of Cognition. In B. Biddle, T.L. Good, & I.F. Goodson (Eds.), *International handbook of teachers and teaching* (pp. 1223–1296). Dordrecht: Kluwer Academic Publishers.

Ramussen, C., Zandieh, M., King, K., & Teppo, A. (2005). Advancing mathematical activity: A practice-oriented view of advanced mathematical thinking. *Mathematical Thinking and Learning*, 7(1), 51–73.
Resnick, L. B. (1991). Shared cognition: Thinking as social practice. In L. Resnick, J.M Levine, & S.D. Teasley (Eds.), *Socially shared cognition* (pp. 1–20). Erlbaum, Hillsdale, NJ.
Sánchez, V. (2003). An approach to collaboration in elementary pre-service teacher education. In A. Peter-Koop, V. Santos-Wagner, C. Breen, & Andy Begg (Eds.), *Collaboration in teacher education* (pp. 57–68). Dordrecht: Kluwer Academic Publishers.
Sánchez, V., García, M., & Escudero, I. (2006). Primary preservice teacher learning levels. In J. Novotná, H. Moraová, M. Krátká, & N. Stehlíková (Eds.), *Proceedings of the 30th Conference of the International group for the Psychology of Mathematics Education* (Vol. 5, pp. 33–40). Prague, Czech Republic: Faculty of Education, Charles University.
Schrire, S. (2006). Knowledge building in asynchronous discussion groups: Going beyond quantitative analysis. *Computers & Education, 46*, 49–70.
Shulman, L. (1986). Those who understand: Knowledge growth in teaching. *Educational Researcher*, February, 4–14.
Shulman, L. (1987). Knowledge and teaching: Foundations of the new reform. *Harvard Educational Review, 57*(1), 1–22.
Simon, M. (1994). Learning mathematics and learning to teach: Learning cycles in mathematics teacher education. *Educational Studies in Mathematics, 26*(1), 71–94.
Simon, M. (1995). Reconstructing mathematics pedagogy from a constructivist perspective. *Journal for Research in Mathematics Education, 26*(2), 114–145.
Simon, M. A., & Tzur, R. (2004). Explicating the role of mathematical tasks in conceptual learning: An elaboration of the hypothetical learning trajectory. *Mathematical Thinking and Learning, 6*(2), 91–104.
Sowder, J. (1992). Estimation and number sense. In D. A. Grouws (Ed.), *Handbook of research on mathematics teaching and learning* (pp. 371–389). New York: Macmillan.
Sowder, J. (2007). The mathematical education and development of teachers. In F. K. Lester (Ed.), *Second handbook of research on mathematics teaching and learning* (pp. 157–224). New York: Macmillan.
Steinbring, H. (2005). *The construction of new mathematical knowledge in classroom interaction: An epistemological perspective.* New York, USA: Springer.
Tillema, H.H., & Kremer-Hayon, L. (2002). Practising what we preach – Teacher educators' dilemmas in promoting self-regulated learning; A cross case comparison. *Teaching & Teacher Education, 18*(5), 593–607.
Tillema, H., & Kremer-Hayon, L. (2005). Facing dilemmas: Teacher-educators' ways of constructing a pedagogy of teacher education. *Teaching in Higher Education, 10*(2), 203–217.
Von Glasersfeld, E. (1991). *Radical constructivism in mathematics education.* Dordrecht: Kluwer Academic Publishers.
Vygotsky, L. (1986). *Thought and mind.* Cambridge, MA: MIT.
Wenger, E. (1998). *Communities of practice: Learning, meaning and identity.* CUP.
Wenger, E. (2007). Communities of practice, A brief introduction. www.ewenger.com/theory/index.htm (June, 2007).
Wilson, S. M., Shulman, L., & Richert, A.E. (1987). 150 different ways of knowing: Representations of knowledge in teaching. In J. Calderhead (Ed.), *Exploring teachers' thinking* (pp. 104–124). London: Cassell.
Windschitl, M. (2002). Framing constructivism in practice as the negotiation of dilemmas: An analysis of the conceptual, pedagogical, cultural and political challenges facing teachers. *Review of Educational Research, 72*(2), 131–175.

Wittmann, E. (1995). Mathematics education as a 'design science'. *Educational Studies in Mathematics,* *29*(4), 355–374.

Victoria Sánchez and Mercedes García
Departamento de Didáctica de las Matemáticas
Facultad de Ciencias de la Educación,
University of Seville

PAOLA SZTAJN

15. CARING RELATIONS IN THE EDUCATION OF PRACTISING MATHEMATICS TEACHERS

The purpose of this chapter is to challenge mathematics teacher educators, particularly those working with practising teachers, to care. Noddings' caring theory and Hackenberg's notion of a mathematical caring relation are used to analyze mathematics professional development episodes. The claim made in the chapter is that beyond content and format, the education of practising mathematics teachers requires attention to relations, particularly those established between mathematics teacher educators (as the one-caring) and mathematics teachers (as cared-for).

THE CHALLENGE TO CARE

In 1992, in the United States, Nel Noddings challenged the education community to care. "The need to care in our present culture is acute," Noddings (1992) claimed, adding "patients feel uncared for in our medical system; clients feel uncared for in our welfare system; old people feel uncared for in facilities provided for them; and children, especially adolescents, feel uncared for in schools" (p. xi). Over a decade later, the need to care is still acute – perhaps even more so than before. In this chapter, I propose that besides caring for children in schools, educators need to care for teachers. More specifically, as I have learned in my own work with practising K-8 mathematics teachers, teacher educators need to establish caring relations with practising teachers when working together to improve mathematics instruction.

Mathematics teachers in the U.S. are again under the spotlight as a new wave of reports points to highly qualified mathematics and science teachers as a major factor in the successful preparation of a skillful new generation (e.g., American Association of Colleges for Teacher Education, 2007; Business Higher Education Forum, 2007; National Academies of Science, 2006). But the current recognition of the teachers' role in mathematics and science education has come tied to an increase in demand for more rigorous evaluation of teaching (Guskey, 2005; U.S. Department of Education, 2007) and accountability accomplished through scientific measures (Cochran-Smith, 2003, 2004). At a time like this, it becomes more likely for teachers to become angry at what may be perceived as the unjust ways (Nieto, 2003) in which they and their students are treated. Therefore, it is timely to add teachers to the list of the uncared for. My purpose in this chapter is to challenge the mathematics teacher education community, particularly those working with practising teachers – to care.

MATHEMATICS PROFESSIONAL DEVELOPMENT: FORMAT, CONTENT, AND RELATIONS

After over a decade of increased attention to the format of professional development initiatives (ways of working with teachers), research has reminded teacher educators that the content of the professional development (what is being worked on) is of utmost importance (Hill & Ball, 2004; Kennedy, 1998). In the mathematics education of teachers, format and content are closely related in a complex interrelationship (Jaworski & Wood, 1999). Therefore, current lists that point to characteristics of high quality or effective professional development include recommendations for both the format and the content of the initiative (see Sowder, 2007, for a comprehensive review of such lists). For example, one empirically established list of key features for effective professional development is organized around structural (reform type, duration, and collective participation) and core (active learning, coherence, and content focus) features (Desimone et al., 2002). Lists such as this, show that both format and content of professional development programmes are at the forefront of mathematics teacher educators' concerns.

Mathematics teaching, nonetheless, is not organized solely around content and format. Teaching is a moral enterprise as much as it is a knowledge endeavor (Ball & Wilson, 1996). It requires a commitment to both knowledge and students, and encompasses a struggle to be both "responsible and responsive" (p. 180) to the learner. Thus, mathematics teacher educators ought to be both responsible and responsive to teachers, attending to both teachers' knowledge and to teachers' needs. Mathematics teacher educators who work with practising teachers also need to consider that their students are adults who come to a learning situation with their own sets of goals (Knowles, Holton, & Swanson, 1998).

One assumption of my work in teacher education is that when mathematics teacher educators and mathematics teachers come together in a professional development encounter, they share one important goal – the improvement of mathematics instruction. Teacher educators and teachers might disagree about what it means or what are the means to improve mathematics instruction. They may think differently about mathematics and mathematics learning in schools, and may play different and perhaps asymmetrical roles in professional development activities. However, in a professional development initiative, they agree to come together to work toward better mathematics teaching. Another assumption of my work is that the burden of initiating and sustaining a successful professional development endeavour is on teacher educators, who need to take upon themselves the responsibility for finding ways to engage teachers in work that leads to better mathematics teaching.

My claim in this chapter is that beyond content and format, teacher educators who work with practising mathematics teachers to improve mathematics instruction need to pay attention to the *relations* established within professional development initiatives. I propose that Nel Noddings' caring theory (Noddings, 1984, 1992) can be used to analyze and promote relations within mathematics professional development situations, which are *mathematical caring relations* (Hackenberg, 2005). Although many relations are established in professional development initiatives, I focus my analysis on the relation established between the mathematics teacher educator and the mathematics teacher to improve mathematics instruction, and I attend to what teacher educators need to do and learn to initiate and sustain mathematical relations that are caring relations. There is also much for the teacher to do and learn in the context of a mathematical caring relation; nonetheless, the focus of this chapter is on the teacher educator whom I believe should be purposefully seeking the establishment of such relations with mathematics teachers. From my perspective, the community of mathematics teacher educators currently lacks concepts to analyze relations in the work they do with practising mathematics teachers. For me, caring theory provides such concepts.

This chapter begins with highlights of important concepts from Noddings' caring theory and a discussion of what makes caring relations in professional development settings mathematical. Three brief narratives of professional development episodes exemplify situations in which caring theory helps the analysis of particular aspects of the work educators of practising mathematics teachers do, the relations they establish with teachers and the relevance of mathematical caring relations in the education of practising mathematics teachers. To conclude, I delineate issues to which I have attended in endeavouring to become a caring mathematics teacher educator of practising teachers.

CARING RELATIONS

In her theory of moral development, Carol Gilligan (1982) proposed a series of stages that go from more to less selfish as a person progresses from paying attention to the self to paying attention to the needs of others and to a combination of self and others. Her ideas inspired discussions about an ethic of care, which has impacted fields such as nursing that have caring as an essential characteristic of practice. In education, this ethic of care is most influential in Noddings' caring theory. While a person who cares can be described as being gentle, kind, thoughtful, or concerned about others, in Noddings' theory caring "requires actual encounters with specific individuals; it cannot be accomplished through good intentions alone" (Tong & Williams, 2006).

Although the word caring is often used to refer to someone's attitude, Noddings places emphasis on caring as a relation instead of caring as a virtue or an individual attribute. If caring were about the characteristics of a single person who cared (the one-caring), we could make a list of desirable caring behaviours and look for those who held these attributes. In a relational approach, caring involves not only the one-caring but also the person to whom the one-caring is attending (the cared-for) and the situation in which they find themselves (Noddings, 2001). Thus, a caring person "is not best construed as one who possesses certain stable, desirable traits that might be identified before she steps into a [caring situation]" (p. 100). Rather, a caring person is one who can establish reciprocal caring relations with a variety of cared-fors in a wide range of situations.

Caring as a relational concept indicates that care depends on the cared-for as well as the one-caring, and caring relations involve characteristics of and engage both the one-caring and the cared-for. For a caring relation to be complete, the cared-for needs to recognize the care and the relation that is being established. Thus, a caring relation is reciprocal and by engaging in it, the cared-for nurtures the one-caring. Noddings (2001) claimed that "the reaction of the cared-for is essential in establishing a relation as one of caring. [...] Any relation in which one person claims to care and the recipient of that care denies the claim is one that demands scrutiny. One does not care simply because one claims to do so" (p. 101). Therefore, no matter how hard a person cares, if the caring is not acknowledged and reciprocated by the cared-for a caring relation is not established.

Caring as a relational concept also requires mutuality. In a mature caring relation, there is a string of caring encounters in which the roles of the one-caring and the cared-for are not static and the one-caring in one moment becomes the cared-for in another (Noddings, 1992). Noddings (2001) also pointed out that caring relations involve two people who know each other. Therefore, to develop caring relations the one-caring and the cared-for need to maintain a connection long enough to get to know each other, requiring time for the establishment of such relations.

Two important concepts in caring theory characterize the "state of consciousness" (Noddings, 1992, p. 15) of the one-caring. They are the one-caring's experiences of engrossment and motivational displacement. Engrossment means full receptivity to the cared-for; it means that the one-caring is ready to hear, feel and see what the cared-for claims to hear, feel and see. During a caring situation, the one-caring is attentive to the needs of the cared-for, trying to make sense of the other person's questions or requirements. By attending, the one-caring can be moved by the needs, goals, or feelings of the cared-for, and the one-caring's engrossment can lead to action. At this point, the one-caring experiences motivational displacement: the one-caring lets go of his or her personal interests, being motivated by and engaged in the projects of the cared-for. Just as we carefully think and plan for our own goals, those who experience motivational displacement begin to think carefully about, plan for, and act toward the goals of another person, embracing these goals as their own.

In caring theory, engrossment and motivational displacement do not tell the one-caring what to do and do not preclude the one-caring from attempting to influence the goals of the cared-for in some way. As Noddings (1992) explained, these two aspects of caring relations "characterize our consciousness when we care. But the thinking that we do will now be as careful as it is in our own service" (p. 16). Therefore, the one-caring assesses the goals and needs of the cared-for before acting in a caring relation. The assessment of what is best for another "involves considering both the felt needs of the other and the values of the one-caring" (Noddings, 2001, p. 101) as the one-caring navigates between inferred needs (those the one-caring thinks are in the best interest of the cared-for) and expressed needs (those the cared-for thinks are best) (Noddings, 2002). Recognizing the caring relation, the cared-for can also evaluate her expressed needs in view of those brought forth by the one-caring, so that both the one-caring and the cared-for navigate between each other's goals.

Overall, caring means an on-going search for competence and, in an approach involving relations, this search refers to engaging in situations in which both the one-caring and the cared-for experience growth. Each situation encompasses many caring encounters in which the roles of the one-caring and the cared-for are intertwined. While in each caring encounter the one-caring and the cared-for may play quite different roles (the one-caring attends, assesses, and acts on the goals of the cared-for while the cared-for recognizes the care, assesses her own needs, and cares back), in a sequence of encounters that involve reciprocity and mutuality both participants in the relation act as one-caring and cared-for. Caring for each other, in a mature caring relation, both participants analyze and revise their own goals while experiencing engrossment and motivational displacement to act on the goals of the other.

CARING RELATIONS AND PROFESSIONAL DEVELOPMENT OF MATHEMATICS TEACHERS

"What is mathematical about caring relations?" asked Amy Hackenberg (2005). She explained that a "mathematical caring relation" is inseparable from mathematics learning and occurs in the context of aiming for mathematical cognitive changes. In such a situation, the teacher as the one-caring takes on "the students' mathematical realities […] as if they were the teacher's own" (p. 47) and moves from the teacher's own ways of operating mathematically to consider the students' mathematical thinking. Combining caring theory with a constructivist perspective on mathematics learning, Hackenberg proposed that the process of decentring was essential to promoting both mathematics knowledge and caring relations, permitting the establishment of caring relations that had the construction of mathematics knowledge at their core.

Taking this idea to the context of mathematics professional development, a teacher educator trying to establish mathematical caring relations becomes engrossed by the participating teachers' mathematical needs and goals. Activities in this context address not only mathematics content, as Hackenberg proposed, but mathematics pedagogy as well, with the aim of improving both teachers' mathematics learning and instruction. Thus, mathematical caring relations in professional development settings are those caring relations that revolve around mathematics teaching and learning. These relations are established around the learning of both mathematics content and pedagogy as teacher educators de-centre from their perspectives on mathematics instruction and take the teachers' mathematical realities as their own.

To situate the discussion of caring theory and mathematical caring relations better within the context of practising teachers' mathematics education, I present three episodes from my experiences working in the mathematics education of practising elementary school teachers over the past decades. This work was conducted in the United States and in Brazil, and it has many similarities across both countries, despite cultural differences. These episodes provide a context for examining the role mathematical caring relations play in mathematics professional development that aims at improving the mathematics instruction of practising teachers. After presenting the episodes, I offer and analysis of them to clarify further how mathematical caring relations were related to the education of these practising teachers.

Episode One

I had been working with a group of elementary teachers in a school in Brazil for about a month. One meeting, we were all complaining about mathematics textbooks and the use of set theory to teach children to count – an approach to learning to count that remained from the New Math movement in that country during the late 1960's (Sztajn, Ortigão & Carvalho, 2000). After listening to teachers for a while, paying attention to their concerns and doubts, I asked: "Why do we have to teach this?" I went beyond and suggested "Let's cut set theory from the curriculum and approach counting from another perspective. Children do not need sets to learn to count."

I made this suggestion based on what I knew about mathematics education in Brazil. For example, I knew that set theory was not assessed in national tests or any other tests and would not make a difference in students' performance in middle school. I was well informed about the Brazilian mathematics curriculum and I cared deeply about children's mathematical learning. I also knew I had full support from the school administrators to propose changes in the school mathematics curriculum and the approaches used to teach mathematics – they hired me to do exactly this. However, all my knowledge and concern for children were not enough to convince teachers to consider, let alone try, what I suggested. My ideas were dismissed during that meeting, and many of my suggestions after this were not welcome.

A few weeks later, scheduling conflicts required me to change my times at the school, but the teachers and I could not find new times to work together. This scheduling problem forced me to stop working at that school. To this day I wonder whether the teachers and I could have overcome the schedule difficulty if they had not stopped valuing my contributions to their mathematics instruction.

Episode Two

This episode is a composite of similar situations I have experienced with many teachers in different settings. I always have a goal for teachers in professional development discussions to talk about what children in their classrooms know. However, it is often the case that one or two teachers in a group keep coming back to what their students do not know. I encourage teachers to focus on what children can do, and only then to address what children cannot do and why. A few teachers, however, find this focus unworkable.

In particular, one such teacher comes to mind. She taught fourth grade and always insisted that her children did not know the basic facts for any of the four operations. Students didn't know anything, as far as this teacher was concerned and she often talked about how her fourth graders could not multiply because they "just" hadn't memorized their facts. "And they can't do anything without those facts," the teacher would add. So she would talk about all the tricks she showed the students to help them remember the basic facts. Because of the teacher's interest in helping students learn the basic multiplication facts – a topic she considered essential to fourth graders – we began to talk about fact strategies, the order in which children usually learn multiplication facts, and some of the harder multiplication facts (Van de Walle, 2004). From these conversations, the teacher began to pay attention to how students were learning their facts and she started sharing more detailed descriptions about what facts children knew, which ones were easier to her students. As I focused my work on how children learn basic facts, the teacher increased her awareness of students' knowledge of basic facts. We then began to question what it meant to know multiplication and the teacher generated new ways of thinking about learning mathematics. Eventually, the teacher and I began to talk about children's multiplicative knowledge – a topic that was on my agenda for the professional development of fourth-grade teachers in that school.

Episode Three

It is often the case that in my work with practising teachers there is no grading or advancement on the salary schedule associated with any assessment of teachers' efforts or knowledge. As a teacher educator of practising teachers, I often wonder about giving teachers extra work to do such as homework or extra activities to complete in between classes, not only because I know teachers are very busy people, but because I ask myself what would motivate teachers to complete these assignments?

I remember when I was first working with a group of teachers from a slum area close to the university in Rio de Janeiro. I was not a professor at that time and I was much younger than all the teachers with whom I was working; I felt overwhelmed by the many challenging teaching situations the teachers faced. In one class I asked the teachers to complete a problem before the next meeting "or else" there would not be a class the following week. The "or else" threat was a slip of the tongue, and I spent the week between classes stressed in anticipation, wondering what I would do if teachers showed up empty handed for our next meeting. It made me remember how I felt when I realized, during the first year I taught seventh-grade mathematics, that I could not threaten students with anything I could not deliver. That week went by slowly. When the teachers and I met again, some explained there was too much going on to complete the given assignment. But some of the teachers had completed their assignment and I was relieved to learn that I could build on their work to continue our classroom discussion. To my surprise, none of the teachers claimed they just did not do what I had asked them because I was in no position to ask them for anything – which would have been true as we were all volunteers in this work. Also to my surprise, the teachers shared the work they did, helping those who could not complete the homework to engage in the learning opportunities designed for that particular lesson.

Using Caring Theory to Analyze the Education of Practising Teachers

In what follows I use caring theory to analyze each of the three episodes and to consider what it means to establish mathematical caring relations in the context of practising teacher education. Of course these episodes can be considered from a variety of viewpoints or through different theories, and some of my claims may seem obvious or already known if considered from a different point of view. My goal is not to contrast caring theory with other theories; rather, my goal is to show how I learned to analyze mathematics professional development from a caring perspective, and how Noddings' caring theory involves a set of concepts that can support the work mathematics teacher educators do.

At the time of Episode One, I attributed the Brazilian teachers' reluctance to accept my suggestions to the culture of schooling and to teachers' attachments to their consecrated ways of doing things. I also thought there could be power struggles involved in teachers' ignoring of my suggestions because I had been hired by, and in a way represented, the administration – not the teachers. Therefore, I attributed the problems that emerged in my work with the teachers to the teachers, the school, and possible struggles between administrators and faculty. I felt I just happened to be in the middle of all "their" problems.

Caring theory, however, brings a very different perspective to that episode. From this perspective, because a relation involves both the one-caring and the cared-for, I am forced to examine and question my role as the teacher educator and the one-caring in that particular situation. Looking back, I believe the reciprocity essential to a caring relation had actually not been established because the teachers who were the cared-fors did not acknowledge me, the teacher educator, as the one-caring; they did not care back. Although my suggestions for changes in the curriculum were based on knowledge of mathematics education and curriculum, I had very little knowledge about the teachers with whom I was working. The teachers did not know me either. They probably thought that while I was informed about the mathematics curriculum and children's learning of mathematics, I did not know much about their work in classrooms. Perhaps the teachers understood from our discussions how much I cared about children and children's learning. What the teachers did not know was that I also cared about them, as teachers, and their work. I do not remember whether I did not show or did not engage in engrossment and motivational displacement – the acts of carefully listening to the teachers and acting on their expressed goals. I thought I cared, but the teachers did not recognize my care and they did not perceive my suggestion to change the curriculum as part of a caring relation. Without feeling cared-for, the teachers were not ready to take such a big leap to a suggestion that did not come from them and did not reflect the goals they perceived as important (such as improving students' understanding of set theory rather than not teaching it). Teachers were not prepared to take up the challenge of questioning their school curriculum themselves and they did not see me as a one-caring who could support them in facing such challenge.

More time was needed to establish a caring relation in which teachers learned that I was listening to them. My premature proposal interfered with the development of a caring relation between the teachers and me as it fostered teachers' perception that I did not care about them, what their work entailed, and the resulting challenge they would face if they accepted my suggestion. I appeared to them as one caring only about my own agenda.

In contrast, Episode Two shows how engrossment and motivational displacement can help in professional development settings. By being attentive to the needs of the fourth-grade teacher, I was able to focus on what she wanted to learn. I de-centred the professional development initiative from my own agenda to talk about basic facts – a topic I had not planned to address at that school because of my concerns about an already existing emphasis on memorization and drill-and-practice to the exclusion of problem solving and strategic reasoning in mathematics classrooms. However, as I set my concerns to the side to listen to and understand the teacher's arguments, I engaged in the teacher's search, putting into practice motivational displacement. At the same time, I did not abandon my responsibility for what I saw as the teacher's professional growth, and I continuously asked the teacher to think beyond basic facts. Eventually, our conversations got to a point that they were professionally satisfying to both of us in that we both learned and thought our agendas were being addressed.

However, at the time of Episode Two, when I began talking to the fourth-grade teacher about basic facts, I did not analyze it from the perspective of caring theory. Therefore, I questioned whether I was helping this teacher, because I thought I was "giving in" to her desire to work on basic facts. I saw talking about basic facts as a very low priority conversation to have with a teacher who already emphasized rote learning of mathematics. From my perspective at the time, I needed to challenge the teacher, bring her new ideas to think about, propose to her other ways of looking at mathematics. Talking about basic facts felt like a missed opportunity, a waste of our time together, and I struggled with the issue of who should set the agenda (Richardson, 1992) for the professional development conversations.

Through caring theory, I now realize that it was the (what I perceived at the time as) "giving in" to the topic of basic facts that opened up the conversations between the teacher and me, allowing for the establishment of a caring relation. Such

307

relation eventually led us to talk about issues in mathematics education that I considered important. More significantly, "giving in" on basic facts led us to a point where we could share interests and engage in discussions that offered learning opportunities for both of us. If I had dismissed the teacher's question because it did not interest me or fit my agenda, we would not have established the caring relation that proved so fruitful in the long run. Through the topic of basic facts, the teacher saw that I was interested in her problems, that her goals engaged me beyond my own goals, and that I was there to help her with her needs. This perception allowed the teacher to care back by also demonstrating interest in what I considered important.

Episode Three, further, shows some of the ways in which I think teachers care back to teacher educators. At the time of this episode, and many other times when I gave teachers some assignment to complete in between professional development meetings, I wondered why would teachers engage in extra work? My typical response to this question was that it must be that the teachers were internally motivated and interested in improving their mathematics instruction. So they would engage in the activities I suggested because these activities could help them become better teachers. However, there were many other activities from which the teachers could choose to engage with, so, why the one I suggested?

Using caring theory, I realized that when teachers engaged in the activities I proposed for their classrooms, which might not seem to have an immediate value to them, they were caring back for me. In this case, teachers were the ones-caring – they were actually experiencing engrossment and motivational displacement. They were listening to my goals and needs, and they were acting upon them as if they were their own. They were willing to try what I suggested, engage in what I thought was a need, and implement something different in their classroom to bring back for discussion in the professional development setting. When the teachers returned to the professional development setting having, for example, experimented with a new open-ended mathematics problem in their classroom, ready to talk about what students did and how students solved the problem, they were the ones-caring.

There is one more aspect from Episode Three that I consider compelling. The teachers from that group who completed the homework for that class worked with their colleagues who did not do the assigned task in completing the activities proposed for that particular professional development class. In this case, I think these teachers were establishing caring relations among themselves. They were listening and attending to each other's needs and working to fulfil such needs as the ones-caring.

Together, these three episodes reveal important aspects of caring theory in mathematics professional development. They present the one-caring and the cared-for as they each engage in their own state of consciousness – the one-caring experiencing engrossment and motivational displacement, the cared-for recognizing (or not!) the care and caring back. They show the importance of both reciprocity and mutuality. They indicate that although a caring relation is not necessarily symmetrical in the sense that the one-caring and the cared-for are not always engaged as equals in the relation, when such relation is established, both one-caring and cared-for engage in learning opportunities. These episodes indicate that it is not enough for teacher educators to say they 'care'; the teachers need to recognize this care through their interactive experiences. These episodes also exemplify how caring relations can be mathematical as they have the improvement of mathematics instruction at their core. Finally, the episodes show that caring relations require the teacher educator and the teacher to know each other, which necessitates time.

One can argue that some of these ideas are not new when one analyzes learning or professional development. As I mentioned before, one can look at the episodes I presented with other lenses and come to similar conclusions or suggestions. For example, many educational theories support the notion that learning begins from where the learners are (e.g., constructivists would claim so, those working with meaningful organizers or concern-based adoptive models would claim so, etc.). Bransford, Brown, and Cocking (2000) refer to this as the leaner-centred perspective of learning. The difference from considering where the teachers are (teachers are the learners in the case of professional development) due to a learning perspective or a caring-relation perspective is that, in the first case, the teacher educator is often focused on his or her goals, thinking about finding out where the teacher is to decide how to go from there to where the teacher ought to be. From a caring theory perspective, considering where the teacher is represents the beginning of a caring relation in which the one-caring is ready to pursue the goals of the teacher-as-cared-for. And in the context of this caring relation, the one-caring and the cared-for increase their knowledge of each other as they steer between their own goals and the goals of the other.

When highlighting relations, not only content or format, mathematics professional development requires a caring encounter between teacher educator as one-caring and teacher as cared-for in the context of a mathematics-learning situation. The theory of caring relations provides a fertile approach to analyze the episodes I presented. The concepts of engrossment and motivational displacement, the need to know the people with whom you work, and the need to provide time and continuity for caring relations to be established are all important aspects of a caring mathematics professional development. When teacher educators engage in professional development activities with practising mathematics teachers, attention to caring theory can facilitate the establishment of mathematical caring relations that lead to teachers' and teacher educators' professional learning.

CONCEPTUALIZING CARING PROFESSIONAL DEVELOPMENT

In *Learning to Trust*, Marilyn Watson (2003) observed one teacher in an inner-city school and extracted key elements the teacher considered when establishing caring relations. For example, Watson talked about the caring teacher learning to like all students, helping all students see that she liked them, and continuously examining her working models of the children with whom she worked. In talking about caring school leaders, Noddings (2006) noted that caring leaders listened, asked probing questions, and made it comfortable for teachers to discuss their difficulties and doubts instead of hiding them. These recommendations from Watson and Noddings can be adapted to the context of teacher education. Thus, I conclude this chapter by considering what caring mathematics teacher educators do, that is, what is important for mathematics teacher educators to establish mathematical caring relations with a variety of mathematics teachers in diverse professional development contexts.

Some of the issues I attend to in becoming a caring mathematics teacher educator are the establishment of caring relations with teachers, teachers' need, and other caring relations that are established in the context of the professional development work. These do not exhaust the issues to which a caring mathematics teacher educator pays attention. Rather, they indicate ways in which I attempt to actualize the implementation of mathematical caring relations in the work I do with practising teachers, and I discuss what I had to learn in order to engage with these issues within caring mathematics professional development settings.

Relations are the essence of care and the establishment of mathematical caring relations needs to be at the centre of caring professional development initiatives. But to engage in such relations as a mathematics teacher educator, I had to learn to listen and to hold what I knew and believed about teachers as working models that were open to change. When a teacher educator has fixed models of what teachers are, or of what teachers know and believe, it becomes difficult to actually listen to what teachers are saying instead of listening to what the teacher educator thinks the teachers are going to say. To practice engrossment and motivational displacement, I learned to gather information about the practising teachers, the children the teachers worked with, and the school environment in which the teachers operated; I learned to use these data to continuously analyze and revise previously held models I had about practising teachers so that I could listen to them better.

As a caring mathematics teacher educator, I learned to search for the unique qualities of each teacher, valuing the needs and the contributions all teachers offer. It is not enough for a teacher educator to learn to listen to a handful of motivated teachers. Rather, it is necessary that the teacher educator also finds a way to communicate with those who seem disengaged. While I believe all teachers are "alike in their need for autonomy, belonging, and competence", each teacher is "unique in skills, intelligence, temperament, [...] and experience" (Watson, 2003, p. 53). Teachers are also different in their mathematical content, pedagogical, and affective knowledge. Thus, by listening to every teacher and by searching for the individual contribution each teacher offers, or the needs they present, the mathematics teacher educator begins to establish mathematical caring relations with each teacher.

At the same time, because a mature mathematical caring relation requires reciprocity and mutuality, I learned to see when teachers were caring back and to let teachers get to know me, my interests, my work, my talents, my difficulties and my needs when it came to mathematics teaching and learning. Although both teacher-educator as one-caring and teachers as cared-for are essential parts in the establishment of new mathematical caring relation, I believe it is the teacher educator as one-caring who has to initiate and nurture a mathematical caring relation until it becomes a mature relation in which the teacher educator and the practising teacher act as both one-caring and cared-for. Therefore, as a caring mathematics teacher educator I had to learn to allow teachers to take the role of the one-caring, also becoming the one who is cared-for in mathematics professional development initiatives.

Teacher educators need to meet teachers' needs throughout the duration of the professional development. They adjust planned activities to match teachers' needs, providing extra scaffolding for those who need it. When teacher educators attend to what teachers want, support learning experiences that are interesting and meaningful to the teachers, and connect to what teachers do in their classrooms, teachers can become more engaged. Learning to attend to teachers' needs as an ongoing professional development activity that goes beyond an initial needs assessment was a skill I had to develop to establish caring mathematical relations. Because teachers' needs change as professional development and mathematical caring relations unfold, as a caring mathematics teacher educator I learned to continuously adjust my goals in response to teachers' newly perceived needs. Teacher educators have to recognize when teachers' needs are changing, being attentive to and understanding of shifting needs in teachers' professional lives.

This chapter has focused on the relation between the teacher educator and the teacher. However, caring professional development also includes other caring relations such as teachers acting as the one-caring and cared-for among themselves (as alluded to in Episode Three). As a teacher educator I learned to use inclusive language (e.g., we instead of I) to encourage caring relations among participating teachers. Teacher educators who work with groups of teachers in professional development settings have to help teachers get to know each other, value their differences, and learn from their diversity. The creation and design of shared experiences is an important aspect of the work of the group for it to become a caring professional development environment. As a caring teacher educator, I needed to learn to support teachers in helping one another and the group as a whole.

CONCLUDING REMARKS

Caring theory offers teacher educators concepts to use in analyzing relations established in professional development settings. Ideas and proposals available for caring teachers and leaders can be adapted to caring professional development and teacher educators. As mathematics teacher educators begin to work on the establishment of caring mathematical relations, research can help the community further understand the role of caring in the education of practising teachers and develop caring theory in the context of mathematics teachers' professional development. How do caring professional development initiatives unfold? How are they implemented and promoted? What are some of the outcomes of such initiatives? As mathematical caring relations become valued in professional development, knowledge about them can begin to be built. The concepts from caring theory can help the teacher education community begin to address these questions, while at the same time providing support for those who are facing the challenge to care for the so many uncared-for teachers currently working in schools.

The challenge to care in professional development will be addressed when those who work with practising teachers start paying attention not only to the format and the content of what they do but also to the relations in which they engage. These three aspects of professional development (format, content, and relations) are equally important and it is only when all three of these are equally valued that the professional development of practising mathematics teachers becomes of high quality.

ACKNOWLEDGEMENTS

I am thankful to Amy Hackenberg for bringing Caring Theory to my attention, to Barbara Jaworski for challenging me to write an analytical chapter about care, to Martha Allexsaht-Snider and Sally Hudson-Ross for their comments on an earlier version of this chapter, and to Dorothy Y. White who helped me learn to care in many professional development situations.

Time for the preparation of this chapter was supported by a grant from the National Science Foundation (ESI-054453543). Any opinions, findings, and conclusions or recommendations expressed in this materials are those of the author and do not necessarily reflect the views of the Foundation.

REFERENCES

American Association of Colleges for Teacher Education (2007). *Preparing STEM teachers: The key to global competitiveness. Selected profiles of teacher preparation programs.* Washington, DC: Author.

Ball, D. L., & Wilson, S. M. (1996). Integrity in teaching: Recognizing the fusion of moral and intellectual. *American Educational Research Journal*, 33, 155–192.

Bransford, J. D., Brown, A. L., & Cocking, R. R. (2000). *How people learn. Brain, mind, experience, and school.* Washington, DC: National Academy of Science.

Business Higher Education Forum (2007). *An American imperative: Transforming the recruitment, retention, and renewal of our nation's mathematics and science teaching workforce.* Washington, DC: Author.

Cochran-Smith, M. (2003). The unforgiving complexity of teaching: Avoiding simplicity in the age of accountability. *Journal of Teacher Education*, 54(1), 3–5.

Cochran-Smith, M. (2004). Taking stock in 2004: Teacher education in dangerous times. *Journal of Teacher Education*, 55(1), 3–7.

Desimone, L. M., Porter, A. C., Garet, M. S., Yoon, K. S., Birman, B. F. (2002). Effects of professional development on teachers' instruction: Results from a three-year longitudinal Study. *Educational Evaluation and Policy Analysis*, 24(2), 81–112.

Gilligan, C. (1982). *In a different voice. Psychological theory and women's development.* Cambridge, MA: Harvard University Press.

Guskey, T. R. (2005). Taking a second look at accountability. *Journal of Staff Development*, 26(1) 10–18.

Hackenberg, A. (2005). A model of mathematical learning and caring relation. *For the learning of Mathematics, 25*(1), 45–51.

Hill, H. C., & Ball, D. L. (2004). Learning mathematics for teaching: Results from California's Mathematics Professional Development Institutes. *Journal of Research in Mathematics Education, 35,* 330–351.

Jaworski, B., & Wood, T. (1999). Themes and issues in inservice programmes. In B. Jaworski, T. Wood, & A. J. Dawson (Eds.), *Mathematics teacher education: Critical international perspectives* (pp. 125–147). London: Falmer Press.

Kennedy, M. (1998). *Form and substance in inservice teacher education. Research Monograph, 13.* Madison, WI: National Institute for Science Education.

Knowles, M. S., Holton, E. F., & Richard A., Swanson, R. A. (1998). *The adult learner.* Houston, TX: Gulf Publishing.

National Academy of Sciences (2006). *Rising above the gathering storm: Energizing and employing America for a brighter economic future.* Washington, DC: Author.

Nieto, S. (2003). *What keeps teachers going?* New York: Teachers College Press.

Noddings, N. (1984). *Caring: A feminine approach to ethics and moral education.* Berkeley, CA: University of California Press.

Noddings, N. (1992). *The challenge to care in schools.* New York: Teachers College Press.

Noddings, N. (2001). The caring teacher. In V. Richardson (Ed.), *Handbook of research on teaching* (4th ed., pp. 99–105). Washington, DC: American Educational Research Association.

Noddings, N (2002). *Caring and social policy: Starting at home.* Berkeley, CA: University of California Press.

Noddings, N. (2006). Educational leaders as caring teachers. *School Leadership and Management, 26*(4), 339–345.

Richardson, V. (1992). The agenda-setting dilemma in a constructivist staff development process. *Teaching and Teacher Education, 8,* 287–300.

Sowder, J. T. (2007). The mathematical education and development of teachers. In F. K. Lester, Jr. (Ed.), *Second handbook of research on mathematics teaching and learning,* (pp. 157–223) Charlotte, NC: Information Age Publishing and National Council of Teachers of Mathematics.

Sztajn, P., Ortigão, M. I., Carvalho, J. B. P. (2000). *E agora? O que fazer sem os conjuntos?* [What to do without set theory?]. *Revista Presença Pedagógica, 6*(3), 37–47.

Tong. R., & Williams, N. (2006). *Feminist ethics. Stanford encyclopedia of philosophy.* Retrieved on January 2008 from http://plato.stanford.edu/entries/feminism-ethics/#2.

U.S. Department of Education (2007). *Report of the Academic Competitiveness Council.* Washington, DC: Author.

Van de Walle, J.. (2003). *Elementary and middle school mathematics: Teaching developmentally* (5th Edition). New York: Allyn & Bacon.

Watson, M. (2003). *Learning to trust.* San Francisco: Jossey-Bass.

Paola Sztajn
University of Georgia
USA

A. J. (SANDY) DAWSON

16. TRUST AND RESPECT: A PATH LAID WHILE WALKING[1]

This is a story about trust and respect, how it is essential for meaningful professional development to occur, how it is built/learnt/developed, and about some of the things that one must do in order to maintain it. This particular story takes place across a vast region of the western Pacific Ocean. The story is told through a series of vignettes interlaced with observations about lessons learnt, conclusions re-examined, theories formulated, tested and then rejected or maintained for future testing. It is a story about how peoples with at least nine different languages from nine distinct cultures, many operating at subsistence levels, introduced me to their ways of working, their world views, and taught me about the pervasive power and influence of family, clan, or as it is called in Hawai'i, o'hana.

INTRODUCTION

This is a story about trust and respect, how it is essential for meaningful professional development to occur, how it is built/learnt/developed, and about some of the things that one must do in order to maintain it. This particular story takes place across a vast region of the Pacific Ocean. It involves mathematics teacher educators from the State of Hawai'i as well as current and former American trust territories: the Republic of Palau, the Federated States of Micronesia (the FSM which consists of the states of Yap, Chuuk, Pohnpei, and Kosrae), the Commonwealth of the Northern Mariana Islands (CNMI), the Republic of the Marshall Islands (RMI), and the territories of American Samoa and Guam. The region encompasses a population of approximately 1.7 million people living on 110 islands spread across 4.9 million square miles of the Pacific Ocean (see Figure 1), presenting unique challenges for professional development initiatives.

---

[1] The title of the chapter is derived from an earlier paper I wrote titled "the Enactive Perspective on Teacher Development: 'A path laid while walking'" contained in Jaworski, Wood and Dawson (1999). This material is based upon work supported by the National Science Foundation under Grant numbers ESI 9819630 & ESI0138916. The content does not necessarily reflect the views of the NSF or any other agency of the U.S. government.

B. Jaworski and T. Wood (eds.), *The Mathematics Teacher Educator as a Developing Professional*, 315–332.
© 2008 Sense Publishers. All rights reserved.

The story is told through a series of vignettes interlaced with observations about lessons learnt, conclusions re-examined, theories formulated, tested and then rejected or maintained for future testing.

February 1999 was a watershed month for me. I received, unsolicited, an invitation[2] to apply for the position as director of a National Science Foundation (NSF) Project that would create and assist with the education of a cadre of mathematics educators from across the Pacific. Other than for American Samoa, I had never heard of the countries where the Project would occur, so during the

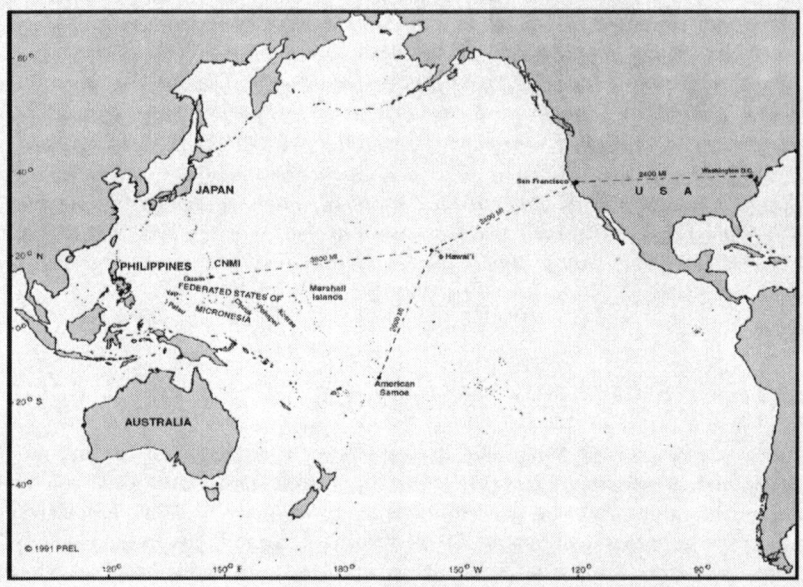

Figure 1. Map Showing Distances across the Region (not to scale)

telephone interview arranged after I submitted an application for the job, I used my computer and the Internet to bring up images and information about Micronesia as I attempted to deal with questions about cultural awareness and sensitivity. I must have done okay, because I was offered the job and moved to Honolulu four months later to embark on this life expanding experience. Some nine years later I am now in a position to reflect on what occurred, how it challenged, refined, and eventually affirmed, though in modified form, long held beliefs about the professional development of mathematics educators. This is a story about how peoples with at least nine different languages from nine distinct cultures, many operating at

---

[2] The invitation came from Pacific Resources for Education and Learning (PREL), a non-profit organization with head offices in Honolulu, Hawai'i that was awarded the two NSF grants reported on in this chapter.

subsistence levels, introduced me to their ways of working, their world views, and taught me about the pervasive power and influence of family, clan, or as it is called in Hawai'i, o'hana.

IN THE BEGINNING

> Lorenzo and Christian (pseudonyms) were surprised when they met each other waiting to board the Island Hopper[3] plane bound for Honolulu. They had played tennis earlier that morning – as they did before work most every day – and neither had mentioned leaving the island. "Where you goin?" Lorenzo asked. "I'm goin to Honolulu for some kind of work meeting. I will be gone for three weeks. What about you? Where you goin?" "I go Honolulu too," said Lorenzo, "also for work and three weeks as well. Where you staying? Maybe we get a taxi together." Christian pulled out some papers to check the name of the place he was staying. "Tokai University," he said in answer to Lorenzo's question. "Me too!" Lorenzo remarked with amazement at the coincidence. "What you do?" he asked. Christian, who was also very surprised that they would be staying at the same place, said, "I'm the math specialist at the DOE. What you do?" Lorenzo gave off his characteristic little laugh, and said, "I teach the math for teachers course at COM/FSM!" "No!" exclaimed Christian, "We both in math? I never knew that. Ain't that crazy – we goin to the same meeting! Maybe we get to play some tennis."

This, as with all the vignettes in this chapter, is a true story. Lorenzo and Christian were two of four mentors chosen to be part of the ten NSF sponsored mathematic mentor teams selected from the Pacific island nations/states noted earlier. The main purpose of the NSF grant was to create and develop local teams of educators capable of providing leadership for elementary and secondary teachers of mathematics. The fact that Lorenzo, based at the College of Micronesia (COM), and Christian, hired as a math specialist by the local Department of Education (DOE), didn't know one another professionally, and had never talked about the teaching and learning of mathematics, is not surprising. This has been true in many jurisdictions world wide, not just across the Pacific. It did highlight the need, however, for a mechanism where such individuals could interact professionally. The NSF Project structured the teams so that there were two individuals each from both the local college and DOE. All ten teams met for the first time in Honolulu, January 2000, and took part in a three-week orientation program that launched the Project. Forty mathematics educators attended, twenty with college affiliations, and twenty with DOE affiliations.

---

[3] So called because the Continental Airlines flight, east to west on Monday, Wednesday, and Friday, and west to east on Tuesday, Thursday, and Saturday, hops from island to island as it makes its way across the Pacific between Guam and Honolulu with stops on Majuro, Kwajalein, Kosrae, Pohnpei, and Chuuk.

FIRST ATTEMPTS: DEVELOPING TRUST AND RESPECT

Previous experiences with First Nations peoples of Canada introduced me to the mechanism of wisdom circle, and the use of a talking stick. I introduced circling with some trepidation that first morning as a tool for opening dialogue between and among the mentors. I was not sure that this form of communication would be accepted by Pacific island cultures. As laid out by Baldwin,[4] the three circle principles and practices as used in the Project were the following:

Rotating Leadership: Each person helps the circle function by assuming small increments of leadership. In ... circling, leadership shifts moment-by-moment and task-by-task. Rotating leadership trusts that the resources to accomplish the circle's purpose exist within the group.

Sharing responsibility: Each person pays attention to what needs doing or saying next and does his/her share. In ... circling, responsibility also shifts moment-by-moment and task-by-task. Shared responsibility is based on the trust that someone will come forward to provide whatever the circle needs: helping each other take action, calling for silence, or offering the next meeting place.

Relying on spirit: Each person places ultimate reliance in the centre and takes his/her place at the rim. Through simple ritual and consistent refocusing, the centre, literally and symbolically, becomes sacred space – a place where everyone's willingness to listen dwells.

The participants adopted the following circle agreements – about what was said in circle and how the interactions were handled.

What is said in circle belongs in the circle. Confidentiality allows people to speak their minds knowing that they will not be gossiped about. Confidentiality allows people to take verbal risks, to experiment with ideas, to keep changing their minds as their understanding grows.

The circle is a practice in discernment. Discernment is the ability to listen, sort, and speak without having to be 'right' or in total agreement before other people's opinions and views can matter. Someone else's view doesn't have to be right or wrong; it may simply be different.

Each person takes responsibility for asking the circle for the support s/he wants and needs. Asking for what you need next allows you to stay at the rim and avoid power struggles and personal drama as ways of getting attention. There are times when your request may not be responded to in the manner you expected; this is not failure, it is learning. Generally, if a request fits the task and orientation of the group, someone in the circle will have both the willingness and resources to respond. If a request doesn't fit task and orientation, there will be a lack of interest. Circle members will learn to negotiate what they can and cannot do, and hold intention for the direction of group energy.

Each person takes responsibility for agreeing or not agreeing to participate in specific requests. This is the corresponding half of the above agreement. Circle

---

[4] This description draws heavily from Christina Baldwin's (1998) book.

requires a fine degree of attention to how time and energy are used in group process. We can support someone and not take direct part in what they need. We may choose not to support a request. We may challenge whether or not a request serves group purposes. If a request fits our task, others are likely to help carry it through. If a request doesn't fit, we need to work together to hold the focus.

Anyone in the circle may call for silence, time-out, or ritual to re-establish focus, to re-center, or to remember the need for...guidance. We're are using this form because it promotes communication, and communication promotes both creativity and efficiency. If communication breaks down, we need to stop and restore it.

Agreements are adaptable. If something is not working, revise the agreements and maintain the process. Agreements are like our mini-constitution. If trouble develops, we search together for an agreement that better supports how we're treating each other and what we're getting done.

Practice listening without interrupting. This can be an important reinforcement of the *talking piece* idea even when the form of council currently in use is open discussion.

These practices were introduced during the first meeting in Honolulu. The talking 'stick' used was a sacred eagle's feather presented to me by a First Nations band of Northern Canada. Whoever is holding the talking stick is the only one allowed to speak, and all others in the circle listen attentively to what is being said. Circle was used once a day during meeting time, perhaps to open proceedings in the morning or to bring everyone back together at the end of the day. Subsequent use by mentor teams across the region varied widely, but it was particularly strong on those islands where circle is a central component of the manner in which tribal councils conduct business. Circle agreements and principles contributed significantly to the development of respect and understanding among the mentors.

> During the circle to close the initial Project meeting, Sandy related how nervous he had been when introducing circling to the group that first day. He asked how others had felt that day, and what they thought now about circling. One of the Samoan men responded with a laugh! "This is how we operate at home. I felt like I was back in my fale (meeting house) at a tribal council meeting. We don't use a talking stick, but each person speaks in turn as we move round the circle. The talk keeps going round-and-round till everyone has said all they wish to say. When we reach that point the decision we need to make is usually very clear. You need not to have been worried at using circle with us." Others around the circle were nodding their heads in agreement indicating that their cultures also used forms for circle.

Clearly, my worries about using circle were unfounded. Use of the circling technique sent the message that each person in the circle would be listened to respectfully, and this contributed to the building of trust amongst mentors and Project staff.

A. J. (SANDY) DAWSON

The 40 men and women who were the mentors for the 10 communities where the Project is located speak at least 10 different languages. Though their cultures are similar in many ways there were fascinating differences among them. From these beginning days onward, the practice was initiated by setting time each day for one or two of the mentor teams to share some aspect of their culture. During the first days together, mentors and Project staff learnt the words and/or phrases to greet and to thank each other in the language of the Pacific island community that was sharing aspects of their culture. This too contributed to the slowly developing sense of trust and respect within the group.

ISLAND SMART: WORLD SMART

The Director of the DOE was sitting at the head of the conference table quietly preparing his 'chew'. Others around the table were already chewing their previously prepared betel nut wraps occasionally spitting into an empty soda can held inconspicuously below the table edge. This was my first meeting with this island's DOE personnel and the betel nut ritual was foreign to my experience as was spitting openly into a soda can! The Director smiled at me, and said, "We believe there is wisdom in the bag," pointing to the woven bag sitting in front of him on the table. It contained the ingredients needed to prepare the betel nut for chewing: lime, the betel nut, the leaf of a betel nut tree, and perhaps a cigarette. After a short pause, the Director said (speaking in English), "This is a custom in our culture. It allows time before a meeting for everyone to pause and think about the issues we will discuss. It is a reflection time, and allows us to settle into the meeting gracefully and gradually." With that being said, he turned and smiled at each person round the table, and then opened the meeting by proclaiming, "We need to have a curriculum, in mathematics as in all subjects, that allows for our children to be Island Smart *and* World Smart. To be one without the other will either create children who are ignorant of the outside world, hence at the mercy of outside forces, or drive the children from their homeland because they have not learnt about their culture and its history."

Over the past 15 years, Pacific island communities worked to implement national and regional standards for mathematics education that would prepare their children for a life at home and/or a life in the outside world. The vision of teaching and learning articulated in these documents required that teachers have significant mathematics pedagogical and content knowledge. It also required that they adopt a problem solving/inquiry stance to teaching and learning, that they be responsive to students' ideas and organize subsequent instruction in ways that required students to extend or reconsider their initial thinking, that they gather informal and formal data about students' conceptions of mathematical ideas, and that they use this information to inform classroom instruction. This type of teaching required a significant re-conceptualization on the part of many teachers of the nature of mathematics learning and teaching. Moreover, it was not an approach to education that was found to exist in many Pacific island schools.

An over arching requirement when interacting with Pacific island cultures is respect: respect for each individual, respect for the culture's values, respect for the social structure, respect for the elders, and respect for the spirituality of the culture. The question facing Project leaders was how to assist with the re-conceptualization process in a manner that honoured the respect expected from Project leaders who came from outside the culture.

For many years prior to moving to the Pacific, I had steeped myself in an orientation developed by Caleb Gattegno[5] called subtle (which stands for the Subordination of Teaching to Learning), which is a way of fostering learners' engagement with mathematics. This approach was chosen as a central core of the Project because of the respect and honour it has for learners. The central aspects of the subtle approach include a deep respect for and acceptance of the capabilities of learners. A subtle teacher always expects learners to be able to grasp the concepts being presented. At the same time, a subtle teacher does not expect the learner to do that every time, in every circumstance. an acknowledgement that in the teacher/student dyad, the learner is central; that is, the learning of the student is of paramount importance, and the teacher's performance, the teacher's lesson, must be subordinated to the learning of the student. the recognition that it is the learner who must do the learning, and that the teacher's function is to create situations and experiences that focus the learner's attention on the key concepts of the mathematics being presented. A teacher's role is to constantly seek activities that attract a learner's awareness. The subtle teacher realizes that not all learners will gain the same understandings during any particular activity so the subtle teacher requires an ever developing, rich collection of activities that directs learners' awareness to the key ideas being presented. The discipline to provide the learner with the minimal essentials for understanding to occur, to not 'tell' the learner everything, or almost everything, in the belief that 'telling' fosters learning. In order for knowledge to be a permanent, accessible, and useable aspect of learners' repertoire, they have to play with it, mould it, modify it, and finally make it their own. the further recognition that conversation among and between learners is a valuable tool in a teacher's instructional repertoire, because often one's peers can ask a question, or provide a focus of attention that enables the learner to 'see' something not previously seen through teacher designed activities. Small group work among peers is a valuable method for fostering learning. the understanding that teaching is subtle work, in terms of it being subtle, but also in terms of it being delicate, restrained, and finely grained. An instructional strategy successfully used on one occasion may not work in a different situation with a different set of learners. An appreciation that, in the words of Caleb Gattegno, "only awareness is educable" by which is meant that learners can only acquire knowledge of that which they are aware. For example, a young learner only can become aware of the

---

[5] Gattegno (1970).

commutative principle by experiencing both cases where the principle does not hold (putting on one's socks and then one's shoes versus putting on one's shoes and then one's socks), and instances where it does hold (adding 3 to 5 versus adding 5 to 3).

These ideas pervaded the workshops and discussions and social activities held with mentors throughout the eight years that I worked with them. Coupled with the use of circling, and the ideas behind that mechanism, the subtle approach turned out to be a 'good fit' with the cultures of the Pacific. These were the basis for establishing the first level of trust and respect that allowed the Project to move ahead successfully. But it would take something else to really get things moving: this was embedded in the experiences that the mentors themselves had.

HAVING INFLUENCE VERSUS HAVING POWER[6]

Grace was a classroom teacher, the youngest to participate in our eight-year Project. There were several other classroom teachers in the Project, but there were also participants whose positions as curriculum coordinators or mathematics specialists allowed them to regularly engage in professional development opportunities with teachers. Over weeklong summer sessions these mentors participated in activities to increase their own understanding of mathematics content and skills in mentoring. Then the mentors were expected to conduct similar professional development activities for teachers in their local communities.

Grace anguished over this last expectation. She was one of the youngest teachers in her community. Her community was relatively small and she was biologically related to many of the other teachers in her school. It was culturally inappropriate for her to approach these older teachers with new ideas. As one of the youngest teachers at the school the expectation was that she was to learn from the elder teachers and not the other way round. She was concerned that she would be perceived as arrogant if she offered a workshop or even took the initiative to approach the other teachers with the ideas she was learning.

Throughout the summer Project activities, the 40 mentors shared various options they had for engaging their local teachers in the Project activities. Some would conduct summer institutes, and other would convene sessions after school. Almost all of these involved the mentor's initiative in recruiting their local teachers. Grace knew these options for inviting or requiring her local colleagues to participate would not work for her. She shared her concerns with the Project leaders since she was sure that she would not have an opportunity to share what she had learned if it meant approaching the teachers.

---

[6] Dr. Joseph Zilliox (professor, College of Education, University of Hawai'i) who was a consultant to and active participant in the Project wrote this vignette.

The following summer when the Project convened once again, mentors were asked to share the work they had done locally. Grace had one of the best experiences to share. While she had no power to initiate a workshop with her local teachers she did have influence through the work she could do with her own students. During the school year Grace did not approach the teachers. Instead she engaged the children in her classroom in the activities she had learned. The students responded enthusiastically and talked to the students in the others classes about the interesting things they were doing in mathematics. Upon hearing about the different activities in Grace's class the teachers approached her and asked if she would share the activities. Grace's influence on the practice of the other teachers came not through imposition, but through an invitation from them. While she did not have power to impose change, she had influence to invite change.

Finding ways to respect the social structure of some island communities was a significant challenge, as Grace's vignette illustrates. It was a challenge that had to be met by the mentors themselves because Project staff, myself included, had no direct knowledge, experience, or place in the culture. It was a learning experience for me and required that I suspend judgment about some cultural practices, just as I had to overcome my aversion to the spitting associated with chewing betel nut. Because of the hierarchal social structure on some islands, a child from a lower social level could not speak in the presence of a child from a higher level. The cultural conflict between being *community* smart and *world* smart is rather clear as is the conflict between the ways of teaching being advocated by mentors, and the cultural mandate found in classrooms. Over time, and after much subtle work by the mentors and others in their educational communities, certain social structure mandates were lifted from classroom situations. When in teaching and learning situations, it was agreed: hierarchal structures were to be suspended, and all children would be free to speak regardless of who else was in the room. This was a slow change process that is still going on. Social structures did not always outweigh the effect of age structures, however, although Grace was the daughter of a high ranking chief yet she was restricted from approaching older more experienced teachers. Professional development efforts across the Pacific region must be cognizant of these and other cultural constraints and operate within them.

TALKING THE TALK: WALKING THE WALK

The Pacific region professional development history is replete with instances of outside experts flying onto the islands, delivering short-term workshops, and flying off again leaving little or no permanent trace of a positive impact. Even though the Project seemed to be having a positive impact on mentors during the first NSF 3-year grant, it was not until the second 5-year NSF grant that a significant change occurred in the level of trust and respect among the mentors and Project staff. The new grant seemed to signal that initial efforts of the mentors would be supported, maintained, and expanded to include their work with upwards of four hundred prospective teachers of mathematics. The mentors were entrusted with the task of assisting prospective mathematics teachers during their first years of teaching. The seeds sown along the pathways laid down during first grant were bearing fruit as new pathways were explored in the second grant.

The change was evident during the mentor meeting held on Guam immediately after the announcement of the second grant. Cultural sharing had always taken place during mentor meetings, but the level of revelation and involvement was taken to a higher level. Mathematical details regarding the construction of ocean going canoes previously not available to those outside the home island were now being shared. The mentors from one of the islands enacted rituals of greeting, typically reserved for welcoming a visiting family or clan, for mentors from all the other islands. It was clear that mentors felt that their fellow mentors were respected and could be trusted with privileged cultural knowledge and rituals. Previously the Project participants had talked about trust and respect, but now they were also 'walking the walk' by sharing more intimate details of their cultural practices. Focus group discussions held just prior to the beginning of the second grant yielded comments regarding subtle changes regarding the hierarchal roles of men and women on some of the islands, information that was offered because the mentors felt others respected their cultural ways even though perhaps not agreeing with them. Finally, the level of fondness that mentors felt and displayed for each other became palpable. The confirmation of a second NSF grant meant they knew that they would be working together and seeing each for at least another five years.

MATHEMATICS AND NATIVE LANGUAGES

We had used the tangram lesson several times since a mentor designed it during her time as a PEIR in Honolulu working for PREL.[7] The grade six children each had a set of tangrams and followed along as *Grandfather Tang's Story*[8] was read to them. The children gained more experience with this manipulative as they used all seven tangram pieces to fill in four animal shaped outlines. They rotated, slid, flipped, and rearranged the pieces until finding the one way to fill the outline completely. Their next task was to make as many square shapes as possible using in turn 2, 3, 5, 6 and finally all 7 pieces. This was the point at which we discovered that in the native language on this particular Pacific island there was no distinction between rectangle and square. Given the challenge of making squares the children constructed both rectangles and squares not ever knowing that there was a differentiation between the two concepts. The teacher informed us that there was only one word in their language to describe both squares and rectangles.

---

[7] The Pacific Educator In Residence (PEIR) program developed by PREL supported educators from the Pacific region served by bringing them and their families to Honolulu for a year of professional development.
[8] Tompert and Parker (1990).

The yearly week-long institutes led by Project staff had to accommodate the languages and mathematical experiences of the 40 mentors who brought with them 10 different mother tongues and who typically spoke at least two languages in addition to English. One of the reasons that previous professional activities across the region met with little success was due to a lack of recognition of the challenges presented by translating western mathematics into Polynesian languages and life experiences. Part of mentor institute time was devoted to making those translations; i.e., after mathematical topics and/or pedagogical approaches were introduced to the mentors, the follow up task was to discuss within each mentor team how that particular topic or approach could be modified, refined, or even redefined so that it could be used by teachers back on the mentors' home islands.

Though English was the official language of instruction on all islands – at least after the third grade – in fact most instruction was carried out in a mixture of the children's mother tongue and English. The mentors had to first understand the western version of the mathematical concept and language used to describe it, not an easy task either since many times the concepts were new to the mentors. When they gained some understanding of the concept, they then had to work on translating their own understanding into language understandable to the prospective teachers of mathematics with whom they worked back home. The activities and approaches offered to the mentors were in turn offered by the mentors to their prospective teachers during workshops held when they returned to their homes islands. It was when observing the mentors teaching the prospective teachers that Project staff were alerted to the difficulties mentors had to overcome in learning the mathematics initially and then translating that knowledge into different languages in order to share it with local teachers. It is not a task that Project staff, or any outside consultant, could provide to local teachers. This highlights what a crucial role the mentors played in furthering their islands' goal of producing *Island Smart, World Smart* children.

SEIZING THE OPPORTUNITY

When on one island during a fact-finding trip May 2007, Project staff were invited by a former mentor to visit the new high school where he was the school's first principal.[9] We had heard wonderful stories about this school, so we were eager to see for ourselves what was special about it. It was when we were escorted into the mathematics lab that we got a major surprise. There were 20 grade 10/11 students working in small groups around the room huddled in front of computers using *Geometer's Sketchpad* (GSP) to complete their end of term mathematics assignments. The walls were decorated with posters that provided instructions how to perform various

---

[9] It has not been unusual for mentors to be promoted to administrative positions within their home island's DOE. On this island alone, one mentor became a school principal and another became the superintendent of schools for the DOE. The mentors' reputations grew significantly as they moved into the community working with prospective teachers.

geometric functions using GSP. We asked the principal, who was a mathematics teacher, where he had learnt about GSP. He told us that one of the other mentors, an elementary teacher, had come back from an institute the previous summer and reported she had a copy of GSP but did not think she could use it, due to a lack of computers, at her school. The principal said, "I'll take it. I know exactly where we can use it." The result of his action was what we witnessed in the mathematics lab. We recently learnt that the principal has now convinced that mentor to switch from the elementary level to be a mathematics teacher at his secondary school.

During the eight year history of the Project, two Scholars in Residence were selected and brought to Honolulu under another of PREL's programs designed to *build capacity through education* by bringing quality professional development to the Pacific region. One of those scholars, Dr. Eric Muller from Brock University in Canada, was selected to provide leadership in the area of computers in mathematics education. During an annual mentor summer institute, Eric introduced each mentor team to *Geometer's Sketchpad*, copies of which, through Eric's contacts, had been generously donated by Key Curriculum Press, the developer of GSP.

Each summer institute had a focus on one school mathematics strand (number and number operations, probability and data analysis, algebra and pre-algebra, geometry and measurement) as well as addressing and pedagogically demonstrating a variety of mathematics process standards as identified by the National Council of Teachers of Mathematics (NCTM): problem solving, connections, reasoning and proof, communications, and representations. Number and number operations were the aspects of the mathematics curriculum most focused on across the region. In many instances, geometry was seldom taught, and probability and data analysis was an area essentially new to both mentors and prospective teachers across the region. The newness of these topics when combined with the translation difficulties presented by languages that did not contain words and phrases congruent with the mathematical topics found in mainland USA textbooks illustrates the major challenges mentors faced in carrying out their mandate of assisting prospective teachers to teach western mathematical ideas to Pacific island children. But these challenges and others were surmountable provided there was trust and respect between those presenting and those receiving the instruction.

SURMOUNTABLE CHALLENGES

During a visit to his home island, one of the mentors, Century, invited me to the school where he formally taught so that I could see how well he taught fractions to the fifth-grade class. Although he would mainly speak in his mother tongue, he felt that I would be able to understand what was happening and could give him feedback about the lesson in the same way he was supposed to observe and hold a post conference with a prospective teacher. I gladly agreed.

A PATH LAID WHILE WALKING

The school consisted of two single floor buildings arranged in an L-shape with six classrooms arranged end-to-end running the length of the building. All rooms were the width of the building and had mesh screened windows on both sides. I arrived at the classroom where Century was to teach. There were 45 children in the room. There were no desks, no chairs, and no tables in the room. The blackboards were unusable except for a small area on the lower corner of the board. It really did not matter that the boards were in such poor condition because there had not been any chalk for at least a year. Century called the children to order, and each took up a small square of space on the floor. It was as if there were 6 rows of 8 desks arranged in an array, yet there were no desks and no chairs. The children were smiling and attentive as Century began his lesson. He used pantomime, gesturing with his arms to draw imaginary figures in the air. He had students stand up at the front of the class and share a collection of coconuts he extracted from the bag he always carried on his shoulder. Near the end of the lesson, Century invited me to query the students about what they had learnt. In halting English the students explained to me what they knew as a result of the lesson. It had lasted approximately 45 minutes and I confess that I found it most ingenious and highly successful.

By any standards, the instructional space and equipment available was unacceptable, yet despite these conditions teaching and learning took place. Century knew these children, they respected and trusted him (and he them), so despite the circumstances together teacher and children forged a learning community in a manner that even an outsider like myself could see that growth in the understanding children had about fractions.

As Century and I were leaving the school, the principal, who had been meeting with another Project staff member, stopped us to issue an invitation for me to engage parents of the school's children in a workshop the next day where, he said, "You could demonstrate the kind of teaching of mathematics you are encouraging the teachers to use with their children." I protested, "I don't have the language to lead a workshop with parents," but my other Project colleague, Marcus, interjected to say, "I will be your 'voice'." Though his mother tongue was not that of this island, he could speak it well enough to communicate what I had to say. Marcus reminded me that, "You often say the mentors should find ways around insurmountable challenges. Here you have one where you can learn to instruct when the path you are walking doesn't contain English!" Marcus had skewered me with my own words. I had little choice but to agree to this request.

The next morning one short announcement was made over the island radio station that there would a meeting that afternoon for parents of XYZ School. Just after the 2:00 p.m. dismissal time at the school, approximately 70 parents were waiting in the schoolyard for the meeting to begin. From somewhere about 50 dilapidated desks appeared and were set up in two adjoining classrooms that had one usable blackboard. Several short pieces of chalk also appeared. Talking excitedly, the parents squeezed into the room, some took the chairs but many stood next to the screened windows trying to catch what little breeze there was in the stifling 85F-degree heat. The smell of the pig farm next to the school wafted by on the breeze!

The principal and then Marcus welcomed the parents to the meeting and explained what was to happen: they were to be taught math by someone who could not speak their language in the fashion teachers and children in the school were teaching and learning mathematics. That comment brought forth loud laughter and the clapping of hands. First I divided the people into small groups, and said they were to help each other with the mathematics questions I was going to give them. So I began the mathematics lesson – using short sentences and long pauses so that Marcus could translate – by inviting them to find a way to arrange the nine digits (1, 2, 3, 4, 5, 6, 7, 8, 9) in the following scheme such that each of the digits was used once and only once so that a true statement was produced:

Many of the parents had very little schooling themselves. What many of them knew about adding was buried deep in memory, if it was there at all. Yet they enjoined the task with enthusiasm, much laughter, and friendly competition between the groups. As solutions were found, I challenged them, through gestures as the conversational noise in the room was too great for Marcus or me to get their attention, to find other solutions, to find patterns, and finally to make up a new task using the nine digits and challenge the group next to them to solve this new task. Two hours later the last of the parents left the school to walk home. Marcus, Century, and the principal said parents told them what a great time they had, that they never thought they could do their children's mathematics and what a great feeling it was to leave the school feeling so successful.

Trust and respect was the ingredient common to these two tales of surmounting challenges. Just as Century and the children had the trust and respect of each other, the parents trusted and respected Marcus and the principal. When they discovered they could do the mathematics tasks set for them, their confidence that the instructor was not going to embarrass them, or make fun of them, was increased, a basic first step in building trust. I trusted them and respected their abilities as human beings and my confidence in them was well rewarded.

CULTURAL SHARING: AN EXAMPLE FROM ONE ISLAND

The centrality of building cultural awareness and sensitivity among the mentors, and the crucial role of fostering of trust and respect among and between mentors and prospective teachers was a goal of the Project from its inception. The decision to set aside time during each institute for cultural sharing began with the first grant.

The 2006 institute on Yap was one of the most memorable of all the meetings because of the uniqueness of Yapese culture, the fact that most of the mentors had not previously been to Yap, and the impact of the cultural sharing arranged by the Yap mentor team. The official currency in Yap is the U.S. dollar. However, Yap is famous for its stone money, which is still in use for traditional exchanges such as the purchase of land or in village ceremonies. The picture on the left shows the large circular discs that are Yap stone money. The cultural event hosted by the Yapese mentors involved a trip to a rural village where the mentors were treated to a display of Yapese craft making and stick dancing. In preparation for travelling to and living on Yap for the institute, mentors were provided with the following information about 'expectations' while on Yap.

In Yap female toplessness is common and socially acceptable, but it is considered highly offensive for women to bare their thighs in public. Short shorts, bikinis, and miniskirts are a definite no-no. These dress restrictions do not apply to women when diving or sunbathing on some private beaches, but bring a wraparound with you for when you get out of the water or leave the beach.

For the cultural event institute participants were bussed to a rural village. Disembarking from the bus, the mentors were greeted by a village elder who led them on a half-hour walk through the jungle to the village. When they emerged from the jungle, they saw a stone money path and craftsmen demonstrating their practice, as shown in the pictures below.

These activities were a precursor to the "stick dance" that was the climax of the mentor cultural experience on Yap. The dance was performed during a very heavy rainstorm, but that didn't seem to deter the dancers from giving an enthusiastic and spirited display. The water rose to a depth of six inches during the storm

Village's Stone Money Path

These dances are crucial to the maintenance of the Yapese culture.

Dances belong to particular communities and can only be performed if the village grants permission. Educating young people to perform the dances are one means that villages use to maintain their language and culture. Once a year during Yap Days, the villages come together to dance and to

Village Elder

Villagers gather to display their handicrafts

Young boys juggle prior to dancing

celebrate their culture. This presentation to mentors was the first the village had undertaken since rebuilding the village meetinghouse that had been destroyed by a typhoon fourteen months previously. It was an honour and a sign of respect for the mentors to be invited into the village to witness the dance and engage with the villagers in their craft display.

These activities were a precursor to the "stick dance" that was the climax of the mentor cultural experience on Yap. The dance was performed during a very heavy rainstorm, but that didn't seem to deter the dancers from giving an enthusiastic and spirited display. The water rose to a depth of six inches during the storm. These dances are crucial to the maintenance of the Yapese culture. Dances belong to particular communities and can only be performed if the village grants permission.

Educating young people to perform the dances are one means that villages use to maintain their language and culture. Once a year during Yap Days, the villages come together to dance and to celebrate their culture. This presentation to mentors was the first the village had undertaken since rebuilding the village meetinghouse that had been destroyed by a typhoon fourteen months previously. It was an honour and a sign of respect for the mentors to be invited into the village to witness the dance and engage with the villagers in their craft display

CENTRALITY OF FAMILY

To say that family is central to Pacific islands cultures is to under estimate the impact and influence that family has on all that is done on the islands. Family events and relationships have first priority for an individual's involvement in any activity. To disrespect a family member or some aspect of the family culture would be a severe and unforgivable affront. Project arrangements and mentor involvement had to constantly be cognizant of the constraints of family demands, but sensitive to the opportunities that respect for family values, priorities, and connections provided.

> Marcus was born and raised on the islands, a man who provided guidance and direction in the early years of the Project as to the ins and outs of working compassionately and sensitively within island cultures. He informed Project staff about, among many things, inter-island family connections, about island foods, about chewing betel nut, and about the etiquette of drinking sakau.[10] He introduced Project staff to influential people on the islands, took us to and educated us about remote culture sites (e.g., Nan Madol[11] on Pohnpei) and influential educational institutions (e.g., Xavier High School[12] on Chuuk), and drew me into and made me part of his extended family. The latter experience allowed me to have a brief insider's view of the functioning of an island family, the pressures and responsibilities experienced by family members, and most crucially the importance of building a trusting and respectful relationship with island peoples before attempting to work with them on professional development activities.

When combined with my previous educational experience in Canadian aboriginal settings, the introduction to Micronesian cultures I received by being accepted into Marcus' family proved invaluable. It reinforced my belief about the absolute necessity of building trust and respect in order for the Project and the work of the

---

[10] Sakau, which is extracted from the pounded roots of a pepper shrub, contains up to 14 naturally occurring painkillers that, unlike alcohol, sedate the drinker making the ritual of consumption a quietly social act.

[11] Nan Madol is a 200-acre stone complex comprised of approximately 92 man-made islets located on the southeast shore of Temwen island off the coast of Pohnpei.

[12] When Xavier High School opened in 1952, it was the first high school in the Trust Territory of the Pacific Islands. Xavier has an enrollment of 151 students (96 boys, 65 girls), serving the islands nations of the Federated States of Micronesia, the Republic of Palau, and the Republic of the Marshall Islands.

mentors to have an impact on the prospective teachers of mathematics across the region.

PACIFIC ISLAND GENEROSITY

The generosity of Pacific island people is notorious in the region. Early in the life of the Project, we had to learn not express admiration about, for example, a shirt one of the male mentors was wearing, or the jewellery worn by a female mentor, because by island custom and tradition the mentor was honour bound to give the object to the person who admired it; literally give the shirt off one's back. Pacific islanders in professional working relationships also expressed generosity so long as they felt there was respect for their cultural ways and relationships. One did not have to necessarily agree with their cultural practices, but one did have to respect them *and* when on their islands abide by those practices. The mentors brought that respect and knowledge with them to the Project and to the meetings. Over eight years of working together, the Project staff forged its own sense of family, its own unique pathways for creating and implementing professional development activities while honouring the cultural practices of their colleagues. As trust and respect among the mentors grew, their conversations deepened, and the bonds among the mentors strengthened. Their Pacific island family now included colleagues from other islands, who brought with them different cultures and languages that enriched and educated all involved with the Project. These gifts of relationship were given in true Pacific island fashion, without reservation and with, as we would say in Hawai'i, *much aloha*.

REFERENCES

Baldwin, C. (1998). *Calling the circle: The first and future culture*. New York, NY: Bantam Books.
Gattegno, C. (1970). *What we owe children: The subordination of teaching to learning*. London: Outerbridge & Dienstfrey
Jaworski, B. J., Wood, T., & Dawson, S. (Eds.) (1999). *Mathematics teacher education: Critical international perspectives*. London: Falmer Press.
Tompert, A. & Parker, R. A. (1990). *Grandfather Tang's story*. New York, NY: Dragonfly Books, Crown Publishers.

A. J. (Sandy) Dawson
Pacific Resources for Education and Learning (PREL) &
College of Education
University of Hawai'i at Mānoa (UHM)

SECTION 4

SYNTHESIS

BARBARA JAWORSKI

17. DEVELOPMENT OF THE MATHEMATICS TEACHER EDUCATOR AND ITS RELATION TO TEACHING DEVELOPMENT

Volume 4 of this Handbook has dealt uniquely with the mathematics teacher educator as a developing professional. This chapter brings Volume 4 to a close by drawing out themes and issues running through the 16 chapters. These include forms of knowledge, theoretical frames, modes of activity and mediational structures in mathematics teacher and teacher educator development, critical inquiry in development and research, and interactions and inter-relationships between teachers and educators. Collaborative partnerships between teachers and educators are seen to provide powerful potential for mutual development.

INTRODUCTION

Mathematics teaching is complex. Teachers need to know mathematics, pedagogy related to mathematics, mathematical didactics in transforming mathematics into activity for learners in classrooms, elements of educational systems in which teachers work including curriculum and assessment, and social systems and cultural settings with respect to which education is located. In addition, teachers know intimately the students with whom they work and the particularities of the schools where teaching takes place. Each of these elements can be discussed in its own right as if it is independent of the others. For example, *pedagogy related to mathematics* can be considered separately from *social systems and cultural settings with respect to which education is located*. In fact what happens in practice is a complicated, dependent interweaving of these elements related to the particularities of context in which teaching takes place. Each teacher, and teachers as local or global communities, work(s) within this complexity and experience(s) the interweaving as part of their professional identity drawing on knowledge tacit or explicit.[1]

---

[1] Teachers' knowledge in teaching is the focus of the first volume of this handbook: Sullivan, P., & Wood, T. (Eds.). (2008). *International handbook of mathematics teacher education: Vol. 1, Knowledge and beliefs in mathematics teaching and teaching development.* Rotterdam, the Netherlands: Sense Publishers.

BARBARA JAWORSKI

Mathematics teacher educators work with teachers to develop teaching. Both educators and teachers have a common aim to provide better learning opportunities for students learning mathematics. Educators provide courses, summer institutes, professional events of various kinds to enable practising teachers to develop knowledge in the areas indicated. Educators work with teachers in school settings, encouraging a professional dialogue about teaching and using their own knowledge to promote forms of practice conducive to students' learning. Educators work with prospective teachers, both in and out of the school setting, to develop knowledge of teaching and to enable the newcomers to grow into professional practice. Thus, the kinds of knowledge needed by teachers are also fundamental to the work of teacher educators. While they cannot know the students and context of each school in any depth, educators bring a profound understanding of mathematics, didactics, pedagogy and systemic factors related to a range of settings. Their pedagogic knowledge extends to creation of opportunity for teachers to learn and develop mathematics teaching. In addition, educators need a knowledge of the professional and research literature relating to the learning and teaching of mathematics, knowledge of theories of learning and teaching, and knowledge of methodologies of research that inquires into learning and teaching in schools and educational systems. I used the diagram in Figure 1 in my introductory chapter to this volume.

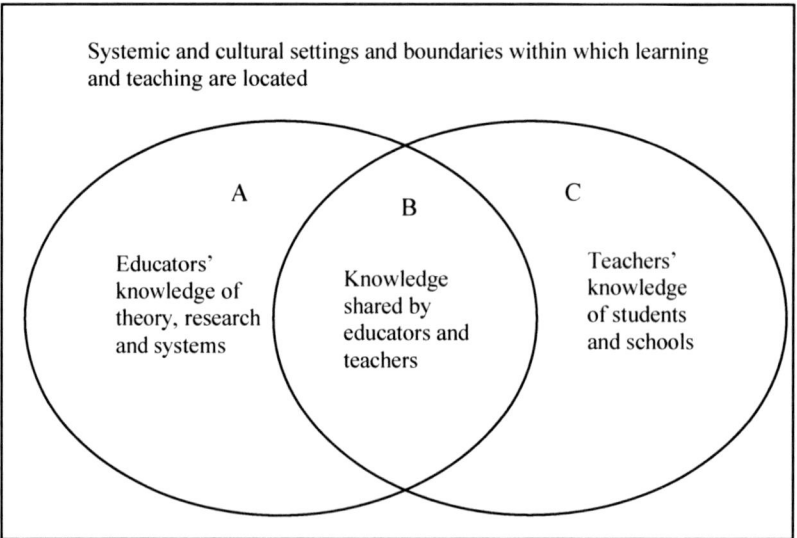

Figure 1. Knowledge in teacher education

A diagram such as we see in Figure 1 apparently simplifies complexity. I use it for three main reasons:

1. To emphasise that knowledge of mathematics teachers and teacher educators is deeply related, although there are specialist areas and of course differences between what is known by individuals or groups in the various areas.
2. To emphasise a common enterprise: all are concerned with the mathematical education of students and all seek to know more about how to provide for the diversity of learning needs.
3. To emphasise that teacher educators are not all-knowing. They bring knowledge to the enterprise, as teachers do, and this knowledge also has to be tested in practical settings and developed through practice.

Traditionally, teacher education has been seen as a transfer of knowledge from educator to teacher. Educators, in out of school settings, pass on knowledge about mathematics, pedagogy etc. in courses and other programmes and teachers use this knowledge in the arena of classrooms. Or, educators work with teachers in schools, supporting teachers in developing their expertise in the classroom. The traditional view sets up a hierarchy in which educators are superior in knowledge and teachers learn from the educators. Implicit in this perspective is that there will be a similar hierarchy between teachers and their students.

In a *constructivist* frame, common particularly in teacher education programmes in the 1980s and 1990s (Jaworski & Wood, 1999), educators recognise teachers as independent cognisers, constructing knowledge of mathematics teaching through their experiences whether in courses or classrooms and through processes of assimilation, accommodation and reflective abstraction (Piaget, 1950). As a consequence, educators seek to provide relevant experiences from which teachers can construct the knowledge they need. Unlike the traditional view there is no expectation that teachers will develop knowledge as conceived by educators and this creates a source of issues as educators seek access to teachers' conceptions to perceive outcomes from teacher education initiatives. Explicit in this perspective on teacher education is that teachers see students also as constructors of knowledge: in their professional learning settings educators use well documented strategies to foster teachers' perspectives relating to students' growth of mathematics knowledge[2]. The constructivist perspective raises many issues for teaching and teacher education (e.g. Carter & Richards, 1999; Irwin & Britt, 1999). Not least, both teachers and educators grapple with the creation of opportunity for learners' construction of relevant knowledge and the issues that arise when the provider of the experience is not satisfied with the apparent

---

[2] Examples include practices evident in current literature such as CGI (Cognitively Guided Instruction) use of examples of student thinking as a basis for developing teachers thinking (e.g., Carpenter, Fennema, Peterson & Carey, 1988); lesson study where teachers plan lessons together and learn individually from the joint process (e.g., Fernandez & Yoshida, 2004); use of video cases in which video acts as a tool to promote new ways of seeing and support individuals in developing teaching (Sherin, 2002; Nemirovsky, Dimattia, Ribiero & Lara-Meloy, 2005). Volume 2 of this handbook describes many, related tools and processes, including using narratives, cases, lesson studies, videos, CGI: Tirosh, D & Wood, T. (Eds.). (2008). *International handbook of mathematics teacher education: Vol. 2. Tools and processes in mathematics teacher education.* Rotterdam, the Netherlands: Sense Publishers.

constructions that emerge. Here also the educator/teacher is positioned as provider and evaluator, superior to the learner.

The recognition of issues in traditional and constructivist frames begs some reframing of perspective, and I see the chapters in this volume to an extent, collectively, providing this reframing. In Chapter 1, I hinted at a "shift"

> in tone and nuance in the ways educators write about educating teachers. There is less of a surety of models of practice that educators promote with teachers and much more a sense of uncertainty in inquiry. With this uncertainty comes, almost paradoxically, a strength of purpose, new ways of speaking about mathematics teacher education, and new paradigms of practice. These build on notions of reflection, for both teachers and teacher educators, on teacher-as-researcher and simultaneously educator-as-researcher models, and on growing recognitions of epistemology, of complexity and the importance of not trying to oversimplify.

A multiplicity of paradigms has become evident in the literature; particularly, sociocultural approaches have now become more widely used and discussed. However, the shift is not from the constructivist to the sociocultural, but rather to a recognition that different lenses on practice can afford different ways of seeing and doing, and that we can learn as a community from them all. In this volume, with the teacher educator in focus, we have looked particularly at ways in which teacher educators view the enterprise of mathematics teacher education, and moreover, how teacher educator knowledge and understanding grows *alongside* the concomitant growth of knowledge for teachers. Coming back to Figure 1, educators have a responsibility for the development of knowledge in areas B and C, and this includes their own knowledge in B. They use knowledge in A, which also grows through their engagement in research. I return to these areas from a perspective on teachers' knowledge later in the chapter.

The various chapters in this volume have inspected and analysed the nature of mathematics teachers educators' knowledge and its growth as part of MTE's educative practice in working with teachers. Chapters take different theoretical perspectives and epistemological positions relating to teacher education programmes and practices. In my next main section I offer a synthesis of themes and issues distilled from these chapters.

MATHEMATICS TEACHER EDUCATOR (MTE) KNOWLEDGE AND ITS GROWTH

The chapters of Volume 4 tell many stories, and readers will see their own stories in the chapters and in the volume as a whole. In working with the authors on the chapters, I also have my own stories which will permeate my synthesis here. I have organised this under a series of headings as follows:
- Learning in and from practice – mediation and synthesis
- Categorisations of teacher knowledge leading to MTEs' structuring of programmes
- Perception, conception, attention and awareness

- The teaching problem
- Sociocultural views on teacher and MTE learning
- Ethical and moral perspectives in mathematics teacher education: caring and reciprocity
- Growth of knowledge through reflection, inquiry, research and writing

I start from the practice of being a teacher educator and refer to theoretical models which teacher educators have themselves synthesised from reflection on and analyses of their own practice. This leads to a consideration of teacher educator knowledge and ways in which knowledge is conceptualised and made available for teacher education programmes and courses. In such courses teacher educators have goals for teachers' learning, but achieving those goals is not straightforward. Different theoretical perspectives offer alternative conceptualisations of the problems and their resolution. This brings me to what I call "The teaching problem": it is the problem of how any one person or group of people can promote or achieve the learning of others and how different theories address this problem. In such promotion there are many ethical and moral issues, particularly in the kinds of relationships that are fostered between teacher educators and teachers. Collaboration between educators and teachers, especially in programmes where both are engaged as researchers is seen as one fruitful way forwards. The activities of a teacher educator research community, particularly in terms of reading the literature and writing for publication, are analysed as important mediators in teacher educator learning and their relation to teacher learning.

Learning in and from Practice – Mediation and Synthesis

Two models relating to growth of knowledge through practice are offered respectively by Pat Perks and Stephanie Prestage [13][3] and by Orit Zaslazvsky [5]. The first model, illustrated in Figures 2a and 2b, relates teacher and MTE knowledge. It suggests a knowledge base for teachers followed by an expansion of this for teacher educators who, like the authors of the model, have themselves formerly been mathematics teachers. In this model teacher knowledge is grounded in knowledge from existing school practices and research (professional traditions); knowledge from being in the classroom (practical wisdom); and the prospective teacher's knowledge from being a learner of mathematics (learner knowledge). We see similar categories for the teacher educator. The *classroom events* (for the teacher) translate into *maths ed sessions*, the classroom settings in which teacher educators meet teachers; *professional traditions* here relate to knowledge from systems, institutions and culture, e.g., expectations to do research and publish; *practical wisdom* relates to knowledge from engagement with teachers in education-focused events. Here, teacher educators' *learner knowledge* is the

---

[3] When I refer to authors of chapters in this volume, I include their chapter number as written here [13].

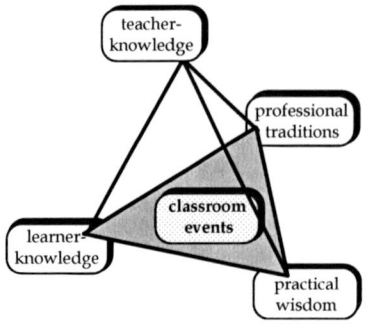 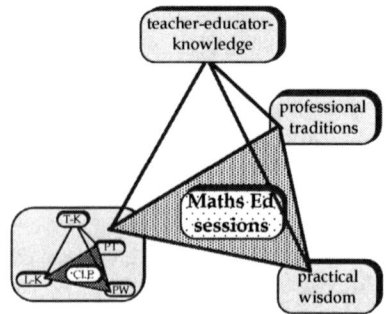

Figure 2a. Teacher knowledge   *Figure 2b. Teacher educator knowledge*

knowledge they bring from having been teachers themselves and we see in the figure that this strand of knowledge brings with it the full complexity of teacher knowledge.

In the second model, from Orit Zaslavsky (illustrated in Figure 3), we see two learning cycles: teachers as learners from tasks designed by teacher educators who facilitate teachers' learning and learn themselves in this process. The two cycles are interwoven and interdependent.

In the first model (Perks and Prestage), we see teacher knowledge as a mediating factor for the development of teacher educator knowledge; in the second (Zaslavsky), the design and use of tasks for teachers' learning (itself drawing on specialist MTE knowledge) mediates the process of teacher educator learning. In expressing growth of knowledge as a *mediational* process, I am drawing here on a sociocultural theoretical frame rooted in Vygotsky (1978) and Leont'ev (1979) in which tools and signs are seen to mediate the learning process. *Teacher knowledge* and *design and use of tasks* can be seen as "intellectual tools" (Wartofsky, 1979, cited in Wells, 1999, p. 69) through which the MTE knowledge grows. Other models of learning can be theorised similarly: for example Victoria Sánchez and Mercedes García [14] discuss dilemmas they face in their practice as teacher educators. The dilemmas arose from choices which had to be made from differing alternatives of equally perceived values. Through engaging with the arguments supporting each of the alternatives, the teacher educators deepened their own conceptions of the teacher education issues with which they were concerned and enhanced their own professional expertise. Thus the dilemmas acted as mediators for their learning. Razia Fakir Mohammad [8] shows how recognition of contradictory forces in her work with teachers led to her own enhanced knowledge in practice. Understanding the nature of these contradictions opened up possibilities to deal with the challenges in her practice. In each of these cases, we

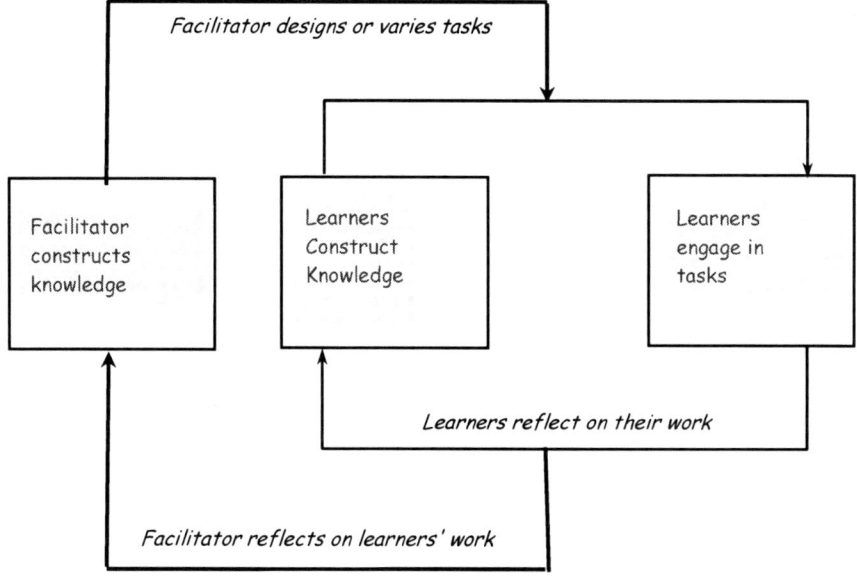

Figure 3. Task-based knowledge of teachers and teacher educators

see teacher educators learning from their engagement in activity and their grappling with the issues it raised for them. Through participative activity with teachers, or in preparing for such activity, teacher educators begin to extract elements of generality which contribute to a synthesis of action and a theorising of action; in some cases a grappling with tensions and issues. The emergent theoretical models, as synthesised in the two examples above, offer a vision of practice in which complexity of practice can be seen holistically and from which new forms of practice can emerge.

Categorisations of Teacher Knowledge Leading to MTEs' Structuring of Programmes for Teacher or for MTE Learning

While we might seek to identify teacher and teacher educator learning as growth of knowledge in practice in which holistic complexity is preserved, we desire nevertheless to make sense as teacher educators of the elements and processes of teacher learning in order that teacher learning can be fostered in our programmes and projects. Research and theorising over two decades and in recent times have sought to define or reveal elements of teacher knowledge which can lead to formulations of teacher education practice facilitating development of teachers' knowledge. Examples include Shulman's (1987) *seven* categories of teacher knowledge from which researchers in mathematics education have distilled examples of *pedagogic content knowledge* in distinction from either knowledge of mathematics or knowledge of pedagogy in mathematics teaching (e.g., Graeber & Tirosh, in Volume 1 of this handbook), of *mathematical knowledge for teaching* (e.g., Ball & Bass, 2004) in distinction from mathematics knowledge per se, and of *didactical knowledge* (e.g., Winsløw & Durrand Guerrier, 2007) to emphasise the didactical transposition from mathematical knowledge to knowledge for teaching mathematics.

Christer Bergsten and Barbro Grevholm [11] trace analytic approaches to distinguishing elements of teachers' knowledge in the MTE literature: for example the competency approach taken in the Danish KOM project in which *eight* specific competencies are identified (Niss, 2004). They comment on the report "Adding it up" (Kilpatrick, Swafford & Findell, 2001) in which mathematical proficiency is defined through *five* major strands. An aim in producing such distinctions is to develop in prospective or practising teachers each of these competencies or strands of proficiency, and a major question for mathematics educators is how to design programmes through which such competency or proficiency can emerge.

Bergsten and Grevholm point to "the didactic divide": a gap between disciplinary and pedagogic knowledge, between theory and practice, observed in such programmes. It raises a question of how theory and practice can be linked in helpful and realistic ways for prospective teachers who rely on their programme to provide such links. Jaworski and Gellert (2003) point to four modes programmes showing varying combinations of theory and practice with the ideal being a fully integrated model. In such a model, the timing of theoretically focused events and work in classrooms is such as to allow two-way reflections in which theory can be understood through practice and practice be informed by theory. However, such models are internationally much less common and harder to achieve from institutional perspectives than those which separate theory and practice.

Of important consideration in addressing the theory-practice divide in educational programmes is where and how teachers learn *mathematics*, or *mathematics for teaching*. In Volume 1 of this handbook, issues of teachers' knowledge for teaching mathematics are discussed, including *what* mathematics is needed for different levels of teaching. MTEs, must also address *how* their design of courses enables teachers to learn this mathematics. For example, there might be a distinction between courses which are primarily focused on mathematics and others with strong elements of pedagogy and didactics. Chapters from Amy Roth McDuffie, Corey Drake and Beth Herbel-Eisenmann [12] and Victoria Sánchez and Mercedes García [14] make explicit the thinking and choices of MTEs in distinguishing mathematical content from, or relating it to aspects of pedagogy and didactics for prospective elementary mathematics teachers. Sánchez and García distinguish a mathematics course for teachers from a mathematics methods course, discussing in detail the factors influencing what distinguishes such courses. Perks and Prestage [13], working with prospective secondary teachers, in a programme that is only one year in length, show how tasks that focus on methods of mathematics teaching can also promote deeper understanding of mathematical concepts. In each of these programmes, we see that the kinds of tasks designed by

MTEs for teachers are often multipurpose tasks that address the deepening of mathematical knowledge alongside focuses on pedagogy and didactics. MTEs emphasise the importance for teachers of developing *reflective* attitudes to such distinctions through which complexity can be acknowledged and of providing opportunities for linking with school-based elements of the programme.

So far in this section, I have been focusing on elements of teacher knowledge relating to programmes designed by teacher educators for teachers. Designing of such programmes could be seen as expressed above as a mediating tool in development of MTE knowledge. However, can we see a parallel formulation in designating elements of MTE knowledge, leading to programmes designed for MTE development? And who should organise such programmes? Orit Zaslavsky [5] offers seven themes that reflect goals for mathematics teacher education, not based on the conventional content topics of teacher education such as teaching decimals or grouping practices, but which suggest the kinds of competency needed by MTEs in promoting teaching development. They include, *coping with conflict, dilemmas and problem situations*; *selecting and using (appropriate) tools and resources for teaching;* and *sharing and revealing self, peer and student dispositions*.

Such areas of competency form the content basis of a programme specially designed for the education of MTEs of practising teachers in Israel, the MANOR programme. In introducing this programme, Ruhama Even [3] points to a gap in the literature involving *missing research on the education of teacher educators*, the *ill-defined nature of the field of educating practising teachers*, and a *lack of information on the practice of mathematics teacher educators*. The MANOR programme (deriving from the Hebrew word MANHIM, meaning "guides") addressed this gap through its design, delivery and evaluation. It was led by experienced MTE researchers and drew its participants from a range of professional areas including teacher mentors, mathematics coordinators in schools and staff of curriculum implementation projects. The curriculum for the programme included three main areas of study captured under the broad headings of *mathematics education, mathematics* and *teacher education*. Ruhama Even coined a new word "knowtice" to designate the integration of *knowledge* and *practice* in the MANOR programme. It seems to me that the programme as described not only addresses the key areas of knowledge indicated in Figure 1, but comes close to the *integrated* form of programme suggested in Jaworski and Gellerts' four modes mentioned above. Important to this integration are elements of theory and research as part of which participants read the research literature, address theoretical concepts and produce their own written materials – itself a formative process about which I say more below.

In this section, I have drawn attention to categorisations of teacher knowledge in teaching mathematics and issues raised for MTEs in promoting teacher knowledge leading to MTEs' own learning. Categorisations offer a multiplicity of ways of cutting through the complexity of teachers' knowledge and of knowledge in teaching mathematics. MTEs' awareness of categorisations from reading the literature, in resonance with their own experiential knowledge can form a powerful knowledge base for the construction of programmes and projects in teacher education. The MANOR programme offers just one example, but as yet a unique example in terms of education of MTEs. In any such programme, choices have been made as to the topics to be addressed and modes of addressing them. Such choices are not straightforward, as Sánchez and García have pointed out, and it is clear that MTEs learn particularly from grappling with dilemmas presented. However, despite considerable evidence of such learning in the research literature, Olive Chapman [6] points out that MTEs rarely speak about their own learning in reporting from research in their programmes – the focus being rather on what teachers have learned.

We see an exception here in the chapter of Amy Roth McDuffie, Corey Drake and Beth Herbal-Eisenmann [12] who together reflect on their teacher education programmes for prospective teachers and recognise ways in which they themselves have learned from these programmes. Making MTE learning in a teacher education programme more overt seems important to addressing how mathematics teacher education develops. An important question for MTEs seems to concern how their own growth of knowledge through engaging with literature on categorisation and subsequent grappling with the dilemmas of choice relates to the experiences and learning of teachers in the resulting programmes or projects. A further question, to which I return later in the chapter, is whether teachers might learn in similar ways given appropriate contexts.

Perception, Conception, Attention and Awareness

Martin Simon [1] makes a distinction between teacher education programmes which have process goals only and those which have both content and process goals. He focuses particularly on the second category, programmes which have both content and process goals, and speaks of "courses or workshops for teachers in which teacher educators aim to promote particular mathematical and pedagogical concepts, skills and dispositions", perhaps based around categorisations such as are mentioned above. Simon suggests four areas of research-based knowledge that are currently *insufficient* for promoting desired (by the reform movement) concepts skills and dispositions. Drawing strong parallels between teachers' teaching of mathematics and teacher educators' teaching of mathematics teaching, he suggests that the insufficiency is one of understanding conceptualisation processes. Thus teachers, in seeking to promote mathematical understanding in pupils *perceive* what is required from pupils in terms of the mathematics to be understood, and they structure pupils' activity around such *perception*. Simon calls this a *perception-based* mode of thinking. It is teacher-focused rather than seeking out the conceptual position of the pupil. If a teacher alternatively takes a conceptual focus – seeking to understanding pupils' current conceptions – then attention can be given to how pupils can move to a higher conceptual level and what forms of activity can support and promote such a conceptual shift. This is described as a *conception-based* mode of thinking. Similarly, teacher educators may be caught up in a perception-based mode of designing teacher learning activities seeking to promote what the educator perceives as being required, rather than exploring first teachers' current conceptions and basing their programme on promoting conceptual shift. Simon grounds his conception-based perspective within a Piagetian theory of assimilation: the issue for teacher educators is how to promote a change in teachers' assimilatory structures. A first step is seen to involve research that illuminates these structures.

These issues and this challenge are addressed by Ron Tzur [7] who refers us to *the learning paradox* (LP) (e.g., Bereiter, 1985), an issue deriving from a constructivist notion of assimilation in reaching an understanding of a new concept. The issue concerns what (mental) mechanisms exist to promote construction of a concept that is more complex that those already assimilated. It appears that a student needs already to have some understanding of a concept prior to learning it, in order to allow concept construction. The LP presents a dilemma for teachers and educators taking a conception-based approach to promoting learning of declared concepts. Tzur, having himself grappled with this dilemma over many years, distinguishes *intuitive* and *non-intuitive* teaching: intuitive referring to following ones own intuitions to promote learning in others (a perception-based perspective), and non-intuitive involving an awareness of the LP and developing strategies to overcome it. Tzur traces his own learning as a teacher educator and offers a nested model of developing awareness: *learning mathematics through teaching it*; *learning pedagogy through teaching teachers*; *learning teacher education through teaching prospective mathematics teacher educators (MTEs)* and *learning to mentor MTEs through scholarly collaboration and continual teaching*. In each case the theoretical model is one of mental construction, the teacher or MTE's own mental construction of the concepts being taught and the strategies for teaching them, related to the mental constructions of the learners who are the focus of the teaching. The model has resonances of that offered by Perks and Prestage, although couched in constructivist theory while Perks and Prestage ground their work socioculturally. I return to such theoretical differences shortly.

Tzur relates his rationalization of these complex processes to Mason's psychology of awareness and attention. John Mason [2] starts from what might seem an obvious statement: "teachers cannot make learners learn, nor can they do the learning for their learners". Elsewhere (Mason, 2002, p. v), Mason makes the point " I cannot change others. I can work at changing myself". So, as teachers or MTEs, *all* we can do is "direct learners' attention", through tasks, activity and interaction designed to promote the learning we seek. *Awareness* develops from what we attend to at any time, and "if you put yourself in an unfamiliar situation, you may become aware that experts are attending to details unnoticed by novices". Thus, as teachers or MTEs, as a result of being aware of what contributes to our own learning, we can create situations finely attuned to learners' attention – "not only 'walking the talk' but 'talking the walk'".

Mason emphasises the necessity of teachers and MTEs challenging themselves by working on elements of their practice – working on their own mathematics, or working on their own pedagogy and so on. In the spirit of this overt learning through experience, they become aware of tasks that can be especially fruitful in generating attention and promoting awareness. Such tasks can be offered to others with the possibility of generating learning outcomes. Orit Zaslavsky's chapter [5] takes these ideas further through an analysis of task development.

The Teaching Problem

Issues related to perception and conception, attention and awareness focus on the thinking of individuals and on ways in which attention is focused and awareness develops. Attention and awareness relate to the construing individual (although they can be seen as more socially based as I address below) whereas perception and conception relate to the complex process of conceptualising the construal of *other* individuals, the learners to whom teaching is directed by teachers and MTEs.

This brings into focus the relationship between teaching and learning from constructivist perspectives. Constructivism is a theoretical perspective addressing mental processing in coming to know, that is in the *learning* of the individual. Alternative ways of conceptualising learning are offered through a range of sociocultural theories in which knowledge is seen to grow through participation in social practice. Vygotsky (1978, p. 57) has written that learning is first of all in the social plane before being internalised to the mental plane. James Wertsch (1991) takes this further suggesting that rather than seeing mental functioning as *deriving* from participation in social life (a dualistic separation), "the specific structures and processes of intramental processing can be traced to their genetic precursors on the intermental plane (p. 27)". In other words, individual mental thought processes and functionings have their origins fundamentally in social interaction. Here we talk of learning, and not of teaching. Although scholars often extrapolate from theories of learning to consequences for teachers and teaching (in the term *constructivist teaching* for example), the links are not obvious.

Within a social frame, from a position of situated cognition and social practice theory (e.g., Lave & Wenger, 1991), Jean Lave challenges us as follows:

> People who have attended school for many years may well assume that teaching is necessary if learning is to occur. Here I take the view that teaching is neither necessary nor sufficient to produce learning, and that the socio-cultural categories that divide teachers from learners in schools mystify the crucial ways in which learning is fundamental to participation and all participants in social practice. (Lave, 1996, p. 157)

We have seen above (with reference to the learning paradox) that the shift from theorising learning in constructivist terms to conceptualising teaching in such terms leads to problems. There are also problems in conceptualising teaching in sociocultural terms: if a teacher is supposed to enable others to learn, how can a teacher enable *learning through participation*? The issue here, relating to *participation*, a concept in sociocultural theory, parallels that of conceptualising *construal* of other individuals, those we would teach. In whichever theory we locate our conceptualisation, we propose models of *learning,* and experience shows us that *learners do learn*. However, we cannot readily use these theories to conceptualise *the promotion* of learning. Learners may not learn what teachers or MTEs want them to learn. Here is the crux of the teaching problem: *what does it mean to get other people to learn what we want them to learn*? Can learning theories address this (deceptively simple) question, and will any answers provided, according to different learning theories, differ?

I see this question as fundamental not only to Volume 4, but to this handbook as a whole. It concerns the *agency* of both teachers and MTEs in the learning processes respectively of students learning mathematics and teachers learning teaching mathematics. Three chapters in Volume 4 address this question from sociocultural perspectives.

Sociocultural Views on Teacher and MTE Learning

Perks and Prestage root their model, discussed above, in Vygotskian theory. They see expertise not as located in the individual but as *distributed* across systems involving other people and the artefacts they use to *mediate* learning of *scientific concepts* – concepts developed through the context of instruction, including those in both mathematics and mathematics teaching. According to Vygotsky (1986), a scientific concept involves from the first a *mediated* attitude towards its object: mediation being provided by artefacts or social dynamics within a system. *Distributed knowing* refers to knowledge distributed across the people within a system, rather than knowledge located within the individual, or knowledge that is common to all. Different forms of expertise can be seen to be rooted in the practices in which experts are engaged and to which newcomers gain access. Here we think of practices of mathematics learning and teaching within systems of education and schooling. Such ways of thinking offer a sense of holding complexity through focusing on community and system in which activity takes place. Knowledge can be seen as situated in the practice with which activity is associated: students, teachers and MTEs learn as part of their engagement in communities of practice (Wenger, 1998) and the associated systems of which they are a part. Perks and Prestage write,

> ... not only do we as tutors have to develop as teacher educators, we have a limited time frame to work with prospective teachers on developing, in Vygotsky's sense, scientific concepts about teaching. Thus a major part of our development lies in looking for efficient and effective experiences to promote the learning of our prospective teachers.

They use activity theory (Cole & Engeström 1993; Engeström, 1999), conceptualising the activity system as a *learning zone*, a system that allows for the learning of participants to contribute to the development of the system. This allows them to "trace connections between the acts that make up the action of learning about teaching in the activity systems of our current practices". I see here a sociocultural framing of the cognitive position articulated by John Mason regarding the drawing of a learner's attention to key concepts through suitably designed tasks which an expert recognises can educate awareness.

How sociocultural perspectives can contribute to our understanding of how learning takes place is also addressed by Merrilyn Goos [4] who draws on sociocultural theory to inform research and development interventions that aim to improve teachers' (and MTEs') opportunities to learn. Goos uses Valsiner's *zone theory*, extending Vygotsky's zone of proximal development (ZPD) to include two other zones: zone of free movement (ZFM) and zone of promoted action (ZPA). Together these zones, defined from the perspective of teacher (or MTE) as learner, include also indicators for intervention by those wishing to promote learning (the teachers of teachers, or of MTEs). Briefly, ZPD offers a set of possibilities for development influenced by learners' knowledge and beliefs in practice; ZFA can be interpreted as constraints within professional contexts, and ZPA as approaches that can engage with possibilities for development and promote actions that can be seen as viable within a given ZFM. Goos analyses initiatives in mathematics teacher education with respect to interaction between these zones, and extends her analyses to consider her own learning in relation to these perspectives on teacher learning. Thus, both Perks and Prestage, and Goos address and theorise the question of agency of the teacher or MTE promoting the learning of others and learning themselves as part of this process. In my next main section, I examine further the inter-relationship of these processes as I see it.

In the third chapter reporting research based in sociocultural theory, Simon Goodchild [10] extends Wenger's theory of *community of practice* to conceptualise teacher and MTE (didactician) learning as taking place reflexively within *communities of inquiry*. Teachers and didacticians together form a project community. Through participation in established communities of practice (school or university) teachers and didacticians use inquiry as a critical tool to promote learning within the project. Inquiry results in critical alignment with the norms of established practice, allowing teachers and didacticians to act within their practice while at the same time questioning its dynamics and exploring new ideas (Jaworski, 2006). An important aim of the project is the development of teaching to promote enhanced learning environments for pupils in mathematics, and didacticians aim to support teachers in this development. Goodchild draws on Paolo Freire's (1972) notion of *conscientization*, from a critical theory paradigm, to support the notion of critical alignment. Freire's conscientization, addressed as a challenge against poverty and oppression in the global human condition, is a metaphor for the condition of teachers (and MTEs) 'oppressed' in their practice by historic, economic, social and cultural contradictions over which they have little or no control. Inquiry as a tool, developing into inquiry as a way of being provides a means of critical action and associated growth of knowledge while working from within the conditions that impede action. The ideas relate strongly to the position on dilemmas articulated by Sánchez and García.

A difference between the programme discussed by Goodchild and the two earlier programmes in this section can be seen in relation to Martin Simon's distinction mentioned earlier. The project discussed by Goodchild would probably be located within Simon's first category of programmes that are process oriented, while the others have both content and process goals for prospective and/or practising teachers. The degree of agency exercised by the MTEs in these programmes is related to their responsibilities within their programme: in the first two programmes MTEs' would have a higher responsibility related to institutional settings, and their courses, curricula and assessment, than in the third, a research and developmental partnership between university and schools where goals of the programme could be mutually negotiated and defined. However, in all three we can see agency interpreted through a model of co-learning within institutional settings. I will return to this idea shortly. Next, however, I focus on moral and ethical questions for MTEs as they work within programmes and exercise agency in promoting learning for teachers and for themselves.

Ethical and Moral Perspectives in Mathematics Teacher Education: Caring and Reciprocity

The conceptualisation of teaching in relation to learning, and the practical realisations of such conceptualisation beg questions about the nature of relationships between learners and teachers, between teachers and MTEs, and between MTEs less or more experienced. Such questions can lead to moral and ethical considerations. What does it mean to take a moral or ethical approach to promoting mathematics learning and teaching development? The chapters in Volume 4 have offered a variety of perspectives relating to this question.

Paola Sztajn [15] discusses *caring theory*, delineating issues she attends to in becoming a *caring* mathematics teacher educator of practising teachers. Her caring theory derives from Nel Noddings work (e.g., 1992) and extends beyond caring for pupils in schools to caring for teachers in professional settings. Sztajn speaks of caring as a relational concept which plays out in the interactive setting involving carer and cared-for and with a reciprocal dimension. Thus it is not just a case of the MTE caring for the teachers with whom she works, but rather of the MTE taking responsibility to initiate a caring ethic through which mutual caring can result. *Caring*, contrary to popular perceptions of being cosy and comforting, is a cognitive construct with a critical dimension. As the carer seeks to understand the perspective and conceptions of the cared-for, the latter is drawn into a reciprocal perception of the carer's motives in promoting learning and insight into what is to be learned. In constructivist terms this reciprocal positioning addresses the learning paradox. In sociocultural terms it creates a mutual space in which reciprocal participation generates co-learning. Reciprocity allows both participants not only to respect the perspectives and position of the other, but commonly to seek to know how the other sees and thinks.

Relationships between teacher and didacticians in the project described by Simon Goodchild [10] can be seen to exemplify a caring ethic. Goodchild, drawing on Hostetler (2005), speaks of *good research* requiring ongoing attention to human well-being and going beyond the usual ethical consideration of avoiding risk or harm towards research that will be positively 'good' for participants. Beyond doing 'good' however, is the critical dimension mentioned above. A caring relationship affords inquiry and overt probing into normative processes. Through examples from her own practice, Sztajn shows that a caring attitude from one partner not only provides the partner with support, drawing on respect and trust from the other, to risk challenging new actions, but can work also in the other direction when the cared-for really tries to understand and achieve what the carer seeks to promote. In this way the agency of the carer is shared by the cared for. Goodchild, however, offers several examples where reciprocity is not achieved despite the good will of the didacticians and their critical agency.

This reminds us forcibly of the teaching problem and lack of ready solutions. Razia Fakir Mohammed [8], coming overtly from a moral position in which she desired to care for participants in her research was brought up against the lack of reciprocity when a teacher with whom she sought to work in a caring way challenged what she offered. An implication was that she had not understood sufficiently the issues the teacher faced in the institutional environment, and that what she seemed to propose could not work for him. It was clear that the caring relation had not established reciprocity and that the MTE's care for the teacher had not translated into the kind of action that could achieve developmental goals.

In his work in mathematics teacher education in the Western Pacific, Sandy Dawson [16] worked from an ethic of trust and respect in which he sought mutuality with his teacher and MTE colleagues in the Pacific region to promote and sustain development. Dawson describes aspects of culture in the communities of the Pacific very different from those he was familiar with, despite his considerable work with First Nations people in North America. Care showed itself in an overt willingness to know participants' views and feelings on the approaches he used, and to communicate his own nervousness and good will in using them. Reciprocity emerged when, despite considerable cultural differences, commonalities of purpose were recognised and his own aims and values were reflected in the words and actions of his partners.

Thus, we see in these chapters examples of reciprocity in which mutual caring relations seem to result in affective promotion of learning as conceived by the teacher or MTE, and others where reciprocity seems not to be achieved. Konrad Krainer [9] suggests that reflecting on learning is a two-way process: not only should MTEs ask teachers to reflect on their learning as part of a teacher education process, but the MTE should reflect also on his or her own learning: "we do not only demand activities from those for whose growth we are co-responsible, but we do it also ourselves". I infer from these words another aspect of a critical caring relation. As MTEs our agency is not just in the processes and tasks we offer to teachers and the kinds of engagement we seek from them. It rests also in overt demonstration of our own desires to fulfil ourselves the goals we suggest to them and our willingness to grapple with the challenges that arise – walking the talk.

DEVELOPMENT OF THE MATHEMATICS TEACHER EDUCATOR

Growth of Knowledge through Reflection, Inquiry, Research and Writing

I quoted above from the words of Konrad Krainer: "we do not only demand activities from those for whose growth we are co-responsible, but we do it also ourselves". In the mathematics teacher education literature it is now extremely common to see MTEs encouraging teachers to reflect on their practice, and discussing reflection as a mediating artefact (not necessarily using these words) in teachers' learning. In this volume four chapters overtly (in Section II), and several others implicitly, offer MTEs' reflections on their *own* practice as MTEs. The form of what is offered is in some cases autobiographical as in Section II, and in others it is part of a rationale for theoretical perspectives espoused and/or ways of working developed by MTEs in their programmes. This differs from the literature more widely as powerfully emphasised by Olive Chapman [6] in her review of papers reporting from learning of teachers in programmes run by MTEs. Research findings in these papers constituted evidence of MTEs' learning: Chapman writes, "for the most part, this learning was presented as what other teacher educators could learn about the nature of these approaches to instruction rather than the actual learning of the persons conducting the research". When I was editor of JMTE, reading many submissions of this kind, I sometimes suggested to authors that they might like to reflect on how their findings from research had influenced their own thinking and impacted their own practice. In just a few cases, authors incorporated a few paragraphs to address this question, and some expressed that they had found it valuable to address the question which might not otherwise have become overt. When I invited certain potential authors to contribute to this volume of the handbook, a response I received said something like, 'but I have done no research into my own learning/practice as a teacher educator', implying that research into teachers' learning, or into programmes designed to generate teachers' learning, provides no evidence of MTE learning. Yet the findings of such research are manifestly what is learned from the research by the researchers who are MTEs.

I have shifted here from *reflection* to *research*; but is this such a big leap? In the 1980s and even earlier in some cases, in mathematics teacher education, we started to see programmes involving teachers as researchers, investigating, or inquiring into their own practice. In 1987, the Association of Teachers of Mathematics (ATM, 1987) in the UK published a booklet entitled *Teacher is (as) researcher*, emphasising the research-like nature of good teaching. In 1988, a working group was initiated at PME entitled *Teachers as Researchers*, and 9 years later, in 1997, a book *Developing practice: Teachers' inquiry and educational change* was published based on the work of this group (Zack, Mousely & Breen, 1997). Much of the research referred to in these two publications is described as "action research", and might be seen as a part of the wider action research movement that was growing internationally at that time. Stephen Kemmis, a leading member of this movement spoke of reflection as "meta thinking", thinking about thinking. He went on to say:

> We do not pause to reflect in a vacuum. We pause to reflect because some issue arises which demands that we stop and take stock or consider before we

act. ... We are inclined to see reflection as something quiet and personal. My argument here is that reflection is action-oriented, social and political. Its product is praxis (informed committed action) the most eloquent and socially significant form of human action. (Kemmis, 1985, p. 141)

Kemmis conceptualised action research with reference to a critically reflective spiral of plan, act and observe, reflect (Kemmis & McTaggart, 1981; Carr & Kemmis, 1983). Thus, reflection, or *critical reflection* as is sometimes the term used, is central to a process of action research and related to conscientization as quoted by Goodchild from Freire.

MTEs researching their teacher education programmes might not consider they are engaged in action research. Action research is "insider research", that is research or inquiry conducted by practitioners into aspects of their own practice. This contrasts with outsider research in which 'outsider' researchers study the practices of others. It seems to me that MTEs researching their own programmes could be seen to be in either or both of these camps, and perhaps the distinction is not so clear anyway. However, Chapman's review indicates that research reports are usually couched in the language of the outsider rather than that of the insider. So, perhaps a critical factor is how as MTEs we talk about what we do (talking the walk?).

A major difference between insider and outsider research might be seen in terms of the kinds of knowledge it generates: insider research linking to enhanced practitioner knowledge and concomitant development of practice; outsider research linking to academic analyses, publication in research journals and generalised knowledge for the research community. It may be the MTE researchers see themselves largely generating the latter kind of knowledge, and if so this seems a pity. It suggests that the former kind of knowledge is being neither celebrated nor even acknowledged. However, this distinction is becoming blurred as more projects are conceived of as partnerships between teachers and MTEs in which both kinds of knowledge are sought and valued. Chapters from Krainer and Goodchild in this volume speak of such projects, respectively IMST and LCM. Both projects have a fundamental concern to influence development and improve teaching and learning; thus growth of practitioner knowledge is central. Both perform academic analyses of data, systematically collected according to clear research questions, and offer generalised formulations for the research community. A question MTEs might address more overtly concerns the overtly symbiotic nature of these two forms of research and ways in which the resulting forms of knowledge are related (Jaworski, 2003). This might enable to us tackle better the problem of academic research papers sitting on shelves and not reaching the practitioners for whom the findings could be of value (e.g. Hargreaves, 1996).

It is an important responsibility for many MTEs to engage in research, keep up to date with the research literature, produce academic papers and publish in highly rated research journals. Those of us who do this recognise how much we learn from reading and writing: the challenge and indeed the *struggle* to formulate texts that fit the academic mould and do justice to our research and its findings. Several of the authors in this volume have spoken of the formative value of this writing process and of using writing as an educative task in teacher education. Ruhama Even writes,

> Another opportunity to participate in the community of practice was the use of participants writings on the activities they conducted for practising teachers of mathematics a resources for other participants and other teacher educators.

Pat Perks and Stephanie Prestage write

> Writing about the session deepened our ideas as we noted our reactions to the session and the power of the labels [used spontaneously by prospective teachers to identify key ideas that has arisen in the context of programme sessions]. The act of writing extended our learning as we accounted for what had happened in the session. The act of writing, a professional tradition and rule of our activity system, acted also as a tool for learning as teacher educators.

After reading such remarks in early drafts of papers, I asked all authors to what extent they had found writing their chapter, or writing more generally a useful tool. Some of the responses I received said:

- Writing this chapter proved formative for me as it oriented a few changes in focus of attention and thus helped me elucidate the issues I was struggling to put in words so that others understand them.
- We think that the process of writing our chapter has allowed us three things:
  - we have learnt a lot
  - because of that, we have gone ahead in our personal development
  - in addition, we have felt members of an international community of practice/inquiry, since a lot of people have collaborated with us in the discussion of our ideas, improving of our English, etc.
- I find that writing is an opportunity to allow associations to flow more fully than is the case in presentations, where time is short and attention needs to be restrained. What can be done in a live session is to work with the audience's attention. What can be done in or through writing is to explore connections, and then to try to massage that into a coherent narrative which gives at least a flavour of the possibilities availabe to be pursued.

Konrad Krainer [9] points to writing as a valuable process for teachers, who acknowledge it despite the difficulties it caused them.

> Evaluations showed that writing was a tremendous challenge for many teachers; however, at the end, when the studies had been written, mostly positive feelings and views remained. Only recently, a doctoral student

(Schuster, 2008) began to investigate the impact of mathematics and science teachers' writing on their motivation and competences.

Thus, the challenges and mediational structures for MTEs can work similarly also for teachers. We might explore more overtly the parallels between MTE learning and teacher learning and the value of collaborative approaches that encourage recognition of co-learning potential.

MATHEMATICS TEACHER EDUCATORS – WHO ARE WE?

Teaching as Mentoring

The themes and issues discussed above point to the complexity of inter-relationships between the learning of pupils, teachers and teachers educators, with which I have personally grappled for many years (e.g. Jaworski, 2001). In a recent paper, in which I referred to a developmental project in which I have been involved with co-researchers including teachers, I raised three questions about "layers of attention":

> We are interested in developing the teaching and learning of mathematics so that pupils have better opportunities to learn mathematics with understanding and fluency. We see this involving three layers of attention:
>
> - How can teachers *and pupils* create a mathematical environment in their classrooms with suitable opportunity for pupils to learn mathematics with understanding and fluency?
>
> - How can didacticians *and teachers* create a didactical environment in their interactive space (in schools and college) with suitable opportunity for teachers to develop mathematics teaching with understanding and fluency?
>
> - How can xxxxxxx *and didacticians* create a (supra-didactical?) environment ... with suitable opportunity for didacticians to learn (didacting?) with understanding and fluency (Jaworski, 2007, p.6)

The third question is deliberately strange, to draw attention first to its parallels with the two earlier questions, but then to recognise that there is no (obvious) mentor for didacticians (MTEs) in the way that MTEs mentor teachers and teachers mentor students. We MTEs however, do need such mentoring. For example, in writing this chapter for the handbook, I asked my fellow editors to offer me a critical review to help improve the chapter. Our submissions of articles to journals results in a form of mentoring from our colleagues in the field who review papers and produce critical comments. In every case where this has happened for me, the paper produced as a result of working on the critical comments is a better paper than it would have been without this mentoring. The process of writing the paper is formative as expressed above, and the review process provides an important critical dimension. Both are mediating tools in crafting the written product and importantly in the associated learning that takes place. In a critical review of a draft of this chapter, Dina Tirosh made the comment:

> Reading as a critical friend (under four different occasions – reviewing a paper, reviewing as an editor, reviewing as a critical friend and reading the works of my graduate students) differs from reading for the sake of reading. Such processes are important for our own growth. You mentioned in your chapter the contribution of the critical friends to the development of the papers – but, as we know, the readers also develop through doing this type of work.

This comment amplifies further the concept of reciprocity discussed earlier.

Martin Simon points to MTE education through PhD programmes in which PhD students focus their research on aspects of teacher and/or pupils' learning in teacher education programmes with which they are associated. The research process in which they engage and their supervision from more experienced MTE researchers offer forms of mentoring that lead to growth of knowledge in teacher education. The supervisors also learn reciprocally from their activity in this process. In the MANOR programme, Ruhama Even shows that prospective MTEs engage in tasks designed by their mentors (more experienced MTEs) which include reading, writing and critical reviewing. Such tasks are designed to draw new MTEs into MTE culture, from which they become legitimate (less peripheral) participants (Lave & Wenger, 1991).

I am using here the term *mentoring*, rather than the more familiar *teaching* to emphasise the facilitative role of the teacher or MTE. In the above we see three modes of mentoring

1) Actions of human agents, acting as critical friends – the teacher figure – promoting learning of others

2) Actions of human agents in practice influencing their own development through some form of "inner mentor".

3) Actions of human agents working collaboratively together, mentoring for each other – "co-mentoring" (Jaworski & Watson, 1994).

These three modes for the MTE are a part of 'normal' process. The culture of MTE activity is such that we take for granted that we will act in certain ways for our own development – in many respects we see it just as doing our job, not, overtly, as developing as an MTE. The seminars we attend and professional conversations we hold are part of that activity. When we act as mentors for our colleagues, either informally as critical friends or more formally, as reviewers for a journal, or as supervisors of PhD research, we enter into a didactical practice that is an extension of our ordinary norms of practice.

Mentoring versus Instruction

Here I discuss MTE learning and development through engagement in a professional community in which participants are agents for their own and others' development. Part of the practice which develops is their working with teachers for teacher and teaching development. To what extent is this a *mentoring* process in the spirit of the discussion above and to what extent is it *instruction*? And are mentoring and instruction different? For example, are MTEs mentors in the programmes that Martin Simon refers to as having only *process goals*, whereas they are *instructors* in those which have *content and process goals,* when there is a content to be delivered? And, what are the parallels for teachers? Is there a culture in teacher development that parallels that described for MTEs – that is seen as just part of teachers doing their job – developmental agency as part of being a teacher? To what extent are, or can teachers be mentors for teachers? There are many programmes currently in which teachers mentor *prospective* teachers, with or without MTE counterparts (e.g., Jaworski & Gellert, 2003; Van Zoest & Bohl, 2002). In some cases, teachers are "trained" to become mentors as part of the programme. My use of the term "trained" suggests that some form of *instruction* is involved.

Mostly for teachers, the content they teach is *mathematics*, although teachers as mentors for prospective teachers teach "teaching mathematics". MTEs teach "teaching mathematics" rather than, or as well as mathematics per se (e.g., Sánchez and García [14] – see comments above). Does the nature of teaching – as mentoring or instruction – depend on what content is being taught? Particularly, if mathematics is being taught, are we more likely to think in terms of instruction? What would it look like to characterise teaching mathematics (per se) as a mentoring process?

Recall Martin Simon and Ron Tzur's constructs "conception-based" and "non-intuitive" – these deal with the individual mentor who works to overcome the constructivist learning paradox – how the critical friend can enter into the conceptions of the mentee in order to provide support that really interacts with the thinking and needs of the mentee. Caring reciprocity can be similarly conceptualised. Recall John Mason's – I can only do the learning for myself, but through that learning I can notice what tasks are fruitful to generate awareness and offer these to others – and what the expert sees is different from the novice. Both of these perspectives deal with *instruction* in that they have predetermined goals for the learning of others. However, a key element of the associated didactics is that instruction is designed to work with learners' conceptions using tasks or other tools, not to try to take learners to where the instructor wants them to be by the route favoured by the instructor. Nevertheless, in this mode, there is inevitably a factor of "where the instructor wants them to be" and this factor can lead to a direct instruction, or a perception-based approach.

Theory Offering a Frame on Practice

Having worked for many years myself within a constructivist frame (e.g. Jaworski, 1994), I have become increasingly aware that I need an alternative frame to make sense of some of the contradictions that become evident in such theorising, and to take account of culture and context as it is related to the overall complexity of the didactic process. The zone theory of which Merrilyn Goos writes, and activity theory as used by Pat Perks and Stephanie Prestage, both offer such an alternative. Over recent years, along with my colleague Simon Goodchild, I have begun to work with activity theory (e.g., Jaworski & Goodchild, 2006) to make sense of the complexity of learning processes as delineated in the three layers of attention at the start of this section.

Goodchild (2007) has analysed the activity of didacticians in the LCM project of which he writes in his chapter in this volume. Using an activity theory frame, based in the extended mediational triangle of Yrio Engeström (1999; see also Perks & Prestage [13]). Goodchild (2007) suggests there are two different activity systems in the LCM project. One activity system is that of the didacticians, based in university practice, taking an MTE role that includes explicit mentoring of teachers as well as personal development within the didacticians' community. The other system is that of the teachers in the project, based in school practice, working within the established communities of school and educational system to promote learning of pupils. Both teachers and didacticians are committed also to a project community that strives to develop inquiry in designing for and acting in the classroom to promote pupils' mathematical learning. Goodchild's findings suggest that the two systems exist side by side with interaction, but participation in the established communities is a stronger influence on activity than in the fledgling project community.

Taking the notion of two activity systems acting side by side – teachers in established school environments and didacticians in established university environments – with interaction as part of the project, I see mentoring relationships as expressed in Figure 4. Cells 1 and 4 express reciprocal learning (caring?) relationships within each of the two systems. Cells 2 and 3 express cross-system relationships. In Cell 2, we might see a relationship that carries instructional connotations, although there is no direct instructional intention involved. In Cell 3, although we might say there is no expressed intention for teachers to mentor MTEs, the nature of the project and its theoretical grounding in inquiry encourages teachers to speak out critically within the project and MTEs to analyse the meaning and impact of teachers' critical remarks for their development of practice. The table can be extended to include students in classrooms, but I leave that to the reader.

Mentoring relations	Teachers	MTEs
Teachers	Teachers mentoring teachers within their own community and system (1)	Teachers mentoring MTEs across the two activity systems (3)
MTEs	MTEs mentoring teachers across the two activity systems (2)	MTEs mentoring MTEs within their own community and system (4)

Figure 4. Mentoring relations within a developmental research project involving teachers and MTEs

BARBARA JAWORSKI

A very important outcome for the didacticians has been the steadying factor of the established school community and schooling system, informing and putting the brakes on didactic zeal – making clear the limitations on what it is possible to achieve and forcing a more practically-based didactic rationale. Although Goodchild's (2007) paper suggests a deficiency of developmental outcome in an apparent lack of expansive development, it seems to me that the project has achieved a great deal by creating a project community in which teachers can act as described, mentoring MTEs in terms of wider systemic factors that are highly influential on what is possible in classrooms and the interpretation of inquiry-based design and instruction in schools and classrooms. Recalling Figure 1, I talked of MTEs influencing growth of knowledge in B and C, drawing on their knowledge developing through research in A. I want now to express this from a teacher direction: teachers draw on their knowledge in practice in B and C (facilitated by the LCM project community) to act as agents for each other and in their own development in school settings. Their activity within the project impinges on MTEs knowledge in B, leading MTEs to question theoretical knowledge in A and reformulate associated didactical practice. The associated activity systems are both essential to the process described and the project acts as a catalyst, or a mediational tool in promoting learning within the established settings.

The LCM project was a developmental research project with *process* goals (Simon, this volume). It did not have *content* goals, for example in teaching mathematics or in teaching mathematics teaching. It did not involve courses for teachers or aim that teachers would learn specific elements of mathematics, pedagogy or didactics. Thus, it was set up to seek to create a project community in which partnerships between teachers and MTEs could be constituted democratically and principles of equity achieved. Such goals for the project were not achieved in quite the ways they were envisaged in conceptualisation (Jaworski, 2005). However, the resulting project allowed for the kinds of interaction I have described above. In the project that has succeeded LCM (TBM – Teaching Better Mathematics) teachers and their leaders from the school system have an overt part in planning activity together with didacticians[4]. It is hoped that this might lead to elements of "content" to be offered to teachers in ways designed to address systemic factors from the school community and in ways which build on MTEs' goals of practice expressed in the inquiry-based goals of the project community. The ground work in developing community between teachers and didacticians, the growing dialogue and language in which to talk to each other at meaningful levels allows content activity to be situated and take on new meanings. Thus, the process-

---

[4] See also the chapter by Jaworski in Volume 3 of this handbook: Krainer, K. & Wood, T. (Eds.). (2008). *International handbook of mathematics teacher education: Vol. 3 Participants in mathematics teacher education: Individuals, teams, communities and networks*. Rotterdam, the Netherlands: Sense Publishers.

based programme can lead to a fruitful base for more content-related work. I believe that these ideas are generalizable to programmes more widely, providing that we see teaching as more than just an instructive process with perceptual goals.

I used here the word "we". When I asked in my heading of this final section "who are we?" this provided an opportunity to see ourselves, MTEs, as partners with teachers, drawing reciprocally on distributed forms of knowing and providing reciprocally for the needs we identify jointly. For me, this seems a potentially fruitful way ahead.

REFERENCES

Association of Teachers of Mathematics (1987). *Teacher is/as researcher.* Derby: ATM.
Ball, D. L., & Bass, H. (2004). Knowing mathematics for teaching. In R. Stræsser, G. Brandell, B. Grevholm, & O. Helenius (Eds.), *Educating for the future. Proceedings of an international symposium on mathematics teacher education* (pp. 159–178). Stockholm: The Royal Swedish Academy of Sciences.
Bereiter, C. (1985). Toward a solution of the learning paradox. *Review of Educational Research, 55*(2), 201–226.
Carpenter, T., Fennema, E., Peterson, P., & Carey, D. (1988). 'Teachers' pedagogical content knowledge of students' problem-solving in elementary arithmetic.' *Journal for Research in Mathematics Education, 19*, 385–401.
Carr, W., & Kemmis, S. (1986). *Becoming critical: Education, knowledge and action research.* London: Routledge Falmer.
Carter, R., & Richards, J. (1999). Dilemmas of constructivist mathematics teaching: Instances from classroom practice. In B. Jaworski, T Wood, & S. Dawson (Eds.), *Mathematics teacher education: Critical international perspectives* (pp. 69–77). London: Falmer Press.
Cole, M., & Engeström, Y. (1993). A cultural-historical approach to distributed cognition. In G. Salomon (Ed.), *Distributed Cognitions: psychological and educational considerations,* (pp. 1-46) Cambridge, Cambridge University Press.
Engeström, Y. (1999). Activity theory and individual and social transformation. In Y. Engeström, R. Miettinen & R-L Punamäki (Eds.). *Perspectives on activity theory* (pp. 19–38). Cambridge: Cambridge University Press.
Fernandez, C., & Yoshida, M. (2004). *Lesson study: A Japanese approach to improving mathematics teaching and learning.* Mahwah, NJ: Lawrence Erlbaum.
Freire, P. (1972). *Pedagogy of the oppressed.* Harmondsworth UK: Penguin.
Goodchild, S. (2007). Inside the outside: Seeking evidence of didacticians' learning by expansion. In B. Jaworski, A. B. Fuglestad, R. Bjuland, T. Breiteig, S. Goodchild, & B. Grevholm (2007). *Learning communities in mathematics* (pp. 189–204). Bergen, Norway: Caspar Forlag.
Hargreaves, D. (1996). Teaching as a research-based profession: Possibilities and prospects. *Teacher Training Agency Annual Lecture.* London: Teacher Training Agency.
Hostetler, K. (2005). What is "good" educational research? *Educational Researcher, 34*(6), 16–21.
Irwin, K. C., & Britt, M. S. (1999). Teachers' knowledge of mathematics and reflective professional development. In B. Jaworski, T. Wood, & S. Dawson (Eds.), *Mathematics teacher education: Critical international perspectives* (pp. 91–101). London: Falmer Press.
Jaworski, B. (1994). *Investigating mathematics teaching: A constructivist enquiry.* London: Falmer Press.
Jaworski, B. (2001). Developing mathematics teaching: Teachers, teacher-educators, and researchers as co-learners. In F.-L. Lin & T. J. Cooney (Eds.), *Making sense of mathematics teacher education* (pp. 295–320). Dordrecht, the Netherlands: Kluwer.

Jaworski, B. (2003). Research practice into/influencing mathematics teaching and learning development: Towards a theoretical framework based on co-learning partnerships. *Educational Studies in Mathematics 54*, 249–282.

Jaworski, B. (2005). Learning communities in mathematics: Creating an inquiry community between teachers and didacticians. In R. Barwell & A. Noyes (Eds.), *Research in mathematics education: Papers of the British Society for Research into Learning Mathematics* (Vol. 7, pp. 101–119). London: BSRLM

Jaworski, B. (2006). Theory and practice in mathematics teaching development: Critical inquiry as a mode of learning in teaching, *Journal of Mathematics Teacher Education*, *9*(2), 187–211.

Jaworski, B. (2007). Developmental research in mathematics teaching and learning: Developing learning communities based on inquiry and design. In P. Liljedahl (Ed.), *Proceedings of the 2006 Annual Meeting of the Canadian Mathematics Education Study Group* (pp. 3–16). University of Calgary. Burnaby, BC: CMESG

Jaworski, B., & Gellert, U. (2003). Educating new mathematics teachers: Integrating theory and practice, and the role of practising teachers. In A. J. Bishop, M. A. Clements, C. Keitel, J. Kilpatrick, & F. K. S. Leung (Eds.), *Second international handbook of mathematics education* (pp. 829–987). Dordrecht, the Netherlands: Kluwer Academic Publishers.

Jaworski, B., & Goodchild, S. (1996). Inquiry community in an activity theory frame. In *Proceedings of the 30th Conference of the International Group for the Psychology of Mathematics Education* (Vol. 3, pp. 353–360). Prague, Czech Republic: Charles University.

Jaworski, B. & Watson, A. (Eds.) (1994). *Mentoring in mathematics teaching*. London: Falmer Press.

Jaworski, B., & Wood, T. (1999). Themes and issues in inservice programmes. In B. Jaworski, T Wood, & S. Dawson (Eds.), *Mathematics teacher education: Critical international perspectives* (pp. 125–147). London: Falmer Press.

Kemmis, S. (1985). Action research and the politics of reflection. In D. Boud, R. Keogh, and D. Walker (Eds.) *Reflection: Turning experience into learning*. London: Kogan Page.

Kemmis, S. & McTaggart, R. (1981). *The action research planner*. Geelong: Deakin Univesity

Kilpatrick, J., Swafford, J., & Findell, B. (Eds.) (2001). *Adding it up: Helping children learn mathematics*. Washington DC: National Academy Press.

Lave, J. & Wenger, E. (1991). *Situated learning: Legitimate peripheral participation*. Cambridge, MA Cambridge University Press.

Lave, J. (1996). Teaching as learning, in practice. *Mind Culture and Activity*, *3*(3), 149–164.

Leont'ev, A. N. (1979). The problem of activity in psychology. In J. V. Wertsch (Ed.), *The concept of activity in Soviet psychology* (pp. 37–71). New York: M. E. Sharpe.

Mason, J. (2002). *Researching your own classroom practice*. London: Routledge Falmer.

Nemirovsky, R., Dimattia, C., Ribiero, B., & Lara-Meloy, T. (2005). Talking about teaching episodes. *Journal of Mathematics Teacher Education*, *8*, 363–392

Niss, M. (2004). The Danish KOM-project and possible consequences for teacher education. In R. Stræsser, G. Brandell, B. Grevholm, & O. Helenius (Eds.), *Educating for the future. Proceedings of an international symposium on mathematics teacher education* (pp. 179–190). Stockholm: The Royal Swedish Academy of Sciences.

Noddings, N. (1992). *The challenge to care in schools*. New York: Teachers College Press.

Piaget, J. (1950). *The psychology of intelligence*. London: Routledge and Kegan Paul.

Sherin, M. G. (2002). A balancing act: Developing a discourse community in a mathematics classroom. *Journal of Mathematics Teacher Education*, *5*, 205–233.

Shulman, L. (1987). Knowledge and teaching: Foundations of the new reform. *Harvard Educational Review*, *57*(1), 1–22.

Van Zoest, L. R., & Bohl, J. V. (2002). The role of reform curricular materials in an internship: The case of Alice and Gregory. *Journal of Mathematics Teacher Education*, *5*, 265–288.

Vygotsky, L. S. (1978). *Mind in society*. Cambridge MA: Harvard University Press.

Vygotsky, L. (1986). *Thought and language*. Cambridge MA: MIT.

Wartofsky, M. W. (1979). *Models, representation and scientific understanding*. Boston: Reidel.

Wells, G. (1999). *Dialogic inquiry: Toward a sociocultural practice and theory of education.* Cambridge, UK: Cambridge University Press.

Wenger, E. (1998). *Communities of practice: Learning, meaning and identity.* Cambridge, UK: Cambridge University Press.

Wertsch, J. V. (1991). *Voices of the mind.* Cambridge MA.: Harvard University Press.

Winsløw, C., & Durrand-Guerrier, V. (2007). Education of lower secondary mathematics teachers in Denmark and France. *Nordic Studies in Mathematics Education, 12*(2), 5–32.

Zack, V., Mousley, J., & Breen, C. (1997). *Developing practice: Teachers' inquiry and educational change.* Geelong, Australia: Centre for Studies in Mathematics, Science and Environmental Education, Deakin University.

Barbara Jaworski
Mathematics Education Centre
Loughborough University

Lightning Source UK Ltd.
Milton Keynes UK
15 November 2010

162891UK00001B/6/P